Innovations in the Industrial Internet of Things (IIoT) and Smart Factory

Sam Goundar
The University of the South Pacific, Fiji

J. Avanija
Sree Vidyanikethan Engineering College, India

Gurram Sunitha
Sree Vidyanikethan Engineering College, India

K. Reddy Madhavi
Sree Vidyanikethan Engineering College, India

S. Bharath Bhushan
VIT Bhopal University, India

A volume in the Advances in Computer and
Electrical Engineering (ACEE) Book Series

Published in the United States of America by
IGI Global
Engineering Science Reference (an imprint of IGI Global)
701 E. Chocolate Avenue
Hershey PA, USA 17033
Tel: 717-533-8845
Fax: 717-533-8661
E-mail: cust@igi-global.com
Web site: http://www.igi-global.com

Library of Congress Cataloging-in-Publication Data

Names: Goundar, Sam, 1967- editor.
Title: Innovations in the industrial internet of things (IIoT) and smart
  factory / Sam Goundar, J. Avanija, Gurram Sunitha, K Reddy Madhavi, S.
  Bharath Bhushan, editors.
Description: Hershey, PA : Engineering Science Reference, 2020. | Includes
  bibliographical references and index. | Summary: "This book discusses
  the development of models and algorithms for predictive control of
  industrial operations and focuses on optimization of industrial
  operational efficiency, rationalization, automation, and maintenance"--
  Provided by publisher.
Identifiers: LCCN 2019051895 (print) | LCCN 2019051896 (ebook) | ISBN
  9781799833758 (h/c) | ISBN 9781799833765 (s/c) | ISBN 9781799833772
  (eISBN)
Subjects: LCSH: Automation. | Internet of things--Industrial applications.
  | Cloud computing--Industrial applications. | Operations research.
Classification: LCC T59.5 .I48 2020 (print) | LCC T59.5 (ebook) | DDC
  658.5/14--dc23
LC record available at https://lccn.loc.gov/2019051895
LC ebook record available at https://lccn.loc.gov/2019051896

This book is published in the IGI Global book series Advances in Computer and Electrical Engineering (ACEE) (ISSN: 2327-039X; eISSN: 2327-0403)

British Cataloguing in Publication Data
A Cataloguing in Publication record for this book is available from the British Library.

For electronic access to this publication, please contact: eresources@igi-global.com.

Advances in Computer and Electrical Engineering (ACEE) Book Series

Srikanta Patnaik
SOA University, India

ISSN:2327-039X
EISSN:2327-0403

MISSION

The fields of computer engineering and electrical engineering encompass a broad range of interdisciplinary topics allowing for expansive research developments across multiple fields. Research in these areas continues to develop and become increasingly important as computer and electrical systems have become an integral part of everyday life.

The **Advances in Computer and Electrical Engineering (ACEE) Book Series** aims to publish research on diverse topics pertaining to computer engineering and electrical engineering. **ACEE** encourages scholarly discourse on the latest applications, tools, and methodologies being implemented in the field for the design and development of computer and electrical systems.

COVERAGE

- VLSI Fabrication
- Analog Electronics
- Chip Design
- Programming
- Computer Hardware
- Sensor Technologies
- Optical Electronics
- Qualitative Methods
- Circuit Analysis
- Power Electronics

IGI Global is currently accepting manuscripts for publication within this series. To submit a proposal for a volume in this series, please contact our Acquisition Editors at Acquisitions@igi-global.com or visit: http://www.igi-global.com/publish/.

Titles in this Series

For a list of additional titles in this series, please visit:
http://www.igi-global.com/book-series/advances-computer-electrical-engineering/73675

Innovative Applications of Nanowires for Circuit Design
Balwinder Raj (National Institute of Technical Teachers Training and Research, Chandigarh, India)
Engineering Science Reference • © 2021 • 263pp • H/C (ISBN: 9781799864677) • US $225.00

Optimizing and Measuring Smart Grid Operation and Control
Abdelmadjid Recioui (Université M'hamed Bougara de Boumerdes, Algeria) and Hamid Bentarzi (Université M'hamed Bougara de Boumerdes, Algeria)
Engineering Science Reference • © 2021 • 400pp • H/C (ISBN: 9781799840275) • US $225.00

New Methods and Paradigms for Modeling Dynamic Processes Based on Cellular Automata
Stepan Mykolayovych Bilan (State University of Infrastructure and Technology, Ukraine) Mykola Mykolayovych Bilan (Mayakskaya Secondary School, Moldova) and Ruslan Leonidovich Motornyuk (Main Information and Computing Center, Ukraine)
Engineering Science Reference • © 2021 • 326pp • H/C (ISBN: 9781799826491) • US $195.00

Examining Quantum Algorithms for Quantum Image Processing
HaiSheng Li (Guangxi Normal University, China)
Engineering Science Reference • © 2021 • 336pp • H/C (ISBN: 9781799837992) • US $195.00

Design and Investment of High Voltage NanoDielectrics
Ahmed Thabet Mohamed (Aswan University, Egypt & Qassim University, Saudi Arabia)
Engineering Science Reference • © 2021 • 363pp • H/C (ISBN: 9781799838296) • US $195.00

Research Advancements in Smart Technology, Optimization, and Renewable Energy
Pandian Vasant (University of Technology Petronas, Malaysia) Gerhard Weber (Poznan University of Technology, Poland) and Wonsiri Punurai (Mahidol University, Thailand)
Engineering Science Reference • © 2021 • 407pp • H/C (ISBN: 9781799839705) • US $225.00

HCI Solutions for Achieving Sustainable Development Goals
Fariza Hanis Abdul Razak (Universiti Teknologi MARA (UiTM), Malaysia) Masitah Ghazali (Universiti Teknologi Malaysia (UTM), Malaysia) Murni Mahmud (International Islamic University Malaysia, Malaysia) Chui Yin Wong (Multimedia University, Malaysia) and Muhammad Haziq Lim Abdullah (Universiti Teknikal Malaysia, Melaka, Malaysia)
Engineering Science Reference • © 2021 • 300pp • H/C (ISBN: 9781799849360) • US $195.00

701 East Chocolate Avenue, Hershey, PA 17033, USA
Tel: 717-533-8845 x100 • Fax: 717-533-8661
E-Mail: cust@igi-global.com • www.igi-global.com

Table of Contents

Detailed Table of Contents

Chapter 1

Sasikala Chinthakunta, Srinivasa Ramanujan Institute of Technology, Anantapuramu, India
Shoba Bindu Chigarapalle, Jawaharlala Nehru Technological University, Anantapur, India
Sudheer Kumar E., Jawaharlala Nehru Technological University, Anantapur, India

Typically, the analysis of the industrial big data is done at the cloud. If the technology of IIoT is relying on cloud, data from the billions of internet-connected devices are voluminous and demand to be processed within the cloud DCs. Most of the IoT infrastructures—smart driving and car parking systems, smart vehicular traffic management systems, and smart grids—are observed to demand low-latency, real-time services from the service providers. Since cloud includes data storage, processing, and computation only within DCs, huge data traffic generated from the IoT devices probably experience a network bottleneck, high service latency, and poor quality of service (QoS). Hence, the placement of an intermediary node that can perform tasks efficiently and effectively is an unavoidable requirement of IIoT. Fog can be such an intermediary node because of its ability and location to perform tasks at the premise of an industry in a timely manner. This chapter discusses challenges, need, and framework of fog computing, security issues, and solutions of fog computing for IIoT.

Chapter 2

G. Rama Subba Reddy, Mother Theresa Institute of Engineering and Technology, India
K. Rangaswamy, Sai Rajeswari Institute of Technology, India
Malla Sudhakara, VIT University, India
Pole Anjaiah, Institute of Aeronautical Engineering, India
K. Reddy Madhavi, Sree Vidyanikethan Engineering College, India

Internet of things (IoT) has given a promising chance to construct amazing industrial frameworks and applications by utilizing wireless and sensor devices. To support IIoT benefits efficiently, fog computing is typically considered as one of the potential solutions. Be that as it may, IIoT services still experience issues such as high-latency and unreliable connections between cloud and terminals of IIoT. In addition to this, numerous security and privacy issues are raised and affect the users of the distributed computing environment. With an end goal to understand the improvement of IoT in industries, this chapter presents the current research of IoT along with the key enabling technologies. Further, the architecture and features of fog computing towards the fog-assisted IoT applications are presented. In addition to this, security and protection threats along with safety measures towards the IIoT applications are discussed.

Sai Deepthi Bhogaraju, InkforTech, India
Korupalli V Rajesh Kumar, VIT Chennai, India
Anjaiah P., Institute of Aeronautical Engineering, India
Jaffar Hussain Shaik, KSRM College of Engineering, India
Reddy Madhavi K., Sree Vidyanikethan Engineering College, India

The recent evolution of the fourth industrial revolution is Industry 4.0, projecting the enhancement of the technology, development, and trends towards the smart processing of the automation in industries. The advancements in communication and connectivity are the major source for the Industrial IoT (IIoT). It collaborates all the industrial functional units to work under a single control channel, digital quantification analytic methods deployment for the prediction of machinery, sensors, monitoring systems, control systems, products, workers, managers, locations, suppliers, and customers. In addition to IIoT, AI methods are also playing a vital role in predictive modeling and analytic methods for the assessment, control, and development of rapid production, from the industries. Other side security issues are challenging the development, concerning all the factors digitalization processes of the industries need to move forward. This chapter focuses on IIoT core concepts, applications, and key challenges to enhance the industrial automation process.

Seeja G., VIT Chennai, India
Obulakonda Reddy R., Institute of Aeronautical Engineering, India
Korupalli V. Rajesh Kumar, VIT University, India
S. S. L. C. H. Mounika, Jawaharlal Nehru Technological University, Kakinada, India
Reddy Madhavi K., Sree Vidyanikethan Engineering College, India

The recent industrial scenarios project its advancements and developments with the intervention of integrated technologies including internet of things (IoT), robotics, and artificial intelligence (AI) technologies. Industrial 4.0 revolutions have broken the barriers of all restricted industrial boundaries with the act of those interdisciplinary concepts and have taken a keen part in industrial development. Incorporation of these advancements considerably helps in improving product efficiency and in reducing the production cost. Based on categories of production, industrial automation processes may vary. In this regard, robots are playing a vital role to automate the production process at various levels of industrial operations. The combination of IoT, robotics, and AI technologies enhances the industrial productivity towards getting the success rate. This chapter focuses on how robotic technology with IoT and AI methods enhances the limitations of various industrial applications.

Chapter 5

Fog computing, often projected as an extension to cloud, renders its design to deal with challenges of traditional cloud-based IoT. Fog enlightens its features of low latency, real-time interaction, location awareness, mobility support, geo-distribution (smart city), etc. over cloud. Fog by nature does not work on cloud instead on a network edge for facilitating higher speeds. Fog pulls down the risk of security attacks. Industrial sector is revolutionized by ever changing technical advancements and IoT, which is a young discipline embraced by industry thereby bringing in IIoT. Fog computing is viable to Industrial processes. IIoT is well supported by the middleware fog computing as industrial process requires most of the task performed locally and securely at end points with minimum delay. Fog, deployed for industrial processes and entities which are part of internet, is gaining importance in recent times being titles as fog for IIoT. Additionally, as industrial big data is often ill structured, it can be polished before sending it to cloud resulting in an enhanced computing.

Mechanical IoT is proceeding to extend from assembling and keen homes to retail, nourishment bundling, social insurance, and other use cases. For the present makers, modern IoT is quickly turning into an unquestionable requirement on creation lines for reasons of adequacy and hazard relief. Checking sensors must coordinate with generation line gear, including robots. Information streams and checking/announcing modules can give basic way data, trigger reaction conventions, and influence different exercises being followed on different modern systems. Modern IoT can expand perceivability into the states of robots in a production line, ranch, stockroom, or emergency clinic. Data about parts and gear can consistently observe to augment profitability and to decrease line personal time. Voice assistant robot takes the commands of the operator or mentor sets the functionality accordingly which is stored in the database and then execute the given task.

At present, the need for an ultra-high speed and efficient communication through mobile and wireless devices is gaining significant popularity. The users are expecting their network to offer real-time streaming without much latency. In turn, this will result in a considerable rise in network bandwidth utilization. The live streaming has to reach the end users mobile devices after traveling through the base station nodes, core network, routers, switches, and other equipment. Further, this will lead to a scenario of content latency and thereby causing the rejection of the mobile devices users' request due to congestion of the network and mobile service providers' core network witnessing an extreme load. In order to overcome such problems in the contemporary 5G mobile networks, an architectural framework is essential, which offers instantaneous, ultra-low latency, high-bandwidth access to applications that are available at the network edge and also making the task processing in close proximity with the mobile device user.

Chapter 8

Lakshman Narayana Vejendla, Vignan's Nirula Institute of Technology and Science for Women, India
Alapati Naresh, Vignan's Nirula Institute of Technology and Science for Women, India
Peda Gopi Arepalli, Vignan's Nirula Institute of Technology and Science for Women, India

Internet of things can be simply referred to as internet of entirety, which is the network of things enclosed with software, sensors, electronics that allows them to gather and transmit the data. Because of the various and progressively malevolent assaults on PC systems and frameworks, current security apparatuses are frequently insufficient to determine the issues identified with unlawful clients, unwavering quality, and to give vigorous system security. Late research has demonstrated that in spite of the fact that system security has built up, a significant worry about an expansion in illicit interruptions is as yet happening. Addressing security on every occasion or in every place is a really important and sensitive matter for many users, businesses, governments, and enterprises. In this research work, the authors propose a secret IoT architecture for routing in a network. It aims to locate the malicious users in an IoT routing protocols. The proposed mechanism is compared with the state-of-the-art work and compared results show the proposed work performs well.

Chapter 9

V. Shunmughavel, Computer Science & Engineering, SSM Institute of Engineering and Technology, India

The industrial internet of things (IIoT) has made its development within a short span of time. Initially it was considered as a novel idea and currently it is a major driver in industry applications. It has created productivity and efficiency for industries worldwide. This innovative technology can become a practical reality if engineers overcome a variety of challenges. They are connectivity, cost, data integration, trust, privacy, device management, security, interoperability, collaboration, and integration. In this chapter, several facts behind the above-mentioned challenges are being explored and addressed.

Chapter 10

Sam Goundar, British University, Vietnam
Akashdeep Bhardwaj, University of Petroleum and Energy Studies, India
Safiya Shameeza Nur, The University of the South Pacific, Fiji
Shonal S. Kumar, The University of the South Pacific, Fiji
Rajneet Harish, The University of the South Pacific, Fiji

This chapter focused on the importance and influence of industrial internet of things (IIoT) and the way industries operate around the world and the value added for society by the internet-connected technologies. Industry 4.0 and internet of things (IoT)-enabled systems where communication between products, systems, and machinery are used to improve manufacturing efficiency. Human operators' intervention and interaction is significantly reduced by connecting machines and creating intelligent networks along the entire value chain that can communicate and control each other autonomously. The difference between IoT and IIoT is that where consumer IoT often focuses on convenience for individual consumers, industrial IoT is strongly focused on improving the efficiency, safety, and productivity of operations with a focus on return on investment. The possibilities with IIoT is unlimited, for example, smarter and more efficient factories, greener energy generation, self-regulating buildings that optimize energy consumption, smart cities that can adjust traffic patterns to respond to congestion.

Chapter 11

Autonomous robots are being increasingly integrated into manufacturing, supply chain, and retail industries due to the twin advantages of improved throughput and adaptivity. In order to handle complex Industry 4.0 tasks, the autonomous robots require robust action plans that can self-adapt to runtime changes. A further requirement is efficient implementation of knowledge bases that may be queried during planning and execution. In this chapter, the authors propose RoboPlanner, a framework to generate action plans in autonomous robots. In RoboPlanner, they model the knowledge of world models, robotic capabilities, and task templates using knowledge property graphs and graph databases. Design time queries and robotic perception are used to enable intelligent action planning. At runtime, integrity constraints on world model observations are used to update knowledge bases. They demonstrate these solutions on autonomous picker robots deployed in Industry 4.0 warehouses.

Chapter 12

In the era of mechanical digitalization, organizations are progressively putting resources into apparatuses and arrangements that permit their procedures, machines, workers, and even the products themselves to be incorporated into a solitary coordinated system for information assortment, information examination, the assessment of organization advancement, and execution improvement. This chapter presents a reference guide and review for propelling an Industry 4.0 venture from plan to execution, according to base on the economic and scientific policy of European parliament, applying increasingly effective creation forms, and accomplishing better profitability and economies of scale may likewise bring about expanded financial manageability. This chapter present the contextual analysis of a few Industry 4.0 applications. Authors give suggestions coordinating the progression of Industry 4.0. This section briefly portrays the advancement of IIoT 4.0. The change of ubiquitous computing through the internet of things has numerous difficulties related with it.

Chapter 13

Chandramohan Dhasarathan, Department of Computer Science and Engineering,
Madanapalle Institute of Technology and Science, India
Shanmugam M., Department of Computer Science and Engineering, Vignan's Foundation
for Science, Technology, and Research, India
Shailesh Pancham Khapre, ASET-CSE, Amity University, Noida, India
Alok Kumar Shukla, School of Computer Science and Engineering, VIT-AP University,
Amaravati, Andhra Pradesh, India
Achyut Shankar, ASET-CSE, Amity University, Noida, India

The development of wireless communication in the information technological era, collecting data, and transfering it from unmanned systems or devices could be monitored by any application while it is online. Direct and aliveness of countless wireless devices in a cluster of the medium could legitimate unwanted users to interrupt easily in an information flow. It would lead to data loss and security breach. Many traditional algorithms are effectively contributed to the support of cryptography-based encryption to ensure the user's data security. IoT devices with limited transmission power constraints have to communicate with the base station, and the data collected from the zones would need optimal transmission power. There is a need for a machine learning-based algorithm or optimization algorithm to maximize data transfer in a secure and safe transmission.

Chapter 14

Geetha Prahalathan, Sree Vidyanikethan Engineering College, Tirupati, India
Senthil Kumar Babu, Sree Vidyanikethan Engineering College, Tirupati, India
Praveena H. D., Sree Vidyanikethan Engineering College, Tirupati, India

The industrial production has experienced a technological revolution in the recent past decades. The technological revolution influenced the agriculture industry too. The important areas in the change are not limited to innovation in farming, novel production of agriculture-based tools and equipment, transportation and consumption of food across the globe, marketing the agriculture products, and digitalization. Digitalization is the involvement of digital technology in the existing field for easing the mechanism of handling, processing, recording the data. Digitalization enables sustainable farming. It is required desperately to develop this technology because there is a substantial reduction of clean water and depletion of aquifers effects the cultivation. With the technology, the quantity and quality of the food has to be managed to feed the global population. The familiar digitization technology that makes the agri-industrial sector to experience growth are artificial intelligence, machine learning, sensor networks, internet of things, robotics, cloud data.

Chapter 15

Abhilash B. L., Vidyavardhaka College of Engineering, India
Gururaj H. L., Vidyavardhaka College of Engineering, India
Vijayalakshmi Akella, K. S. School of Engineering and Management, India
Sam Goundar, British University, Vietnam

Due to globalization, demand per capita has increased over the decade; in turn, standard of living has been increased. The emission of carbon dioxide is increasing exponentially in construction industries, which affects the global ecological system. To reduce the global warming potential, net zero energy buildings are very essential. With respect to technological advancements in information technology, the internet of things (IoT) plays a vital part in net zero energy buildings. In this chapter, the various issues and challenges of high-performance zero energy buildings are elaborated using different scenarios.

This chapter discusses the implication of predictive maintenance (PM) for industrial marketing companies. Using an illustrative case study from the Indian industrial air compressor market, it shows that predictive maintenance solutions will change the way of conventional sales and marketing. Sellers need to focus on early innovation adopters among its customers. They also need to engage with existing customers early on in the purchase process and highlight how PM can reduce the total cost of ownership. PM can be sold effectively to different types of customers- transactional, value-oriented, and collaborative. Industrial marketers have to position the solution appropriately to gain competitive advantage.

In the area of tank inspections across the industry, robots were introduced to replace human inspectors in selected operations. The technological gap in adoption of similar technologies by Zimbabwe's bulk fuel storage tanks operators motivated this research. The industry's current NDT practices were investigated, costs and inconveniences were identified, and improvements were explored. Operators of bulk fuel facilities and companies providing tank inspection services were engaged to establish the reasons for the gaps in technological assimilation. Emerging global technologies that enable in-service inspections were identified and their applicability to Zimbabwe's bulk fuel facilities was investigated. A combination of crawler based ultrasonic thickness tests for tank shells, and acoustic emission in-service tank bottom testing was observed to be the most convenient and relevant in-service tank inspection method for Zimbabwe's bulk fuel storage tanks industry. Internet-based remote connectivity and control was considered for data compilation, analysis, storage, and reporting.

Preface

INTRODUCTION

Industrial Internet of Things (IIoT) is the enabler of the Fourth Industrial Revolution (4IR) or Industry 4.0. The Fourth Industrial Revolution will have a profound impact on the way we live, work, and socialize. Industry 4.0 is essentially about the digital transformation of our lives, cities, and industries. Industrial Internet of Things will interconnect all the things that we interact with in our ecosystem. This convergence of cyber physical systems will transform our manufacturing and services sector. It will impact how we manage and organise technology, organisation and the environments in which we operate. Billions of people on smart mobile devices, connecting through network technologies like 5G, have access to unlimited processing power and storage capacity via the cloud. This combined with emerging technologies like Artificial Intelligence (AI), robotics, the Internet of Things (IoT), autonomous vehicles, 3-D printing, nanotechnology, biotechnology, materials science, energy storage, and quantum computing will integrate our physical world with the digital one. Equipped with data from everything, big data analytics, machine learning, we will be making better decisions in real time for the advancement of our society.

As we started connecting Things to the Internet (Internet of Things), we realised that this connectivity not only provides remote control over things, but we can also collect data from it, analyse it and do many other things. Internet of Things are many applications in industries and cities. For example, in retail it is used for digital signage, in-store offerings and promotions, supply chain management, vending machines, and smart payments. In healthcare, it is applied in adherence to standards, support for controlling medical equipment, monitoring of clinical appliances, virtual care, prevention of pandemics, control of infections, and wellness. Smart Cities use Internet of Things in construction of energy efficient buildings (green buildings), in online education (access), environmental impact assessment, energy efficiency grids (smart energy), monitoring traffic congestion, diversion of road traffic, traffic lights, bus/train timetabling, water management and waste disposal. Other applications of Internet of Things can be seen in the financial sector for insurance, assisted and autonomous driving cars, smart buildings, smart living, smart navigation, assisted living, social welfare, and security.

Plants, equipment, appliances, devices and everything else in the Industry are getting connected to the Internet (Industrial Internet of Things). Technologies such as 5G Networks, big data analytics, RFID sensors, blockchain, edge computing, additive manufacturing, artificial intelligence, robotics, augmented reality are seamlessly connecting man and machine in manufacturing (smart manufacturing). Connected industries are relying on these technologies for connected farms to connected markets enables with smart supply chains, smart maintenance in factories, smart factories, digital factories, innovative product design, and software defined engineering. Innovations in the Industrial Internet of Things have already been deployed in the following sectors: industrial products, electronics, high tech engineering, automotive, aerospace, defense, retail, customer support, medical devices, software, utilities, oil, gas, and communications. Manufacturing operational intelligence, operational asset monitoring, medical devices monitoring, and manufacturing predictive maintenance are some example use cases of Innovations in Industrial Internet of Things (IIoT).

INTERNET OF THINGS (IOT)

As price of computer chips became affordable and small enough, they were fitted into all devices, appliances, and everything else that was manufactured. At the same time, wireless network connectivity became faster, cheaper, and ubiquitous. Any device with a computer chip has the capability to connect to the Internet. Therefore, billions of devices are now being connected to the Internet because of affordable computer chips and pervasive wireless connectivity. This connectivity means that now it is possible to collect data from these devices, share that data, analyse that data, monitor the devices and control it the way you want. The large amounts of data (Big Data) being collected and analysed (Data Analytics) provides an insight and intelligence in real time that has never been experienced before. This innovation has led to many smart technologies and applications like smart health, smart cities, smart homes, and devices.

According to (Ranger, 2020), "connecting all these different objects and adding sensors to them adds a level of digital intelligence to devices that would be otherwise dumb, enabling them to communicate real-time data without involving a human being. The Internet of Things is making the fabric of the world around us more smarter and more responsive, merging the digital and physical universes." To avoid confusion between existing devices connected devices, he further clarifies IoT as follows "The term IoT is mainly used for devices that wouldn't usually be generally expected to have an internet connection, and that can communicate with the network independently of human action. For this reason, a PC is not generally considered an IoT device and neither is a smartphone -- even though the latter is crammed with sensors. A smartwatch or a fitness band or other wearable device might be counted as an IoT device, however."

There has been a number of innovative applications of Internet of Things that has resulted in a smarter, healthier and secure society. For example, Internet of Things application in healthcare known as Connected Health or Smart Health enables doctors to monitor their patients both inside and outside the hospitals and get vital data about their health. Alerts will get delivered to doctors if a patient is not doing well and alarms will be raised with other healthcare professionals if a patient is heading towards a critical stage. Individuals can also self-monitor with wearables, smart watches, fitbit, and a myriad of devices out there. Every city in the world is aspiring to become a smart city. The smart city concept uses sensors placed all over the city to collect data in real time and take alternative action if needed. For example, diverting traffic in case of accidents and avoiding traffic jams. Switching traffic lights based on actual traffic is another example.

Tech analyst company (IDC, 2019), "predicts that in total there will be 41.6 billion connected IoT devices by 2025, or "things." It also suggests industrial and automotive equipment represent the largest opportunity of connected "things,", but it also sees strong adoption of smart home and wearable devices in the near term. Worldwide spending on the IoT was forecast to reach $745 billion in 2019, an increase of 15.4% over the $646 billion spent in 2018, according to IDC, and pass the $1 trillion mark in 2022." Ranger (2020), estimates "consumer IoT spending was predicted to hit $108 billion, making it the second largest industry segment: smart home, personal wellness, and connected vehicle infotainment will see much of the spending. Manufacturing operations ($100 billion), production asset management ($44.2 billion), smart home ($44.1 billion), and freight monitoring ($41.7 billion) will be the largest areas of investment."

INDUSTRIAL INTERNET OF THINGS (IIOT)

Industries use many devices, plants, and specialised equipment's in their factories and services. Connecting these devices to the Internet will enable better control of the devices and the data collected and shared will provide intelligence and a competitive advantage. Industrial Internet of Things will make factories smarter, green, and their products will have quality while being cheap because of efficient processes. The ability to use sensors in their plants and equipment, collect the data produced by this sensors through wireless networks, analyse the data collected, and use the analytics to measure and optimize industrial processes is an ideal of example of Industrial Internet of Things. Apart from the devices with sensors communicating to the central server and sending data, they can also communicate, exchange information, and send data (machine-to-machine) to other devices. This machine-to-machine (M2M) communication is critical for industry where assembly lines with different machines are used to manufacture the final product.

According to (McLellan, 2019), "technologies such as 5G, IoT sensors and platforms, edge computing, AI and analytics, robotics, blockchain, additive manufacturing and virtual/augmented reality are coalescing into a fertile environment for the Industrial Internet of Things (IIoT), which is set to usher in what's often described as the Fourth Industrial Revolution or Industry 4.0. He further states "supply chains will have end-to-end transparency thanks to sensors, data networks and analytics capabilities at key points. All other things (trade barriers, for example) being equal, parts and raw materials will arrive just in time at highly automated factories, and the fate of the resulting products will be tracked throughout their lifetimes to eventual recycling." "As a result, businesses deploying IIoT systems will see increased operational efficiency, will reduce their environmental impact, and will have better information on which to base their future plans. Similarly, 'smart farms' will combine emerging IIoT-related technologies into integrated high-resolution crop production systems based on robotics, big data, and analytics." (McLellan, 2019)

Industrial Internet of Things is a panacea for the industry. Our industries and factories at the moment are inefficient, fossil fuel operated, with broken supply chains, polluting the environment with carbon and causing climate change. Application of Industrial Internet of Things in our factories are expected to solve many of these issues. Therefore, Industrial Internet of Things are critical in achieving what is required for the Fourth Industrial Revolution. Getting real-time data from our industrial processes, taking corrective action, and making better decisions are the basics of doing it right. With data analytics, machine-to-machine communication, and artificial intelligence, our industrial processed are elevated to its optimum operational efficiency. The entire supply chain, from the raw material to the finished product, or from the farm to the consumer can be coordinated to create further efficiencies and transparency.

How big is the Industrial Internet of Things? According to (Ranger, 2020), "worldwide spending on the Internet of Things (IoT) is forecast to reach $745 billion in 2019 – up from the $646 billion spent in 2018, and likely to hit $1 trillion in 2022. The vast majority of that spending will be by businesses." The IIoT is already big business according to (IDC, 2019), "manufacturers will spend $197 billion on it in 2020, transport companies $71 billion, and utilities $61 billion. In contrast, consumer IoT spending will reach $108 billion, mainly focused on smart devices and homes., personal wellness, and connected vehicle infotainment". (Ranger, 2020)

INNOVATIONS IN THE INDUSTRIAL INTERNET OF THINGS (IIOT)

Internet of Things are bringing life changing improvement to our lives, cities, and societies. Industrial Internet of Things are doing same to our factories, industries, and farms. Living better, working smarter and continuously innovating is the mantra of the Fourth Industrial Revolution. Innovations means not only using robots to do things for us but adding intelligence to these robots so that they become more efficient, are able to identify issues and resolve them or refer to others for solution. The "big data" that we are collecting from all the devices in the industry are driving the innovation. Data analytics of the collected data enable industries to get manufacturing operational intelligence, monitor operational assets, remote monitoring of devices, remote servicing of devices, monitor the health of their machinery, and predictive maintenance.

The "Industrial Internet of Things (IIoT) adds significant value to businesses, and industries such as manufacturing, transportation, utilities and more, are taking full advantage of IIoT's capabilities", writes Wachsman, (2019). As reported by her, "IIoT devices can also deliver valuable data to businesses for innovation. Industries use data gathered from IoT devices for maintenance, collect data on shop floor/warehouse management, sales, marketing, accounting, and C-Suite." She says "it comes to no surprise that majority of them collect data with IoT devices on equipment's wear and tear. They also collect data regarding energy usage, telemetrics, and weather/climate. According to (Wachsman, 2019), "predictive maintenance dominates current or tentative use cases for IIoT devices. Other use cases include quality control, metering, managing inventory, safety, and smart packaging".

Notable Innovations in Industrial Internet of Things have been mostly in manufacturing, retail, utilities, and in the logistics sector. Manufacturers are using IIoT in their production for predictive servicing of the machines on their assembly lines and thus prevent machines from breaking down, reduce downtime and increase production. Technicians from utility companies are now able to check on remote installations, diagnose issues and rectify them remotely rather than having to travel to the transmitting stations on the top of a rainforest mountain. This saves costs, energy, time, and effort. In smart cities, street lights, traffic lights and other essential services are able to send an alert to the city council technicians to indicate that they are down or not working, and the technicians are able to get the exact location via GPS. Retailers can change their digital signage whenever they wish, and transport companies can locate and understand their fleets. More innovations are in the pipeline as the cost of computer chips and wireless goes down.

Anything connected to the Internet can get hacked and thus the devices in Industrial Internet of Things are vulnerable and we should be concerned about their security. Some of the security risks identified by (Ranger, 2020) are: "the IIoT brings with it new security risks because networking objects that would have otherwise been standalone means there's a risk that hackers could snoop on the data being transmitted, or even attempt to gain control of the devices." He adds that "consumer IoT device manufacturers have long been criticised for poor security practices like weak or non-existent passwords and software that cannot be upgraded when bugs become apparent. The IIoT must contend with these worries and more, but the stakes are potentially much higher. Connecting manufacturing lines or power grids to the internet in one form or another (or even the web) increases the potential attack surface for hackers." (Ranger, 2020)

SMART FACTORY

Data driven decisions make a factory smart. With Industrial Internet of Things, factories collect data from different machines along its assembly line and other associated functions and intermediaries along the supply chain (backward and forward integration). These huge amounts of data (Big Data) from different functions when analysed (Data Analytics) and mined can provide much better approaches to monitor, maintain, and manage production lines. A Smart Factory will combine data analytics with artificial intelligence, machine-to-machine communication, and cloud to achieve optimum productivity, efficiency, and automation. Everything is tracked in real-time, machines talk to one another, machines on production lines are repaired before they break down (predictive servicing and maintenance), energy consumption is optimised, and production lines can be changed and customized in the shortest time possible.

According to (Sjödin, 2018), "the development of novel digital technologies connected to the Internet of Things, along with advancements in artificial intelligence and automation, is enabling a new wave of manufacturing innovation. "Smart factories" will leverage industrial equipment that communicates with users and with other machines, automated processes, and mechanisms to facilitate real-time communication between the factory and the market to support dynamic adaptation and maximize efficiency. Smart factories can yield a range of benefits, such as increased process efficiency, product quality, sustainability, and safety and decreased costs. However, companies face immense challenges in implementing smart factories, given the large-scale, systemic transformation the move requires." They used data gathered from in-depth studies of five factories in two leading automotive manufacturers to analyze these challenges and identify the key steps needed to implement the smart factory concept." (Sjödin, 2018)

The manufacturing sector has benefitted prominently from Smart Factories. Some notable innovations in Smart Factory include virtual manufacturing and micro manufacturing. Virtual manufacturing is when a product is manufactured digitally without the use of any physical resources and machines. The product is tweaked and fine tuned virtually till it reaches the perfection and has the required quality. Based on the learning and experience that was acquired by manufacturing it in the digital environment, now the production can be moved into the physical environment. Virtual manufacturing eliminates wastage of resources, time, energy, saves money, and gets product to the market faster than the competitors. Smart factory will be able to engage in micro manufacturing as the production line is agile and can be changed within minutes for the manufacture of customized products. For many customers, "one size does not fit all" and they want their products heavily customised to suit their requirements. It is now possible with smart factory.

Artificial Intelligence and Machine Learning are expected to further innovate design and production in Smart Factory. Artificial Intelligence will enable the machines in an assembly line to take corrective action when faced with errors, without human intervention and retain the learning for future. Machine Learning will teach other machines based on analytics of data that was collected from the entire supply chain. All the machines within the smart factory will communicate with each other (machine-to-machine), share data, gain intelligence, predict and optimise production. The use of the Smart Factory concept is going to guarantee increased productivity, better production, cost savings, and provide and competitive advantage.

ABOUT THIS BOOK

The rise of the new digital industrial technology, popularly known as Industry 4.0, is focused to integrate industrial processes with full-scale process automation and reshaping the future of industrial revolution. Industrial IoT provides leverage and reality of IoT in the context of industrial transformation. According to (Ranger, 2020), the IIoT is already big business according to (IDC, 2019), "manufacturers will spend $197 billion on it in 2020, transport companies $71 billion, and utilities $61 billion. As quoted by Morgan Stanley, "The factory floor is getting upgraded with new sensors, connectivity and big data that are designed to revolutionize automation". As sensors spread across almost every industry, the Internet of Things is going to trigger a massive influx of big data. We delve into an era where Industrial IoT will have the biggest impact and what it means for the future of industrial revolution. Industrial Internet of Things is expected to change the way we work, live, produce, farm and industrialise.

Industrial IoT is changing the face of the industry by completely redefining the way stakeholders, enterprises and machines connect and interact with each other in the industrial digital eco system. Smart and connected factories in which all the machinery transmits real-time data, enables industrial data analytics for improving operational efficiency, productivity, and industrial processes thus creating new business opportunities, asset utilization and connected services. Industrial IoT leads factories to step out of legacy environments and arcane processes towards open digital industrial eco systems. The Smart Factory is a concept for expressing the end goal of digitization in manufacturing. Smart Factory is a highly digitized shop floor that continuously collects and shares data through connected machines, devices, and production systems. With the Smart Factory concept, production lines can be changed instantly for customised products.

This book offers industry, researchers, and government agencies the recent advancements in Industry Digitization and Automation. The objectives of the book is to explore topics on: Development of models and algorithms for predictive control of industrial operations; Methods for collection, management, retrieval and analysis of industrial data; Insights to taxonomy of challenges, issues and research directions in smart factory applications; Focus on optimization of industrial operational efficiency, rationalization, automation and maintenance. Some chapters of this book deal with Industrial Internet of Things innovative practices like using Voice Controlled Biped Walking Robot for Industrial Applications. These humanoids have been programmed using the "Zero Moment Point" for them to be as agile as humans in walking. Advanced Predictive Analytics for Control of Industrial Automation Process is another chapter that highlights how industrial processes can be controlled by analysing data from earlier processes.

Graduate students, academics, and practitioners in the industry will find this book quite useful. The other objective of this book is to be a primary source for students, academics, and practitioners to reference the evolving theory and practice related to Industrial Internet of Things and Smart Factory. It aims to provide a comprehensive coverage and understanding in the management, organisation, and technological use of Industrial Internet of Things and Smart Factory. The book intends to provide opportunities for investigation, discussion, dissemination, and exchange of ideas in relation to Industrial Internet of Things and Smart Factory internationally across the widest spectrum of scholarly and practitioner opinions to promote theoretical and empirical research.

ORGANIZATION OF THE BOOK

The book is organized into 17 chapters. A brief description of each of the chapters follows:

Chapter 1. The Challenges, Technologies, and Role of Fog Computing in the Context of Industrial Internet of Things

Typically, the analysis of the industrial big data is done at the Cloud. If the technology of IIoT is relying on cloud, data from the billions of Internet-connected devices are voluminous and demand to be processed within the cloud DCs. Most of the IoT infrastructures: smart driving and car parking systems, smart vehicular traffic management systems, and smart grids are observed to demand low-latency, real-time services from the service providers. Since cloud includes data storage, processing and computation only within DCs, huge data traffic generated from the IoT devices probably experience a network bottleneck, high service latency and poor Quality of Service (QoS).Hence, the placement of an intermediary node that can perform tasks efficiently and effectively is an unavoidable requirement of IIoT. Fog can be such an intermediary node because of its ability and location to perform tasks at the premise of an industry in a timely manner. This chapter discusses challenges, need and framework of Fog computing, security issues and solutions of Fog Computing for IIoT.

Chapter 2. Towards the Protection and Security in Fog Computing for Industrial Internet of Things

Internet of Things (IoT) has given a promising chance to construct amazing industrial frameworks and applications by utilizing wireless and sensor devices. To support IIoT benefits efficiently, Fog Computing is typically considered as one of the potential solutions. Be that as it may, IIoT services still experience issues such as high-latency and unreliable connections between cloud and terminals of IIoT. In addition to this numerous security and privacy issues are raised and affect the users of the distributed computing environment. With an end goal to understand the improvement of IoT in industries, this chapter presents the current research of IoT along with the key enabling technologies. Further, the architecture and features of fog computing towards the fog assisted IoT applications are presented. In addition to this, security, and protection threats along with safety measures towards the IIoT applications are discussed.

Chapter 3. Advanced Predictive Analytics for Control of Industrial Automation Process

The recent evolution of the fourth industrial revolution that is Industry 4.0, projecting the enhancement of the technology, development, and trends towards the smart processing of the automation in industries. The advancements in communication and connectivity are the major source for the Industrial IoT (IIoT). It collaborates all the industrial functional units to work under a single control channel, here digital quantification analytic methods deployment for the prediction of machinery, sensors, monitoring systems, control systems, products, workers, managers, locations, suppliers, and customers. In addition to IIoT, AI methods are also playing a vital role in predictive modeling and analytic methods for the assessment, control, and development of rapid production, from the industries. Other side security issues are challenging the development, concerning all the factors digitalization processes of the industries need to move forward. This chapter focuses on IIoT core concepts, applications, and key challenges to enhance the industrial automation process.

Chapter 4. Internet of Things and Robotic Applications in the Industrial Automation Process

The recent industrial scenarios project its advancements and developments with the intervention of integrated technologies including Internet of things (IoT), Robotics and Artificial Intelligence (AI) technologies. Industrial 4.0 revolutions have broken the barriers of all restricted industrial boundaries with the act of those interdisciplinary concepts and have taken a keen part in industrial development. Incorporation of these advancements considerably helps in improving product efficiency and in reducing the production cost. Based on categories of production, industrial automation processes may vary. In this regard, Robots are playing a vital role to automate the production process at various levels of industrial operations. The combination of IoT, Robotics and AI technologies enhances the industrial productivity towards getting the success rate. This chapter focuses on how Robotic technology with IoT and AI methods enhances the limitations of various industrial applications.

Chapter 5. Fog Computing in Industrial Internet of Things

Fog Computing often projected as an extension to cloud renders its design to deal with challenges of traditional cloud based IoT. Fog enlightens its features of low latency, real time Interaction, location awareness, mobility support, geo-distribution (Smart city) etc. over Cloud. Fog by nature does not work on cloud instead on a network edge for facilitating higher speeds. Fog pulls down the risk of security attacks. Industrial sector is revolutionized by ever changing technical advancements and IoT, which is a young discipline embraced by Industry thereby bringing in IIoT. Fog computing is viable to Industrial processes. IIoT is well supported by the middleware Fog computing as Industrial process requires most of the task performed locally and securely at end points with minimum delay .Fog, deployed for Industrial processes and entities which are part of Internet is gaining importance in recent times being titles as Fog for IIoT. Additionally, as Industrial big data is often ill structured, it can be polished before sending it to cloud resulting in an enhanced computing.

Chapter 6. Voice-Controlled Biped Walking Robot for Industrial Applications

Mechanical IoT is proceeding to extend from assembling and keen homes to retail, nourishment bundling, social insurance, and other use cases. For the present makers, modern IoT is quick turning into an unquestionable requirement have on creation lines for reasons of adequacy and hazard relief. Checking sensors must coordinate with generation line gear, including robots. Information streams and checking/announcing modules can give basic way data, trigger reaction conventions, and influence different exercises being followed on different modern systems. Modern IoT can expand perceivability into the states of robots in a production line, ranch, stockroom, or emergency clinic. Data about parts and gear can consistently observe to augment profitability and to decrease line personal time. Voice assistant robot takes the commands of the operator or mentor sets the functionality accordingly which is stored in the database and then execute the given task.

Chapter 7. Realizing a Low Latency M-CORD Model for Real-Time Traffic in Smart Cities

At present, the need for an ultra-high speed and efficient communication through mobile and wireless devices is gaining significant popularity. The users are expecting their network to offer real-time streaming without much latency. In turn, this will result in a considerable rise in network bandwidth utilization. As, the live streaming has to reach the end users mobile devices after traveling through the base station nodes, core network, routers, switches, and other equipment. Further, this will lead to a scenario of content latency and thereby causing the rejection of the mobile devices' users' request due to congestion of the network and mobile service providers' core network witnessing an extreme load. In order to overcome such problems in the contemporary 5G mobile networks, an architectural framework is essential, which offers instantaneous, ultra-low latency, high-bandwidth access to applications that are available at the network edge and also making the task processing in close proximity with the mobile device user.

Chapter 8. Traffic Analysis Using Internet of Things for Improving Secured Data Communication

Internet of Things can be simply referred to as Internet of entirety, which is the network of things enclosed with software, sensors, electronics, allows them to gather and transmits the data. Because of the various and progressively malevolent assaults on PC systems and frameworks, current security apparatuses are frequently insufficient to determine the issues identified with unlawful clients, unwavering quality, and to give vigorous system security. Late research has demonstrated that although system security has built up, a significant worry about an expansion in illicit interruptions is yet happening. Addressing security on every occasion or in every place is an important and sensitive matter for many users, businesses, governments, and enterprises. In this research work we are going to propose a secure IoT architecture for routing in a network. It mainly aims to locate the malicious users in a IoT routing protocols. The proposed mechanism is compared with the state of the art work and compared results shows the proposed work performs well.

Chapter 9. Challenges to Industrial Internet of Things Adoption

The Industrial Internet of Things (IIoT) has made its development within a short span of time. Initially it was considered as a novel idea and currently it is a major driver in Industry applications. It has created productivity and efficiency for industries worldwide. This innovative technology can become a practical reality if engineers overcome a variety of challenges. They are connectivity, cost, data integration, trust, privacy, device management, security, interoperability, collaboration, and integration. In this chapter, several facts behind the above-mentioned challenges are being explored and addressed. There are certain prime challenges in the adoption of IIoT. In the absence of a secure and properly encrypted network, the adoption of IIoT could lead to brand new security challenges and vulnerabilities. The prime concern of IIoT is to focus on connecting more and more devices together.

Chapter 10. Industrial Internet of Things: Benefit, Applications, and Challenges

This chapter focused on the importance and influence of Industrial Internet of Things (IIoT) and the way industries operate around the world and the value added for society by the Internet connected technologies. Industry 4.0 & Internet of Things (IoT) enabled systems where communication between products, systems and machinery are used to improve manufacturing efficiency. Human operators' intervention and interaction is significantly reduced by connecting machines and creating intelligent networks along the entire value chain that can communicate and control each other autonomously. The difference between IoT and IIoT is that where consumer IoT often focuses on convenience for individual consumers, Industrial IoT is strongly focused on improving the efficiency, safety, and productivity of operations with a focus on return on investment. The possibilities with IIoT is unlimited, for example: smarter and more efficient factories, greener energy generation, self-regulating buildings that optimize energy consumption, smart cities that can adjust traffic patterns to respond to congestion.

Chapter 11. Knowledge-Driven Autonomous Robotic Action Planning for Industry 4.0

Autonomous robots are being increasingly integrated into manufacturing, supply chain and retail industries due to the twin advantages of improved throughput and adaptivity. A fundamental characteristic required in Industry 4.0 deployments is the ability of autonomous robotic devices to self-configure in dynamic goal and deployment conditions. In order to handle complex Industry 4.0 tasks, the autonomous robots require robust action plans, that can self-adapt to runtime changes. A further requirement is efficient implementation of knowledge bases, that may be queried during planning and execution. In this paper, we propose RoboPlanner, a framework to generate action plans in autonomous robots. In RoboPlanner, we model the knowledge of world models, robotic capabilities and task templates using knowledge property graphs and graph databases. Design time queries and robotic perception are used to enable intelligent action planning. At runtime, integrity constraints on world model observations are used to update knowledge bases. We demonstrate these solutions on autonomous picker robots deployed in Industry 4.0 warehouses.

Chapter 12. Industrial Internet of Things: Foundations, Challenges, and Applications – A Review

In the era of mechanical digitalization, organizations are progressively putting resources into apparatuses and arrangements that permit their procedures, machines, workers, and even the products themselves, to be incorporated into a solitary coordinated system for information assortment, information examination, the assessment of organization advancement, and execution improvement. This chapter present a reference guide and review for propelling an Industry 4.0 venture from plan to execution, according to base on the economic and scientific policy of European parliament, applying increasingly effective creation forms, and accomplishing better profitability and economies of scale, may likewise bring about expanded financial manageability. This chapter present the contextual analysis of a few industry 4.0 applications. Authors give suggestions coordinating the progression of Industry 4.0. This section briefly portrays the advancement of IIoT 4.0. The change of ubiquitous computing through the internet of things has numerous difficulties related with it.

Chapter 13. Blockchain-Enabled Decentralized and Reliable Smart Industrial Internet of Things

The development of wireless communication in the information technological era, collecting data, and transfer it from unmanned systems or devices could be monitored by any application while it is online. Due to direct and aliveness of countless wireless device in a cluster of the medium could legitimate unwanted users to interrupt easily in an information flow. It would lead to data loss and security breach. Many traditional algorithms are effectively contributed to the support of cryptography-based encryption to ensure the user's data security. IoT devices with limited transmission power constraints must communicate with the base station the data collected from the zones would need optimal transmission power. There is a need for a machine learning-based algorithm or optimization algorithm to maximize data transfer in a secure and safe transmission. IIoT attacks are improving in all perspectives to discover the characteristic behavior of optimized cases and pre-analyzing the features for a back entry. Many more protocols and techniques are imposed on the diverse manufacturers to bring them into a centralized control.

Chapter 14. Digitalisation and Automation in the Agriculture Industry

The industrial production has experienced a technological revolution in the recent past decades. The technological revolution influenced the agriculture industry too. The important areas in the change are not limited to: innovation in farming, novel production of agriculture based tools and equipment, transportation and consumption of food across the globe, marketing the agriculture products, and Digitalization. Digitalization is the involvement of digital technology in the existing field for easing the mechanism of handling, processing, recording the data. Digitalization enables sustainable farming. It is required desperately to develop this technology because there is a substantial reduction of clean water and depletion of aquifers effects the cultivation. With the technology, the quantity and quality of the food has to be managed to feed the global population. The familiar digitization technology that makes the agri-industrial sector to experience growth are Artificial Intelligence, machine learning, sensor networks, Internet of things, Robotics, Cloud data.

Chapter 15. Internet of Things for High Performance Net Zero Energy Buildings

Due to globalization demand in per capita has increased over the decade; in turn standard of living has been increased. The emission of carbon dioxide is increasing exponentially in construction industries, which affects the global ecological system. These kinds of building trap the heat inside the building and hence the need of HVAC systems in the building, which causes carbon emission. To reduce the global warming potential, net zero energy buildings are very essential. With respect to technological advancements in Information technology, the Internet of Things (IoT) plays a vital part in net zero energy buildings. The building architecture is commercially based on their functional, aesthetic and luxury needs. In the commercialized world, the engineers and construction work force utilize maximum embodied energy with higher carbon emission. In this chapter, the various issues, and challenges of high-performance zero energy buildings are elaborated using different scenarios.

Chapter 16. Implication of Predictive Maintenance for Industrial Marketing: A Case Study

This chapter discusses the implication of predictive maintenance (PM) for industrial marketing companies. Artificial Intelligence (AI), Machine Learning (ML) and Internet of Things (IoT) will transform the business world. Industrial Internet of Things (IIoT) is a significant subset of the IoT led transformation and explores the industrial application of IoT. Today, there is no doubt about this transformation, often captured by the term Industry 4.0. Using an illustrative case study from the Indian industrial air compressor market, it shows that predictive maintenance solutions will change the way of conventional sales and marketing. Sellers need to focus on early innovation adopters among its customers. They also need to engage with existing customers early-on in the purchase process and highlight how PM can reduce the total cost of ownership. PM can be sold effectively to different types of customers- transactional, value-oriented and collaborative. Industrial marketers have to position the solution appropriately to gain competitive advantage

Chapter 17. Enhancing In-Service Tank Maintenance Through Industrial Internet of Things and Acoustic Emission Testing

In the area of tank inspections across the industry, robots where introduced to replace human inspectors in selected operations. The technological gap in adoption of similar technologies by Zimbabwe's bulk fuel storage tanks operators motivated this research. The industry's current NDT practices were investigated, costs and inconveniences were identified, and improvements were explored. Operators of bulk fuel facilities and companies providing tank inspection services were engaged to establish the reasons for the gaps in technological assimilation. Emerging global technologies that enable in-service inspections were identified and their applicability to Zimbabwe's bulk fuel facilities was investigated. A combination of crawler based ultrasonic thickness tests for tank shells, and acoustic emission in-service tank bottom testing was observed to be the most convenient and relevant in-service tank inspection method for Zimbabwe's bulk fuel storage tanks industry. Internet based remote connectivity and control was considered for data compilation, analysis, storage, and reporting

ACKNOWLEDGMENT

I would like to especially acknowledge my Fellow Editors (Dr. Avanija Jangaraj, Dr. Gurram Sunitha, Dr. Reddy Madhavi, and Dr. Bharath Bhushan) for all the work they did towards this book. Their initial involvement in proposing and getting this book off the ground is much appreciated as are their words of encouragement and support. We are proud to present the book on the *Innovations in the Industrial Internet of Things (IIoT) and Smart Factory*. We would like to thank all the reviewers that peer reviewed all the chapters in this book. We also would like to thank the admin and editorial support staff of IGI Global Publishers that have ably supported us in getting this issue to press and publication. I would like to thank Ms. Jan Travers for continuing with this edited book project even after it was listed for cancellation because of COVID-19. And finally, we would like to humbly thank all the authors that submitted their chapters to this book. Without your submission, your tireless efforts and contribution, we would not have this book.

For any new book, it takes a lot of time and effort in getting the Editorial Team together. Everyone on the Editorial Team, including the Editor-in-Chief is a volunteer and holds an honorary position. No one is paid. Getting people with expertise and specialist knowledge to volunteer is difficult, especially when they have their full-time jobs. Next was selecting the right people with appropriate skills and specialist expertise in different areas of Internet of Things, Industrial Internet of Things and Smart Factory to be part of the Review Team.

Every book and publisher have its own chapter acceptance, review, and publishing process. IGI Global uses an online editorial system for chapter submissions. Authors are able to submit their chapters directly through the e-Editorial Discovery system. The Editor-in-Chief then does his own review and selects reviewers based on their area of expertise and the research topic of the article. After one round of peer review by more than three reviewers, a number of revisions and reviews, a chapter and subsequently all the chapters are ready to be typeset and published.

I hope everyone will enjoy reading the chapters in this book. I hope it will inspire and encourage readers to start their own research on Industrial Internet of Things and Smart Factory. Once again, I congratulate everyone involved in the writing, review, editorial and publication of this book.

Any comments or questions can be emailed to sam.goundar@gmail.com

Sam Goundar
The University of the South Pacific, Fiji

REFERENCES

IDC. (2019). *The Growth in Connected IoT Devices Is Expected to Generate 79.4ZB of Data in 2025, New IDC Forecast.* https://www.idc.com/getdoc.jsp?containerId=prUS45213219

McLellan, C. (2019). *The Industrial Internet of Things: A guide to deployments, vendors and platforms. The Rise of Industrial IoT.* ZDNet. https://www.zdnet.com/article/the-industrial-internet-of-things-a-guide-to-deployments-vendors-and-platforms/

Ranger, S. (2020). *What is the IoT. Everything you need to know about Internet of Things right now.* ZDNet Special Feature. https://www.zdnet.com/article/what-is-the-internet-of-things-everything-you-need-to-know-about-the-iot-right-now/

Sjödin, D. R., Parida, V., Leksell, M., & Petrovic, A. (2018). Smart Factory Implementation and Process Innovation: A Preliminary Maturity Model for Leveraging Digitalization in Manufacturing. *Research Technology Management, 61*(5), 22–31. doi:10.1080/08956308.2018.1471277

Chapter 1

The Challenges, Technologies, and Role of Fog Computing in the Context of Industrial Internet of Things

Sasikala Chinthakunta
Srinivasa Ramanujan Institute of Technology, Anantapuramu, India

Shoba Bindu Chigarapalle
ⓘD https://orcid.org/0000-0002-3637-507X
Jawaharlala Nehru Technological University, Anantapur, India

Sudheer Kumar E.
ⓘD https://orcid.org/0000-0003-2752-0711
Jawaharlala Nehru Technological University, Anantapur, India

ABSTRACT

Typically, the analysis of the industrial big data is done at the cloud. If the technology of IIoT is relying on cloud, data from the billions of internet-connected devices are voluminous and demand to be processed within the cloud DCs. Most of the IoT infrastructures—smart driving and car parking systems, smart vehicular traffic management systems, and smart grids—are observed to demand low-latency, real-time services from the service providers. Since cloud includes data storage, processing, and computation only within DCs, huge data traffic generated from the IoT devices probably experience a network bottleneck, high service latency, and poor quality of service (QoS). Hence, the placement of an intermediary node that can perform tasks efficiently and effectively is an unavoidable requirement of IIoT. Fog can be such an intermediary node because of its ability and location to perform tasks at the premise of an industry in a timely manner. This chapter discusses challenges, need, and framework of fog computing, security issues, and solutions of fog computing for IIoT.

DOI: 10.4018/978-1-7998-3375-8.ch001

INTRODUCTION

The "Internet of Things (IoT) refers to embodiment of the continuous convergence between the physical aspects of human activities and its reflection on the data world " (Puliafito et al., 2019). The fast technological advancements have remodeled the industrial sector. They extended the industrial business to the automation of industrial processes by avoiding man power interaction in the industry. In the framework of IIoT, the operation of complex physical machines are interconnected with networked sensors and software applications. It is the technological enabler of significant improvements in the efficiency of modern industrial processes. It consist of sensor networks, machines, robots, actuators, machines, appliances and personnel. Here, the data acquired from the sensors and machines are analyzed to get the valuable data to run factory operations. Generally, industrial big data analysis is performed at the Cloud end.

Generally, Cloud involves Data-Centric Network's (DCNs), which provides resources for storage and computation to the clients. So every service request and demands of the customers are analyzed and processed within the remotely located DC's . However, with the increased number of devices associated with the internet and growing technological advances of the IoT, the data handled by the cloud DCs is significant.Both cloud computing and IoT have a gratuitous working relationship.The IoT produces large quantities of data while cloud computing provides a way for that data to reach its destination, thus it helps to make the work more effective (Xu, 2012).

Technological development in manufacturing refers to an advanced manufacturing model enabled by IoT,cloud computing, service-oriented technologies and virtualization that convert manufacturing resources into services that can be accessed and distributed extensively. Even though cloud computing offers several advantages for IoT, its approach normally disputes with

the framework of IIoT. Most often cloud DCs are remote, it may leads to undesirable latency on transmission when the network traffic is heavy. Another limitation of cloud computing is implicit dependency, also known as "vendor lock-in" from the book *Enterprise cloud computing for non-engineer*.

If the technology of IoT depends on cloud (Atlam et al., 2018)collected data from the trillions of devices in the network are enormous and the data should be processed within the cloud DCs for analysis or visualization.Because the IoT devices don't have enough storage,compute and networking resourcesand they are battery powered.Hence IoT uses powerful resources provided by the cloud for storage and computation(Ramli et al.,2019).

The heterogeneous devices on the IoT network may produces numerous data traffic, it leads to high service latency, network bottleneck and poor Quality.Because in the cloud environment computation,data storage and processing done within DCs,they are available remotely to the end users.

Moreover, in order to process many user requests, the DCs must be up and working 24x7 without fail, which eventually result in a huge amount of energy being used.The IIOT includes various sensor devicesd and machines, they produce the large volumes of data for analysis.The data may be time-dependent or sensitive. Therefore, to take some decisions very quickly the data should be processed locally rather than DCs. Machines in the IIoT scenario require a timely response, unwanted delays can result in severe catastrophic failures.

Bonomi et al. proposed the Fog Computing (FC) model as a way of expanding cloudbased technologies to the network edge, sharing processing, storage, and networking resources and services along the CloudtoThings continuum, closer to IoT devices' topological proximity.So, the placement of a special node called fog node as an intermediate node to perform tasks efficiently and effectively is a crucial need in IIoT. Due to its placement and ability to do tasks fog is an intermediary node to perform specific timely manner tasks at the premises of an industry.This chapter discusses about fog computing, layered structure of fog computing and key technologies for fog computing, Industrial Internet of Things and its components, role of fog computing in IIoT and its applications, challenges in IIoT, finally conclusion of the work.

LITERATURE SURVEY

Several researchers were published their work in the field of Fog Computing. This section briefly discusses some of the important works related to Fog Computing. Bonomi et al. talked about the importance of FC in the domain of IoT. Yi et al. focused on FC under "seven themes, namely, fog networking, quality of service (QoS), interfacing and programming model, computation offloading, accounting, billing and monitoring, provisioning and resource management, security and privacy". LM Vaquero et al. explained about the "concept of fog computing in terms of emerging trends and enabling technologies and in usage patterns. They also briefly discuss the challenges ahead". Yannuzzi et al. explained " some of the challenges in IoT scenarios and demonstrate that fog computing is a promising enabler for IoT applications". Dasterji et al. discusses "an overview of fog computing along with its characteristics. They introduce various applications that benefit from fog and present several challenges". More recently, Chiang *et al.* given a tutorial on fog computing. "They discuss at a very high level the differences between fog computing, edge computing, and cloud computing. They also present the advantages of fog computing and discuss the research challenges". Many studies have been published on fog computing in depth (Stojmenovic & Wen,2014)and (Chiang et al.,2017) also in the context of particular application domains, i.e., " vehicular Ad-hoc NETworks (VANETs) (Kai et al.,2016), Radio Access Networks (RAN) (Ku et al.,2017,Peng et al.,2016) and Internet of Things (Chiang et al.,2016)".

FOG COMPUTING

FC is a distributed computing paradigm between traditional cloud computing and various end devices with limited capabilities i.e. storing, computing and networking services (Mahmud et al.,2018). This type of computing suitable for IoT applications that are latency-sensitive. FC is an extension of the cloud but much relevant to the things that operates on IoTdata. As depicted in Fig 1, It works as an interface between DCs and end devices for getting networking, storage and processing resources closer to the end devices.Usually, those devices referred as "fog nodes". With a network link, those nodes can be deployed anywhere.Any device with access to the network, storage and processing will act as a fog node.For example, "industrial controllers, embedded servers, switches, routers and video surveillance cameras". Even though fog and cloud uses the same resources and similar methodologies and attributes such as virtualization and multi-tenancy, FC offers several advantages for IoT devices like scalability, Low latency, geographical and large-scale distribution and reduced operating costs.

Layered Architecture of Fog Computing

Fog computing is a computing paradigm, that brings the execution of some of the DCs operations at the "edge" of the network. The main objective of fog computing is to provide low and predictable latency

Figure 1. Fog Computing
(Alrawais et al.,2017)

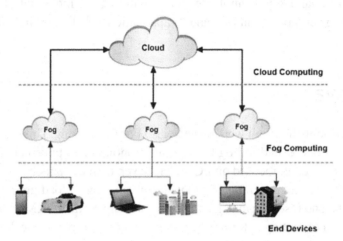

for time-dependent IoT applications(Shi et al.,2015). Figure 2, shows the layered architecture of Fog computing.The architecture of fog computing (Ni et al.,2017)includes six layers from bottom to top: "physical and virtualization layer, monitoring layer, pre-processing layer, temporary storage layer, security and transport layer". The layer of physical and virtualization consists of multiple nodes such as physical nodes,virtual nodes and virtual sensor networks. In this layer, nodes are maintained and managed depending on their service demands and types. The monitoring layer is responsible for monitoring the availability of fog nodes, resource utilization, and sensors nodes and network elements. Hence, through this layer it is possible to moniter the work performed by the particnode in the network. It is also able to know about at what time particular node is executed what task, and what will be needed from it next.

The responsibility for conducting data management activities rests with the pre-processing layer. Throughout this step, the analysis of the collected data, data filtering and trimming will be performed to extract the required information from the collected data.After the data is processed the datais temporarily stored in the temporary storage layer.Data encryption / decryption should be done in the protection layer. This layer can also uses integrity techniques to protect the data from tampering. The preprocessed data is submitted to the cloud in the transport layer, allowing the cloud to retrieve and provide more useful services for the end users (Aazam et al.,2018).

Key Technologies for Fog Computing

The key technologies to implement FC are deterministic virtualization and deterministic networking. Deterministic virtualization enables a fog node to perform functions with various levels of security, time, safety -criticality. For example, "a real-time control function can run on a real-time operating system side by side with a data analytics application on a standard operating system" (Steiner and Poledna,2016).On the other hand, to increase the special distance between physical process and the fog node that controls the process is enabled in deterministic communication. So, control functions are implemented remotely on the fog node.

Figure 2. The layered architecture of Fog Computing
(Aazam et al.,2018)

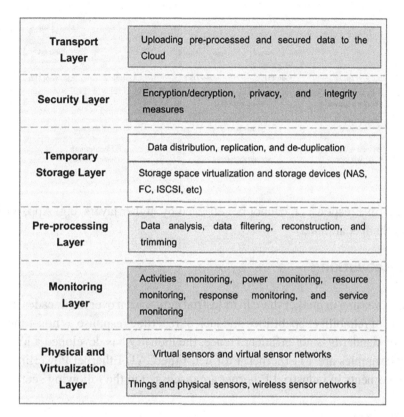

Deterministic Virtualization

Virtualization has been developed as a technique for sharing hardware among many applications, ensuring minimal interference between various applications on each other(Sandstrom et al.,2013) . Fig.3 shows the two types of virtualizations.Generally,Type 1 hypervisors is also known as "bare-metal" or "native" hypervisors. Its implementation allows one or more Operating Systems known as guest OS.Its implementation involves a software layer that is directly executed on the underlyinghardware. It also shares hardware resources, like CPU,memory among different guest OSs, by giving scheduled accessto the resources However, the major drawback with type 1 hypervisor is, it gives limited access to the hardware resources for each guest OS. Hence, in this scenario, each guestOS will not be aware of the existence of the hypervisor. Type 2 hypervisors also known as "hosted" hypervisors.They form a software layer located at the top of an operating system (OS). The structure of a system with type 2 hypervisor is depicted in Figure 3.

Even though Type 1 supervisors are essential for fog computing, type 2 hypervisors also suitable for some applications of FC. However, only the technique of virtualization cannot reduce the limitations of the underlying hardware layer which would depend on the quality of the nonfunctional properties. On the other hand, to know more about the non-functional properties of a layered system that involves virtualization, must know about each layer. For example, Unless the hardware layer provides the capabili-

Figure 3. System structure with and without virtualization
(Barham et al.,2003)

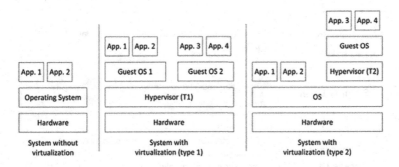

ties to ensure a real-time response, it cannot be retrieved by higher layers, e.g. App, with a comparable level of quality(Barham et al.,2003)

Deterministic Networking

Ethernet has been increasing in market share in industrial automation over the decades. However, because of the consideration of reliability capabilities and lack of real time response as non-functional properties the IEEE 802 set of standards is not efficient.Hence, the industry has developed a lot of real-time variants for Ethernet. Examples are TTEthernet, Profinet, EtherCAT, Ethernet Powerlink and Ethernet/IP. Some of the improvements in industrial Ethernet are discussed in the following sections.

Time Synchronization

The distributed systems includesIEEE has standardized protocols for the synchronization of local clocks in the nodes (e.g., fog nodes). IEEE 1588 was standardized in 2002 with a revision of standards published in 2008.These protocols dynamically elect a grandmaster clock in the distributed system which acts as a time master for other nodes. When the grandmaster is disconnected or fails, the protocols elect a new grandmaster in another way and time synchronization is re-stored.Both IEEE 1588 and IEEE 802.1AS are being updated. The newly added functionality will allow the co-existence of multiple operational grandmasters as well as their time distribution information on standalone network routes.

Schedule Traffic

One way of using synchronized time is to make broadcasting and forwarding decisions over the network. The basic functionality of time-triggered communication is standardized in IEEE 802.1Qbv. Here, the nodes will send the data in accordance with the repetitive communication schedule and the messages will be sent over the network as scheduled.As nodes and switches synchronously process their schedules, they ensure the minimal message queuing delays and guaranteed real-time transmission guarantees are required.

Frame Preemption

IEEE 802.1Qbu standardizes the ability to preempt ongoing frame transmissions. Usually,it is used in networks carrying a limited number of small-complex messages like alarms and long low-critical data messages. In this environment, queuing delays among high-complex messages on each other is acceptable, but queuing delays among high-complex and low-complex messaging may result in the loss of its real-time transmission delay. Therefore,with IEEE 802.1Qbu, the ongoing lower-critical message can preempted by the higher- critical message in the transmission. As a second advantage, when using scheduled traffic (IEEE 802.1Qbv) frame preemption may increase bandwidth utilization of low-complexity messages.

INDUSTRIAL INTERNET OF THINGS

IoT connects essential devices such as sensors in the infrastructure of industries and integrates the instances with current IoT applications.By deploying IIoT, both users and organizations get valuable insights in industrial processes.Therefore,they achieve high IIoT interconnects critical devices and sensors in industries' infrastructure and integrates scenarios with existing IoT applications. With the deployment of IIoT, organizations, as well as users, gain invaluable i nsights into industrial processes. So, they can obtain high productivity along with reliability with reduced cost. In IIoT, machine-to-machine interaction is provided through communication links are. They must satisfy the strict requirements regarding reliability.

Elements of Industrial Internet of Things

The IIoT implementation strategies depends on the components that are required by an IIoT. This section briefly explains the components of IIoT as follows.

Localization of Wireless Sensor and Actuator Networks(WSANs)

With the help of WSANs, different actions are executed depending on sensed data that needs highly reliable and quality data. To convert the data produced from sensors i.e. electrical signal into certain physical action Actuators are used. By using one or more actuators an actor acts on the field. Actor is one of the very important network entity that performs networking responsibilities such as receive,process, transmit and relay the data. In an IIoT environment, WSANs are controlled through over the internet through a remote application over the internet.Actors typically have much more resources compared to sensors such as longer battery life and higher data transmission rate. Nevertheless, the sensed data on the basis of which the actions performed must still be relevant by the time an action is initiated.The actors in IIoT may be Unmanned Aerial Vehicles (UAVs) they are usually known as drones. Furthermore, the sensor and actor environment in IIoT may be complex and hierarchical with the layout of master and slave nodes(Jazdi et al.,2014). A rich middleware can be feasible anchor node like a fog micro-datacenter. Fog can take local actions very fastly and it helps to achieve a cohesive interface with remote application with heterogeneous actors and sensors. In addition, WSANs have many open up privacy and securitychallenges, so there is a need for appropriate solutions for fog environment such as middleware in a strong IIoT environment.

Controlling and Managing Cyber-Physical Systems(CPS)

"A Cyber-Physical System (CPS) allows the networking oftraditional embedded systems and devices in the cyber world" (Gazis et al.,2015).It is a sub-set of IoT that interfaces machines and devices either directly or via remote application. A CPS allows control and remote access to embedded systems and devices.Therefore, it supports numerous flexible services such as remotely turning on the cooling/heating system which is critical to an industrial environment.

Figure 4. CPS Architecture
(Azam et al.,2016)

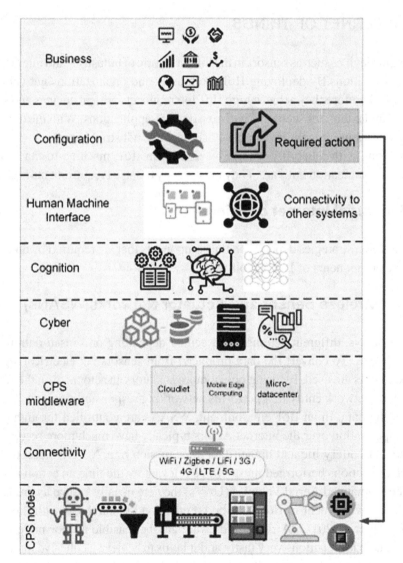

Figure 4 shows the overall layered architecture of CPS.A CPS provides the communications between machines, humans and products. CPS can play vital role in a cost-effective manner in the automation system of industrial process such as maintenance, control, diagnostics, and assistance.Further,business models can also be created using CPS. "A CPS has a control unit which is responsible for controlling sensors, machines, actuators and devices. The entities in a CPS communicate with the real world, gather the data, and process the data inorder to contribute to the industrial process. The CPS nodes requires a communication interface to share the data with several embedded systems, over-the Internet applications, and the cloud. The data sharing is the valuable function of a CPS,since the obtained data may be processed centrally. CPS interacts with machines, actuators and devises,it is extreamely crucial in the

Manufacturing processes processes and thus within the framework of the Industry 4.0 definition. Smart manufacturing, robotic surgery, smart grid, automotive manufacturing are all good examples of CPS based an industrial environment" (Azam et al.,2016).

Industrial Big Data Analytics

When an industrial system is automated and to a certain degree, they may be autonomous, large quantities of data will be involved.By deploying WSANs, WSNs, Virtual Sensor Networks (VSNs), linked machines, devices and appliances continuously generates large volumes of data. The data provide a means to build tailor-made services when the data analytics needed are applied(Salman et al.,2015).

An IIoT without extensive data processing on the sensed data would be incomplete. Information analytics helps in: predictive maintenance, improved fault tolerance, error avoidance, detection and cost-efficiency, etc.

Virtual Sensing and Virtual Sensor Networking(VSN)

A smart system requires multiple sensors to challenge the environment and products to satisfy the the requirements of the system.In this case the deployment of physical sensors is very costly. Virtual Sensors (VSs) are therefore the feasible solution that allows a nearby fog to easily provide the data. Construction of VSNs(Lee et al., 2016)Sensing devices and customisation of virtual sensing devices are used according to the requirements. For example, it is possible to configure VSs for industrial environments,different goods, manufacturing, etc. and machine learning techniques can be applied in the cloud. The program running on the user's mobile device would then provide the service provider with feedback. Different sensors may be needed within an agricultural area to monitor the development of various plants within various environments or maybe even various times of the day.Interactive sensing capabilities by fog can fulfill the standards for customisable and scalable sensornetworking. Through this way, cost savings can be made and more tailor-made programs can be offered.

Web of Things for the Industry

"When actuators and sensors integrated up with the applications and services available on the web, it is called a web ofthings (WoT) - a refinement of IoT" (Bonomi et al.,2012). Many services would be provided over the web in the IIoT environment.Furthermore,the interconnection of sensors, robots and devices needs integration with third-party web-services. For example, data from a recycling firm is integrated with the sensors and devices of a waste management company to build adaptable and enhanced services(Aazam et al.,2016). Therefore in IIoT, WoT would be a major element.

FOG WITH IIOT

This section describes the scope of the current IIoT with the support of FC. Fog is a middleware,that handles the resources, communication among nodes in the network and local processing of the data. The solution provided by an IIoT may have various entities like actuators, devices andsensors etc. Most of the sensors and devices are small in size and have minimal resources example, sensors embedded on the door. So, this type of devices is inefficient to execute large computations like context-awareness and data analytics. In addition to this, the devices are battery powered so energy consumption is also an important challenge. Hence, to do tasks like complex data analysis and management a middleware i.e. fog is useful for IIoT. Figure 5, shows the fog based architecture for IIoT. In the IIoT environment, the fog can act as an edge, cloudlet, an edge device, a nano-datacenter or a micro-datacenter(Tang et al.,2017).. Fog performs complex tasks on behalf of the devices. It also monitors the sensors energy consumption of each sensor and then changes the data generation frequency respectively. It also analyses and maintains other energy sources i.e. thermal, solar etc. Most of the service providers of IIoT possesses proprietary systems, they need better interoperability approaches with multi-protocol translation, functions and API's to solve issues occurred in interoperability. In IIoT, the major task of fog is to design short-range protocols to communicate with the sensor nodes in the network. Example (Sivashanmugam et al.,2003) if a sensor node is working using Bluetooth it may need to communicate to the IIoT node located at long distance in the network. In this environment, the publish/subscribe paradigm (Depuru et al.,2011) is used, here publishers are the information producer's and subscribers are the information receivers. It provides a better way to distribute information among multiple consumers and producers. In the IIoT environment, fog enables pub/sub service provisioning to enhance business processes.

Role of Fog in Industry

Fog computing nowadays plays a critical role in all fields of industry. The whole industry is divided into three sectors: i)extraction, such as mining ii) manufacturing, such as automobiles and iii) services, such as transportation. Each industry have the ability in becoming part of the IoT or the IIoT vision. This section presents some of the areas of industry where IoT can be implemented with the help of fog, to achieve the goals of future smart industries.

Mining

Mining is one of the primary industries that need data analysis. This also increases the scale of the mining sector with the increasing population. Mining entails some risks and is too costly too. Based on IBM surveys, every person needs about 3.11 million pounds of metals, fuel and minerals in his or her lifetime. By using sensors and technologies related to sensors, efficiency would be increased by avoiding excessive waste and costs. In addition, maintenance costs and system failures can be calculated accurately. Before the actual digging process starts, the collected data will save time and money. Therefore, autonomous drilling system or digging, self-driving cars using IIoT standards can be some of the examples of mining

Figure 5. Fog computing in IIoT
(Tang et al.,2017)

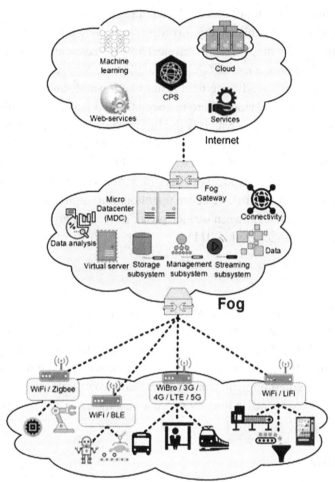

Industrial IoT – robots, actuators, machines, sensors, and devices

industry modernisation. Mining has many challenges and is one of the most dangerous forms of industry. The autonomous drilling method, driverless vehicles may also be some of the examples of modernizing the mining industry using IIoT standards. Mining is one of the toughest kinds of business with many dangers. In the case of mineral and coal mining for example, suffocation, rock sliding and other risks are typical. Besides, certain methods of mining can have dangerous chemical reactions and gas emissions.

Therefore, the use of sensor networks is very useful for capturing data and communication. Additionally, accuracy can also be improved with sensor networking and especially with FC, since extensive processing of data is applied through the co-existing fog. There are also significant concerns for the maintenance and energy efficiency for the mining industry, as it includes heavy machinery and requires a lot of time to manage the entire mining and collection process. With IIoT, mining can lead to better management of machinery and energy-efficiency.

Smart Grid and Power Industry

The smart grid is the latest electrical infrastructure that has grown continuously over the last ten years.

The smart grid comprises renewable energy infrastructure, energy-efficient smart meters (Farhangi, 2010) and smart appliances. In the conventional electric grid scheme, customers are provided with electricity services and billed once a month (Depur et al., 2011). Nowadays, however, the demands are very complex with the technological advancement of automation and autonomous lifestyle, with various electrical machines and appliances. Hence, there is a need for two-way relationship between a consumer and a manufacturer of electrical power, which is the fundamental concept behind the smart grid. "The power resource is distributed in a smart grid to local distribution companies (LDCs), which operate as a micro grid and provide the end-users with electricity" (Chekired et al., 2017). Since the whole concept of a smart grid is not confined to electrical suppliers only, telecommunications operators should also be interested in designing the smart grid. Telecom operators thus sign agreements with local electrical utilities to provide two-way contact between service providers and smart meters through the "Advanced Metering Network" (AMI) (Gungor et al .,2011).

Transportation

"Transport is a key industry and every country's backbone".The transport sector includes commercial public transit buses, metro and subway trains, private cars and cargoes. Intelligent Transportation Systems (ITS) is a transport-related subset of IoT, ITS and the Road Side Unit (RSU) may be fitted with a fog. For example, fog can allow the Internet of Vehicles (IoV), provide location-aware and context-aware services, support in-vehicle entertainment, smart parking and smart traffic lighting management on the basis of road conditions, detours, traffic load these are all examples of ITS allowed by fog computing.

CHALLENGES IN IIOT

To understand the scope of IIoT in the real world, we need to address several challenge, some of them are discussed in this section.

Dynamic Energy Consumption Management

Generally, the industries consume the largest amount of power, so they need to do power management dynamically. The consumption of energy varies based on the industry type even it may vary depending on the season. Thus, consumption of energy affects the lifetime of the network. Hence, energy consumption is a crucial factor in IIoT. It includes not only robotics but also have sensors and actuators. The transition of a large number of data packets continuously in IIoT leads to the consumption of a lot of energy in the network. The energy consumption also affects the time of synchronization. To address this issue, we need algorithms that can solve both time synchronization and energy consumption issues in the IIoT environment. Dynamic management of energy consumption is an important requirement of IIoT.

Interoperability Issues of Devices

Since IIoT involves several subsystems, devices, machines, and external systems, all of them are working together to perform a task. Therefore, integration of the sensors with the system and interoperability techniques becomes a challenge in IIoT.

Security and Privacy

With the progressive increment of the device connectivity to the network large volumes of data are producedwhich may be liable to misuse or theft because most of the industrial applications are deployed on external resources. Hence, IIoT data may be in the risk,that can affect the confidentiality, availability and integrity of the data. So, developing new techniques to ensure security objectives in the IIoT environment is a challenging task.

Fault Detection and Recovery

The IIoT systems highly automated and they have heterogeneous entities. Hence the chance of failures also increases in this environment such as delayed communication, device malfunction, and connectivity failures. An efficient IIoT system must have the capability to detect and rectify common faults in time. Therefore, efficient fault detection algorithms have to be employed at the gateway, middleware or hub or that coordinates different machines and devices.

Service Provisioning Based on Context-Aware

The ability to discover web-services based on the requirement is essential to support a dynamic environment in the industry.In the IIoT environment, the major objective of context awareness is to get contextual information without having existing contextual information or gather contextual information from an existing information. Both cases have different complexities and outcomes, hence context-awareness requires more intelligence and efficiency.

CONCLUSION

The technological advances in the industry introduce the automation of business processes. To provide the support for automation a middleware fog is used in IIoT. This chapter describes the structure of fog computing,IIoT and need of fog in the IIoT environment. Finally, it discusses some of the challenges such as dynamic power management, security and privacy issues, interoperability, and context awareness in the IIoT environment.

REFERENCES

Aazam, M., St-Hilaire, M., Lung, C. H., & Lambadaris, I. (2016, October). Cloud-based smart waste management for smart cities. In *2016 IEEE 21st International Workshop on Computer Aided Modelling and Design of Communication Links and Networks (CAMAD)* (pp. 188-193). IEEE. 10.1109/CAMAD.2016.7790356

Aazam, M., Zeadally, S., & Harras, K. A. (2018). Deploying fog computing in industrial internet of things and industry 4.0. *IEEE Transactions on Industrial Informatics, 14*(10), 4674–4682. doi:10.1109/TII.2018.2855198

Alrawais, A., Alhothaily, A., Hu, C., & Cheng, X. (2017). Fog computing for the internet of things: Security and privacy issues. *IEEE Internet Computing, 21*(2), 34–42. doi:10.1109/MIC.2017.37

Atlam, H. F., Walters, R. J., & Wills, G. B. (2018). Fog computing and the internet of things: a review. *Big Data and Cognitive Computing, 2*(2), 10.

Barham, P., Dragovic, B., Fraser, K., Hand, S., Harris, T., Ho, A., Neugebauer, R., Pratt, I., & Warfield, A. (2003). Xen and the art of virtualization. *Operating Systems Review, 37*(5), 164–177. doi:10.1145/1165389.945462

Bonomi, F., Milito, R., Zhu, J., & Addepalli, S. (2012, August). Fog computing and its role in the internet of things. In *Proceedings of the first edition of the MCC workshop on Mobile cloud computing* (pp. 13-16). 10.1145/2342509.2342513

Chekired, D. A., Khoukhi, L., & Mouftah, H. T. (2017). Decentralized cloud-SDN architecture in smart grid: A dynamic pricing model. *IEEE Transactions on Industrial Informatics, 14*(3), 1220–1231. doi:10.1109/TII.2017.2742147

Chiang, M., Ha, S., Risso, F., Zhang, T., & Chih-Lin, I. (2017). Clarifying fog computing and networking: 10 questions and answers. *IEEE Communications Magazine, 55*(4), 18–20. doi:10.1109/MCOM.2017.7901470

Dastjerdi, A. V., & Buyya, R. (2016). Fog computing: Helping the Internet of Things realize its potential. *Computer, 49*(8), 112–116. doi:10.1109/MC.2016.245

Depuru, S. S. S. R., Wang, L., Devabhaktuni, V., & Gudi, N. (2011, March). *Smart meters for power grid—Challenges, issues, advantages and status. In 2011 IEEE/PES Power Systems Conference and Exposition.* IEEE.

Farhangi, H. (2010). The path of the smart grid. *Power and Energy Magazine, 8*(1), 18-28.

Gazis, V., Leonardi, A., Mathioudakis, K., Sasloglou, K., Kikiras, P., & Sudhaakar, R. (2015, June). Components of fog computing in an industrial internet of things context. In *2015 12th Annual IEEE International Conference on Sensing, Communication, and Networking-Workshops (SECON Workshops)* (pp. 1-6). IEEE. 10.1109/SECONW.2015.7328144

Groom, F. M., & Jones, S. S. (Eds.). (2018). *Enterprise cloud computing for non-engineers.* CRC Press. doi:10.1201/9781351049221

Gungor, V. C., Sahin, D., Kocak, T., Ergut, S., Buccella, C., Cecati, C., & Hancke, G. P. (2011). Smart grid technologies: Communication technologies and standards. *IEEE Transactions on Industrial Informatics, 7*(4), 529–539. doi:10.1109/TII.2011.2166794

Jazdi, N. (2014, May). *Cyber physical systems in the context of Industry 4.0. In 2014 IEEE international conference on automation, quality and testing, robotics.* IEEE.

Kai, K., Cong, W., & Tao, L. (2016). Fog computing for vehicular ad-hoc networks: paradigms, scenarios, and issues. *The Journal of China Universities of Posts and Telecommunications, 23*(2), 56-96.

Lee, W., Nam, K., Roh, H. G., & Kim, S. H. (2016, January). A gateway based fog computing architecture for wireless sensors and actuator networks. In *2016 18th International Conference on Advanced Communication Technology (ICACT)* (pp. 210-213). IEEE.

Mahmud, R., Kotagiri, R., & Buyya, R. (2018). Fog computing: A taxonomy, survey and future directions. In *Internet of everything* (pp. 103–130). Springer. doi:10.1007/978-981-10-5861-5_5

Ni, J., Zhang, K., Lin, X., & Shen, X. S. (2017). Securing fog computing for internet of things applications: Challenges and solutions. *IEEE Communications Surveys and Tutorials, 20*(1), 601–628. doi:10.1109/COMST.2017.2762345

Peng, M., Yan, S., Zhang, K., & Wang, C. (2016). Fog-computing-based radio access networks: Issues and challenges. *IEEE Network, 30*(4), 46–53. doi:10.1109/MNET.2016.7513863

Puliafito, C., Mingozzi, E., Longo, F., Puliafito, A., & Rana, O. (2019). Fog computing for the internet of things: A Survey. *ACM Transactions on Internet Technology, 19*(2), 1–41. doi:10.1145/3301443

Ramli, M. R., Bhardwaj, S., & Kim, D. S. (2019). *Toward Reliable Fog Computing Architecture for Industrial Internet of Things.* Academic Press.

Salman, O., Elhajj, I., Kayssi, A., & Chehab, A. (2015, December). Edge computing enabling the Internet of Things. In *2015 IEEE 2nd World Forum on Internet of Things (WF-IoT)* (pp. 603-608). IEEE. 10.1109/WF-IoT.2015.7389122

Sandström, K., Vulgarakis, A., Lindgren, M., & Nolte, T. (2013): Virtualization technologies in embedded real-time systems. In *2013 IEEE 18th conference on emerging technologiesand factory automation (ETFA)* (pp. 1–8). Los Alamitos: IEEE Press. 10.1109/ETFA.2013.6648012

Shi, Y., Ding, G., Wang, H., Roman, H. E., & Lu, S. (2015, May). The fog computing service for healthcare. In *2015 2nd International Symposium on Future Information and Communication Technologies for Ubiquitous HealthCare (Ubi-HealthTech)* (pp. 1-5). IEEE. 10.1109/Ubi-HealthTech.2015.7203325

Sivashanmugam, K., Verma, K., Sheth, A. P., & Miller, J. (2003). *Adding semantics to web services standards.* Academic Press.

Steiner & Poledna.(2016). Fog computing as an enabler for the Industrial Internet of Things. *e &iElektrotechnik und Informationstechnik, 133*(7), 310-314.

Stojmenovic, I., & Wen, S. (2014, September). *The fog computing paradigm: Scenarios and security issues. In 2014 federated conference on computer science and information systems.* IEEE.

Tang, B., Chen, Z., Hefferman, G., Pei, S., Wei, T., He, H., & Yang, Q. (2017). Incorporating intelligence in fog computing for big data analysis in smart cities. *IEEE Transactions on Industrial Informatics, 13*(5), 2140–2150. doi:10.1109/TII.2017.2679740

Vaquero, L. M., & Rodero-Merino, L. (2014). Finding your way in the fog: Towards a comprehensive definition of fog computing. *Computer Communication Review, 44*(5), 27–32. doi:10.1145/2677046.2677052

Xu, X. (2012). From cloud computing to cloud manufacturing. *Robotics and Computer-integrated Manufacturing, 28*(1), 75–86. doi:10.1016/j.rcim.2011.07.002

Yannuzzi, M., Milito, R., Serral-Gracià, R., Montero, D., & Nemirovsky, M. (2014, December). Key ingredients in an IoT recipe: Fog Computing, Cloud computing, and more Fog Computing. In *2014 IEEE 19th International Workshop on Computer Aided Modeling and Design of Communication Links and Networks (CAMAD)* (pp. 325-329). IEEE.

Yi, S., Li, C., & Li, Q. (2015, June). A survey of fog computing: concepts, applications and issues. In *Proceedings of the 2015 workshop on mobile big data* (pp. 37-42). 10.1145/2757384.2757397

Chapter 2
Towards the Protection and Security in Fog Computing for Industrial Internet of Things

G. Rama Subba Reddy
Mother Theresa Institute of Engineering and Technology, India

K. Rangaswamy
Sai Rajeswari Institute of Technology, India

Malla Sudhakara
iD https://orcid.org/0000-0002-2559-4074
VIT University, India

Pole Anjaiah
Institute of Aeronautical Engineering, India

K. Reddy Madhavi
Sree Vidyanikethan Engineering College, India

ABSTRACT

Internet of things (IoT) has given a promising chance to construct amazing industrial frameworks and applications by utilizing wireless and sensor devices. To support IIoT benefits efficiently, fog computing is typically considered as one of the potential solutions. Be that as it may, IIoT services still experience issues such as high-latency and unreliable connections between cloud and terminals of IIoT. In addition to this, numerous security and privacy issues are raised and affect the users of the distributed computing environment. With an end goal to understand the improvement of IoT in industries, this chapter presents the current research of IoT along with the key enabling technologies. Further, the architecture and features of fog computing towards the fog-assisted IoT applications are presented. In addition to this, security and protection threats along with safety measures towards the IIoT applications are discussed.

DOI: 10.4018/978-1-7998-3375-8.ch002

Table 1. Statistics of Industrial IoT adaption rate

Industrial Adoption	IoT Adoption Rate
Industrial products	26%
Electronic and High tech	24%
Automotive	14%
Federal, aerospace, and defense	12%
Retail and consumer	9%
Medical gadgets	8%
Software	7%
Utilities/oil and gas	4%
Communications	2%
Other	4%

(Source: Statista)

INTRODUCTION

IoT is one of the trending technologies and is predicted to provide challenging results in operational transformations and the role of various industrial systems that are available like systems related to transport and manufacturing. For instance, when IoT is utilized to develop knowledge transportation systems, then transportation authority can keep track of the current location of every vehicle, monitors the motion of that vehicle and forecasts its future place and traffic strength. IoT, the term was first recommended for referring the connected objects that are uniquely recognizable interoperable by Radio-Frequency Identification (RFID) technology (Dedy Irawan et al., 2018). After this, researchers relate this IoT with new technologies, for example, sensors, mobile phones, and many other GPS enabled devices. Today, the straightforward definition of IoT is nothing but, a dynamic worldwide network infrastructure having self-configuration abilities depends on the standard and interoperable interaction protocols in which physical and virtual "objects" will have identities, physical attributes and virtual and so on are integrated into data network (Li et al., 2018). Mainly, the combination of sensors/actuators, labels of RFID and interactive technologies serves as a base of IoT and depicts how distinct objects and gadgets surrounded by are associated with the internet and accepts that objects and devices to assist and communicate with each other to satisfy the primary goals (Li et al., 2012).

Many industries prefer IoT technologies today. They have been conducted different industrial projects based on IoT, specifically in agro-domain, food processing firms, environmental monitoring, security surveillance and the other domains (Zhu et al., 2018). In IIoT (Industrial IoT), numerous devices create massive data that need processing. In the industry 4.0 aspect (Xu et al., 2018), a wide range of IIoT applications like manufacturing of smart meters needs processing of real-time data. Table 1 represents the adoption rate of IoT at the global level as of 2017, and the graphical representation of the corresponding statistics is shown in Figure 1. In the same year, the automotive industry was with 13% adoption rate.

In fact, for attaining the IIoT application requirements, cloud computing (M. Rudra Kumar et al., 2019) is considered a key enabler (Luvisotto et al., 2018). Nevertheless, still, cloud-based IIoT network is facing a few challenges that remained unsolved. Cloud data centres are always deployed remotely, which results in transmission latency that is not bearable. Additionally, surging data created by intel-

Figure 1. Adaption rate of Industrial IoT worldwide as of 2017

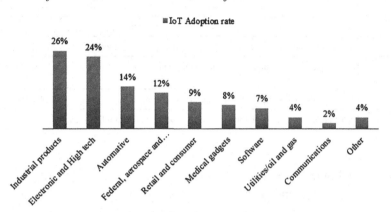

ligent services leads to a heavy burden on the server and also an issue with the network will have an error in the communication network. We observed that fog computing is a challenging solution to solve the earlier challenges in cloud-based IIoT (Iorga et al., 2018; Li, G et al., 2018). Fog Computing (FC) will process the workloads locally on fog nodes that are nearer to the terminals so that to minimize the latency that supports a recent IoT application and services that need less latency, mobility support, and geo-distribution.

As a nontrivial extension of cloud computing, it is not evitable that few of the issues will proceed to persist, particularly issues regarding security and privacy (Alcaraz et al., 2019). The deployment of FC is done by distinct fog service providers even though they are not trustworthy, and the devices are vulnerable to compromise. The fog nodes are confronted with different security and privacy issues (Zhou et al., 2018). Fog nodes can be considered as the proxies for the end-devices to do the operations with secure. In case the devices are lacking with resources (Banerjee et al., 2018). The motto of both the security and also privacy is complicated before the model and simulate on fog-based IoT applications. The literature about security and privacy challenges fog computing for IoT is still at an early phase. In this chapter, the researchers presented IoT technology and fundamental aspects of fog computing in brief, including fog computing in IIoT. Also, an in-depth analysis of Industry 4.0. Later security issues and privacy issues in fog computing with respective to IIoT are described.

Literature Review

IoT is treated as an extensive network infrastructure equipped with several connected components depends on sensors, communication, and networking and data processing technologies (Kamble et al., 2018). The RFID technology is said as the necessary IoT technology that permits microchips for transmitting the identification data to the user via wireless communication. Enhancements in both RFID and WSN will mainly contribute to IoT development. There exist mainly four layers in IoT technology which are interconnected together, each layer has unique functionality, and Table 2 represents a brief description of each layer. For instance, the network layer supports the primary transfer of the data between two devices using a wired or wireless connection, and the service layer is useful for providing various services, etc.

Besides, the rest of the other technologies and gadgets like barcodes, smart mobiles, social media networks, and cloud computing are included to develop a vast network to support IoT and these associated

Table 2. Major Layers in IoT

Layers	Description
Sensing layer	It is integrated with hardware (i.e. RFID) to sense the real world and to capture the information.
Networking layer	This layer provides interconnection support and essential transfer of data between two devices using wired or wireless medium.
Service layer	This layer provides numerous services to meet user expectations and played a key role among other layers.
Interface layer	This layer acts as a mediator between the user and other applications.

technologies are shown in Figure 2 (Pei Breivold et al., 2019). Also, IoT is attracting some industries like logistics, manufacturing retailing, and pharmaceutics. Due to the advances in wireless communication, smart mobiles and smart network technologies are included in IoT. As a result, IoT based technologies made a significant impact on new Information Communications Technology (ICT) and enterprise systems technologies (See Figure 3).

Figure 2. The Major Technologies associated with IoT

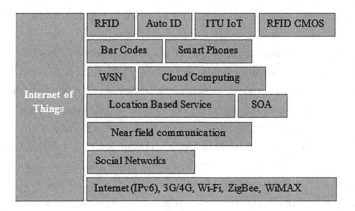

Connecting Devices to Cloud

IoT allows connected devices for gathering information and interact with each other. Cloud Computing (CC) (Rittinghouse et al., 2017) gives a real-time solution to solving the IoT application-related risks. Cloud-based IoT model has segmented under two layers, likely top and bottom layers. The top layer stores information and control layer where a cloud provides an efficient method to manage the services of IoT and deploy the applications of IoT through the exploitation of gadgets and the information gathered from those gadgets. Cloud bridges the interval among the objects and the applications and hides the complexity and the functionalities during implementation. The bottom layer has several IoT objects connected and a cloud. A combination of CC and IoT permits IoT devices with restricted resources for off-loading the data and hard computation on the cloud. It mainly relies on the location where a cloud DC exists where it covers. Such private clouds will rarely affect if not they cover the required areas. In

Figure 3. Impact of IoT-related technology.

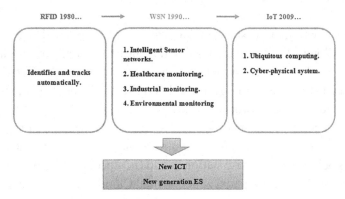

contrast, public clouds concentrate on giving worldwide coverage. Now, let's move on to fog computing context and benefit from it in contrast with cloud computing.

Fog Computing

Few nodes in a network will stretch out a cloud to hold nearer to the things that perform activities on IoT information. In many cases, instead of information retrieval from the cloud, the object can do the same from the nearest fog node, as shown in Figure 4. These nodes are known as fog nodes and executed anywhere using the web, for instance over a factory ground, close through railway track, in a vehicle etc. FC as engineering architectures gives non-functional information to empower IIoT. Fog applications are as assorted as the IoT. The essential thing here is that observing or examining the continuous information from IoT and after that proceeds with further tasks. The activities incorporate machine-to-machine correspondences or human-machine collaboration. Consider a few of the models like bolting an entryway, adjusting hardware settings, zooming a camcorder, opening a valve in light of a weight perusing, making a bar outline.

Principles

In 2012, Cisco introduced FC to clear the restrictions when Cloud DCs combined with IoT (Brennand et al., 2016). Under this section, we described the principles and advantages of FC.

Figure 4. Interaction with IoT devices using Fog nodes

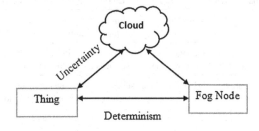

Closer to the Users

Along a Cloud-to-Thing Continuum As stated under Section 2, demerits with the inclusion of Cloud-IoT due to centralization of Cloud DC. FC is worth noting that the expression "towards a network edge" does not mean "only at a network edge," as Fog services can be distributed anywhere including continuum from Cloud to Things that are hosted on Fog Nodes (FNs) (Marin-Tordera et al., 2017). Any device that has enough computing, storage and networking abilities to provide the best services can be an FN (Open Fog Consortium, 2018). Therefore FNs may be: (i) resource-rich end devices (e.g., vehicles, smart traffic lights, video surveillance cameras, industrial controllers); (ii) advanced edge nodes (e.g., switches, gateways, Wi-Fi access points, cellular base stations); and (iii) specialized "core" network routers. Table 3 represents the merits of FC when compared with essential Cloud-IoT combination. It is noticeable that all these are critical benefits of FC (Satyanarayanan, 2017).

System-Level Paradigm

The Fog is a system-level paradigm means it "elaborates from the Things, via network edges, by a Cloud, and across many protocol layers—not just radio systems or specific protocol layer, nor a part of an end-to-end system, but a system spanning between the Things and the Cloud". Therefore, FC will speedups the systems development where an individual computer does not provide complete services with abundant resources. Rather than a service is typical gets decomposed and provided by a hierarchy of FNs such that each of them runs a specific portion of the overall service while cooperating with the other FNs. This pyramid-like a framework which is one of the guiding principles of the Open Fog Reference Architecture (OFRA), as discussed. However, as stated by the OFC in Reference (Open Fog Consortium, 2017), "computational and system hierarchy is not essential to complete Open Fog architectures, but it is still stated in many deployments."

The architecture of FC is divided into three layers similar to Cloud-Fog-Device framework and Fog-Device framework. The Cloud-Fog-Device framework composed of three different layers, such as a device layer, fog layer, and a cloud layer and is represented in Figure 5. The fog-device system has two layers, likely device layer and fog layer. These layers are sorted in ascending order of computing and storage

Table 3. Merits of FC compared with cloud-based IoT

Demerit of Cloud-IoT	Fog Solution
Latency	Fog nodes perform data analytics related to the data collected.
Bandwidth consumption	As some part of the information is used by the surrounded Fog nodes, a minimum quantity of data exchanged with Cloud DC. Also, Fog nodes behave as a medium between the Things and a Cloud, later optimizing data transmitted to a Cloud DC. Overall, FC supports to manage vast quantities of data efficiently by minimizing the consumption of the bandwidth (Dubey et al., 2015).
Privacy and security	Fog node analyzes the sensitive information that is locally stored rather than sending the same to a cloud via the internet. A fog can improve confidentiality and security in recent applications.
Context-awareness	Fog nodes are very closer to the IoT gadgets, increasing the context notion. Exploiting the contextual data allows the best services and reduces the usage of resources.
Hostile environments	FC proves as a base when the service must be available all the time, but the IoT gadgets are disconnected to a cloud. So instead of this, the nearby Fog node must provide the service to IoT devices so that they can connect.

capacity. The existing devices are brought their owners like wearable devices (like smartwatches, mobiles, vehicles and so on) (Ryoo et al., 2018). All devices are under the control of the same owner and can make a group and interact with each other through wireless ad hoc networks. The fog layer has network equipment, like routers, gateways, switches, bridges and base stations, augmented with the computational ability and local servers (e.g., industrial controllers, embedded servers, mobile phones and video surveillance cameras). Those gadgets are called fog nodes in fog computing and can be implemented at any place with the network connections in a smartphone, on a roadside unit and so on. In Fog-gadget system, nodes of it offer many services even in the absence of cloud servers, for instance, decentralized vehicular movement (Mezentsev et al., 2019), indoor floor sketch rebuild (Krieg et al., 2018).

Figure 5. Three-Layer Architecture of Fog Computing for IoT

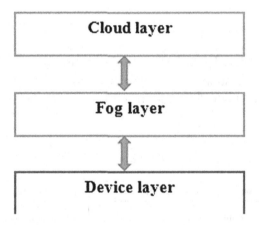

Working Procedure of Fog

Designers either port or compose IoT applications for FNs at a network edge. The fog nodes closer to a system edge utilize the information from the IoT gadgets. At that point, this is significant for the mist IoT application to guide particular sorts of information to the ideal spot for examination, and it is illustrated in Table 4.

1. Mostly private information is examined over a fog node closer to those things which are producing the information. For example, consider the Cisco Smart Grid distribution network, it is fundamental to confirm whether assurance and control circles are working deliberately or not? Subsequently. Closest fog nodes consistently keep track of the issues and attempt to overlook those by sending control directions to actuators.
2. Data that expends seconds or minutes to do an activity is gone through an aggregation node to break down. In a Smart Grid model, each substation may have its self - accumulation hub for showing the operational status of every downstream feeder and horizontal.

3. Minor sensitive information is sent to the cloud for ensuring authentication, big data analytics, and old storage. For instance, every fog node will send occasional outlines of lattice data to a cloud for the correct storage and legacy analysis.

WHAT HAPPENS IN THE FOG AND THE CLOUD?

Fog Nodes

1. Using a well-defined set of protocols, it receives the data from the IoT devices.
2. Execute IoT enabled applications in less duration in real-time.
3. The temporary storage will be in one to two hours.
4. These devices send updated data frequently to the cloud.

The Cloud Platform

1. Receives and adds the retrieved data from many fog nodes.
2. Explore the available information on multiple gadgets to meet the business objective.
3. Send the new job criterion to the fog nodes by observing business insights.

Table 4. Fog node extends the cloud to a network edge.

	Nearest Fog Nodes to IoT Devices	Fog Aggregation Nodes	Cloud
Response time	Milli seconds to seconds	Seconds to minutes	Either minutes, days or weeks
Application	M2M	visualization simple analytics	Big data and Visualization analytics.
Storage Duration	Transient	Less (days to weeks)	More (weeks to years)
Coverage	Local	wider	Global

Industry 4.0

Industrial production faced many revolutions due to its very beginning in the late 18th century. Initially, the method was driven by steam power like entrance. Then the industrial revolution involved in electrification and massive production, whereas the 3rd industrial revolution started in the 1960s with the digital programming of automation systems (Lasi et al., 2014). In today's, we are coping with the 4th industrial revolution, which is also called Industry 4.0, Smart Factory, or Smart Manufacturing. We are in the middle of a critical transformation, considering the way we do produce the products. This transition is known as Industry 4.0, which represents the 4th revolution incurred in manufacturing. Similar to the above mentioned three industrial revolutions which preceded it – steam power, mass production/electricity,

Figure 6. Evolution of Industry 4.0

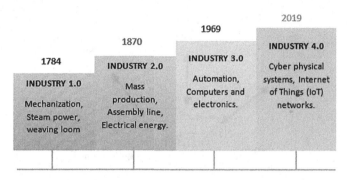

digital age – Industry 4.0 will transform the local and global economies and create a new future for all of us (Gunasekaran et al., 2019). The evolution of industry 4.0 from Industry 1.0 is shown in Figure 6.

Applications

Various industries are still not accepting Industry 4.0. How it influences their operations, many others are already initiated implementation and planned for the future so that smart machines can increase their business levels. The significant applications of Industry 4.0 are shown in Figure 7, i.e. robotics, IoT, 3D mapping, big data analytics, cloud computing and cybersecurity etc. Connected devices can gather a considerable quantity of information that represents the performance levels and most of the issues along with the data analysis to find the patterns which are not possible for a human. Industry 4.0 enables manufacturers to optimize quickly. From sensors collected information, an African gold mine detected issues with levels of O_2 while leaching. Industry 4.0 considers this aspect of completing a new level by IoT assistance and cyber-physical systems. Connecting the floor of the industry to IIoT software enables humans and machines to interact with each other. With Industry 4.0, the manufacturers can collect raw data from every section of the manufacturing process so that the decisions making is done with the quick and efficient manner; and automate and streamline processes, minimizing the waste and improves the productivity and Overall Equipment Effectiveness (OEE). Production can be started from starting to ending obtains that complete raw materials, as well as finished products, achieve the regulatory standards with high quality.

Security and Privacy Challenges in IIoT

IoT performs a crucial role in delivering the best services efficiently to the end-users. It also imposes challenges about security and privacy. Here we specified significant issues with security and privacy of the IoT environment.

Authentication

To keep the IoT devices securely, authentication is the essential requirement. But the drawback here is that there are insufficient storage space and CPU power to perform the cryptographic operations which are essential for an authentication protocol. Those resource-limited gadgets can outsource the cost-oriented

Figure 7. Applications of Industry 4.0

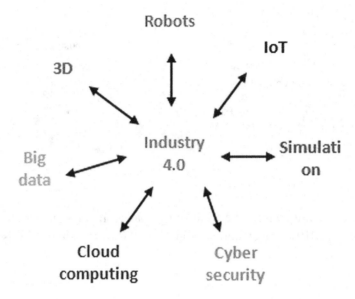

computations and space for fog devices that deploy authentication protocol. Yee Wei Law and colleagues advised a wide-area evaluation framework key management (WAKE) model for the smart grid. This model relies on public-key infrastructure (PKI) using multicast authentication for secure interactions.

Trust

IoT environment combines many gadgets and sensors related to various actuators. No mechanism is good at efficiency. When there is no trust measurement, the users of IoT services must notice whether it is useful in using some vital IoT services. Hence, building the trustworthiness among IoT devices is an essential aspect so that secure platforms can be made to preserve secured and reliable services of IoT.

Privacy

For the research team, violation of privacy regarding user's data is said to be an exciting aspect and grabs their attention towards the clearance of such issue. IoT devices with restricted resources are deficit with the ability to do encryption and decryption over generated information that is vulnerable. One more privacy issue is on the location that infers the location of IoT devices. Many IoT applications are location-related services, particularly the applications of mobile computing. The final privacy issue rises with the security of a user's utilization pattern of generated data. For example, smart meter data represents different usage patterns of IoT users like several persons residing at house, the time they turn Television, etc.

Access Control

Access control is a security approach to procure only authorized users to have access to particular resources, like an IoT device, or gathered data. In IoT, we should have access control to confirm that only authorized users can perform a given action like accessing IoT device information, requesting for an IoT device, or upgrading IoT device software. The IoT suggests new challenges in access control because we're dealing with a wide range of "devices" with restricted resources (that is, power and bandwidth.

Intrusion Detection

An intrusion detection technique identifies the malicious IoT devices and alerts the users in a network to perform relevant actions. Many current approaches in IoT will concentrate on some attacks with the least efficiency. An IoT platform makes it complex to find internal and external attacks. Besides the complex model of intrusion detection techniques meets the restricted resources in IoT is one in the challenging task. The primary challenge is designing and tuning a detection system that can work in large-scale, widely geo-distributed, and highly mobile environments.

Data Protection

The massive amount of data that was generated by the IoT is at rapid growth win increase in the connected components. This data is to be preserved not only at the communication level but also at the processing level. Because of limited resources, it is hard to do data processing over the devices of IoT, and therefore data can be sent to a cloud to perform further process and analyze as well. In such a case, data integrity must be preserved during the processing phase. Since IoT devices are unable to do encryption and decryption, it is hard to authenticate and data integrity.

Security Threats of Fog Computing

It is very vulnerable for cloud computing to get hacked by unauthorized users because of centralized data storage and computing framework. The key vendors of cloud computing are Google, Amazon and Yahoo are facing some privacy leakage issues. A major factor that restricts cloud computing development is Cloud security. Fog computing can be considered as a secured architecture when compared with

Figure 8. Security and Privacy threats in Fog Computing

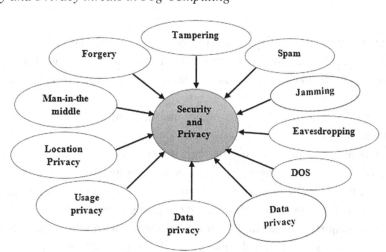

Cloud computing because of some reasons. One of them is, gathered data is maintained and analyzed transiently over nearer fog node surrounded by data sources

Local data storage, exchange, and analysis make it complicated for hackers to get access to users' data. Another reason is, exchanging data among gadgets and a cloud is no longer proceeds in real-time, so it is not so easy for hackers to discern the privacy data of the particular user (Atlam et al., 2018, Thareja et al., 2018). Mainly a hacker can perform the below activities (Roman R et al., 2018) to mislead fog computing. There are many kinds of security and privacy threats encountered in a fog computing network (Ni et al., 2017), among which few of them are illustrated here (See Figure 8).

- **Forgery:** Harmful intruders will generate fake details instead of stealing the data. Besides, network resources like bandwidth, storage, and energy will be highly consumed by fake details.
- **Tampering:** A tampering attacker can maliciously steal, delay or update the transmitting information for disrupting fog computing and desires its efficiency. It's hard to find some tampering behaviours, as wireless communication concerns and user mobility may lead to failure and delay during transmission.
- **Spam:** In short spam, data means unwanted information like redundant data, fake gathered information from the users that are generated by the intruders. Spam data results with the more consumption of network resources, misleading the people via social networks and violation of privacy.
- **Jamming:** An attacker can deliberately generate many fake messages to interrupt the communication channels or resources of computing so that other users are ignored from casual communication and also computation.
- **Eavesdropping:** Malicious attackers grasp the data via communication media to obtain transmitting data and also can retrieve. This network type attacks are more effective if the data is not encrypted.
- **Denial-of-Service:** An attacker disrupts the services provided by fog nodes so that to make them unavailable to its intended users. This attack involves the utilization of network resources to ignore the requests from unauthorized users from being fulfilled.
- **Man-in-the-Middle:** A malicious attacker stands between two parties to modify the exchanging information between these parties, but those parties will feel like they are only interacting with each other.

Privacy is another major problem in fog computing since the user's data is get transmitted, processed or shared. Data owners are not showing their willingness to un-hide their information to others. The privacy of a user includes four aspects, like identity privacy, data privacy, usage privacy, and location privacy.

- **Identity Privacy:** User identity includes name, mobile number, license number and public key certificate with which one can connect with a particular user.
- **Data Privacy:** When users are maintaining their information on the fog nodes, their information may be revealed to others, and transmission can be done between the two parties. From that data, sensible data can be retrieved like name, designation, address, and other vital information. For instance, a vote reveals the intention of a person regarding politics.
- **Usage Privacy:** It majorly represents the pattern utilization with which the user uses the services of fog nodes. Consider, smart meter data will reveal the daily activities of human-like when he/

she will go to bed, reading and other daily activities of human, which vividly violates the privacy of humans.

- **Location Privacy:** At present many mobile applications gather user's location data. It describes what we have to provide our details of location if we wish to enjoy the online services. Nevertheless, location details sharing are must in such cases so that our product will be delivered according to it.

CONCLUSION

In modern IIoT, a vast number of devices are generating a tremendous amount of information, and that required substantial processing for analysis. Also, with regards to industry 4.0, increasing more IIoT applications, for example, smart manufacturing, automation of industry required a robust data centre to process this real-time information. Nonetheless, the cloud-based IIoT system is as yet confronting some unsolved difficulties such as remote deployment and latency. Fog computing, on the other hand, is a promising answer for the address above difficulties by processing locally on the fog nodes closer to the terminals to minimize the latency issues. Since it is essential to understand the necessary background of IoT along with FC, in this chapter, the authors presented a brief idea about the IoT and FC along with its principles, working procedure, architecture in an IIoT environment. Then, the evolution of Industry 4.0 is presented. Further More, the security and privacy threats in fog computing are addressed, which includes security attacks and privacy mechanisms. These issues pose several challenges to overcome the problems encountered by numerous industries using IoT technology.

REFERENCES

Al-Turjman, F., & Alturjman, S. (2018). Context-sensitive access in industrial internet of things (IIoT) healthcare applications. *IEEE Transactions on Industrial Informatics*, *14*(6), 2736–2744. doi:10.1109/TII.2018.2808190

Alcaraz, C. (2019). *Security and Privacy Trends in the Industrial Internet of Things*. Springer. doi:10.1007/978-3-030-12330-7

Atlam, H. F., Walters, R. J., & Wills, G. B. (2018). Fog computing and the internet of things: a review. *Big Data and Cognitive Computing, 2*(2), 10.

Banerjee, U., Juvekar, C., Wright, A., & Chandrakasan, A. P. (2018, February). An energy-efficient reconfigurable DTLS cryptographic engine for End-to-End security in iot applications. In *2018 IEEE International Solid-State Circuits Conference-(ISSCC)* (pp. 42-44). IEEE. 10.1109/ISSCC.2018.8310174

Celso, A. R. L., Domingos, F., Antonio, A. F., & Leandro, A. (2016). Fox: A Traffic Management System Of Computer-based Vehicles Fog. In *2016 IEEE Symposium on Computers and Communication (ISCC)*. IEEE.

Da Xu, L., He, W., & Li, S. (2014). Internet of things in industries: A survey. *IEEE Transactions on Industrial Informatics*, *10*(4), 2233–2243. doi:10.1109/TII.2014.2300753

Dedy Irawan, J., Adriantantri, E., & Farid, A. (2018). Rfid and IoT for attendance monitoring system. In *MATEC Web of Conferences*. 10.1051/matecconf/201816401020

Dubey, H., Yang, J., Constant, N., Amiri, A. M., Yang, Q., & Makodiya, K. (2015, October). Fog data: Enhancing telehealth big data through fog computing. In *Proceedings of the ASE bigdata & socialinformatics 2015* (p. 14). ACM.

Gunasekaran, A., Subramanian, N., & Ngai, W. T. E. (2019). *Quality management in the 21st century enterprises: Research pathway towards Industry 4.0*. Academic Press.

Iorga, M., Feldman, L., Barton, R., Martin, M. J., Goren, N. S., & Mahmoudi, C. (2018). *Fog computing conceptual model* (No. Special Publication (NIST SP)-500-325).

Kamble, S. S., Gunasekaran, A., Parekh, H., & Joshi, S. (2019). Modeling the internet of things adoption barriers in food retail supply chains. *Journal of Retailing and Consumer Services*, *48*, 154–168. doi:10.1016/j.jretconser.2019.02.020

Kaur, K., Garg, S., Aujla, G. S., Kumar, N., Rodrigues, J. J., & Guizani, M. (2018). Edge computing in the industrial internet of things environment: Software-defined-networks-based edge-cloud interplay. *IEEE Communications Magazine*, *56*(2), 44–51. doi:10.1109/MCOM.2018.1700622

Krieg, J. G., Jakllari, G., & Beylot, A. L. (2018, May). InPReSS: INdoor Plan REconstruction Using the Smartphone's Five Senses. In *2018 IEEE International Conference on Communications (ICC)* (pp. 1-6). IEEE. 10.1109/ICC.2018.8422975

Lasi, H., Fettke, P., Kemper, H. G., Feld, T., & Hoffmann, M. (2014). Industry 4.0. *Business & Information Systems Engineering*, *6*(4), 239–242. doi:10.100712599-014-0334-4

Lavanya, R. (2019). Fog Computing and Its Role in the Internet of Things. In *Advancing Consumer-Centric Fog Computing Architectures* (pp. 63–71). IGI Global.

Li, G., Wu, J., Li, J., Wang, K., & Ye, T. (2018). Service popularity-based smart resources partitioning for fog computing-enabled industrial Internet of Things. *IEEE Transactions on Industrial Informatics*, *14*(10), 4702–4711. doi:10.1109/TII.2018.2845844

Li, S., Da Xu, L., & Zhao, S. (2018). 5G Internet of Things: A survey. *Journal of Industrial Information Integration*, *10*, 1–9. doi:10.1016/j.jii.2018.01.005

Li, Y., Hou, M., Liu, H., & Liu, Y. (2012). Towards a theoretical framework of strategic decision, supporting capability and information sharing under the context of Internet of Things. *Information Technology Management*, *13*(4), 205–216. doi:10.100710799-012-0121-1

Luvisotto, M., Tramarin, F., Vangelista, L., & Vitturi, S. (2018). On the use of LoRaWAN for indoor industrial IoT applications. *Wireless Communications and Mobile Computing*, *2018*, 2018. doi:10.1155/2018/3982646

Marín-Tordera, E., Masip-Bruin, X., García-Almiñana, J., Jukan, A., Ren, G. J., & Zhu, J. (2017). Do we all really know what a fog node is? Current trends towards an open definition. *Computer Communications*, *109*, 117–130. doi:10.1016/j.comcom.2017.05.013

Mezentsev, O., & Collin, J. (2019, April). Design and Performance of Wheel-mounted MEMS IMU for Vehicular Navigation. In *2019 IEEE International Symposium on Inertial Sensors and Systems (INERTIAL)* (pp. 1-4). IEEE. 10.1109/ISISS.2019.8739733

Ni, J., Zhang, K., Lin, X., & Shen, X. S. (2017). Securing fog computing for internet of things applications: Challenges and solutions. *IEEE Communications Surveys and Tutorials*, *20*(1), 601–628. doi:10.1109/COMST.2017.2762345

Open Fog Consortium. (2017). *OpenFog Reference Architecture for Fog Computing*. Retrieved from https://www. openfogconsortium.org/wp-content/uploads/OpenFog_Reference_Architecture_2_09_17-FINAL.pdf

Open Fog Consortium. (2018). *Top 10 Myths of Fog Computing*. Retrieved from https://www.openfog-consortium.org/ top-10-myths-of-fog-computing/

Pei Breivold, H. (2019). Towards factories of the future: Migration of industrial legacy automation systems in the cloud computing and Internet-of-things context. *Enterprise Information Systems*, 1–21.

Ramli, M. R., Bhardwaj, S., & Kim, D. S. (2019). *Toward Reliable Fog Computing Architecture for Industrial Internet of Things*. Academic Press.

Rittinghouse, J. W., & Ransome, J. F. (2017). *Cloud computing: implementation, management, and security*. CRC Press. doi:10.1201/9781439806814

Roman, R., Lopez, J., & Mambo, M. (2018). Mobile edge computing, Fog et al.: A survey and analysis of security threats and challenges. *Future Generation Computer Systems*, *78*, 680–698. doi:10.1016/j. future.2016.11.009

Rudra Kumar. (2019). Energy Efficient Scheduling of Cloud Data Center Servers. *International Journal of Innovative Technology and Exploring Engineering, 8*(11), 1769-1772.

Ryoo, I., Sun, K., Lee, J., & Kim, S. (2018). A 3-dimensional group management MAC scheme for mobile IoT devices in wireless sensor networks. *Journal of Ambient Intelligence and Humanized Computing*, *9*(4), 1223–1234. doi:10.100712652-017-0557-6

Sadeghi, A. R., Wachsmann, C., & Waidner, M. (2015, June). Security and privacy challenges in industrial internet of things. In *2015 52nd ACM/EDAC/IEEE Design Automation Conference (DAC)* (pp. 1-6). IEEE. 10.1145/2744769.2747942

Satyanarayanan, M. (2017). The emergence of edge computing. *Computer, 50*(1), 30–39. doi:10.1109/ MC.2017.9

Singh, D., Tripathi, G., & Jara, A. J. (2014, March). A survey of Internet-of-Things: Future vision, architecture, challenges and services. In 2014 IEEE world forum on Internet of Things (WF-IoT) (pp. 287-292). IEEE.

Thareja, C., & Singh, N. P. (2019). Role of Fog Computing in IoT-Based Applications. In *Emerging Research in Computing, Information, Communication and Applications* (pp. 99–112). Springer. doi:10.1007/978-981-13-5953-8_9

Whitmore, A., Agarwal, A., & Da Xu, L. (2015). The Internet of Things—A survey of topics and trends. *Information Systems Frontiers*, *17*(2), 261–274. doi:10.100710796-014-9489-2

Xu, L. D., Xu, E. L., & Li, L. (2018). Industry 4.0: State of the art and future trends. *International Journal of Production Research*, *56*(8), 2941–2962. doi:10.1080/00207543.2018.1444806

Zhou, W., Jia, Y., Peng, A., Zhang, Y., & Liu, P. (2018). The effect of iot new features on security and privacy: New threats, existing solutions, and challenges yet to be solved. *IEEE Internet of Things Journal*, *6*(2), 1606–1616. doi:10.1109/JIOT.2018.2847733

Zhu, C., Rodrigues, J. J., Leung, V. C., Shu, L., & Yang, L. T. (2018). Trust-based communication for the industrial Internet of Things. *IEEE Communications Magazine*, *56*(2), 16–22. doi:10.1109/MCOM.2018.1700592

Chapter 3
Advanced Predictive Analytics for Control of Industrial Automation Process

Sai Deepthi Bhogaraju
InkforTech, India

Korupalli V Rajesh Kumar
iD https://orcid.org/0000-0002-7989-1824
VIT Chennai, India

Anjaiah P.
Institute of Aeronautical Engineering, India

Jaffar Hussain Shaik
KSRM College of Engineering, India

Reddy Madhavi K.
Sree Vidyanikethan Engineering College, India

ABSTRACT

The recent evolution of the fourth industrial revolution is Industry 4.0, projecting the enhancement of the technology, development, and trends towards the smart processing of the automation in industries. The advancements in communication and connectivity are the major source for the Industrial IoT (IIoT). It collaborates all the industrial functional units to work under a single control channel, digital quantification analytic methods deployment for the prediction of machinery, sensors, monitoring systems, control systems, products, workers, managers, locations, suppliers, and customers. In addition to IIoT, AI methods are also playing a vital role in predictive modeling and analytic methods for the assessment, control, and development of rapid production, from the industries. Other side security issues are challenging the development, concerning all the factors digitalization processes of the industries need to move forward. This chapter focuses on IIoT core concepts, applications, and key challenges to enhance the industrial automation process.

DOI: 10.4018/978-1-7998-3375-8.ch003

INTRODUCTION

Internet of Things is a network that connects mechanical, digital, and computing machines with the least human-computer interaction. In 1982, a Coca Cola vending machine at Carnegie Mellon University was the first machine connected via the internet and it was designed to know if the cool drinks in the vending machine are cool without the need for a physical check. Later in 1990, an Internet controlled toaster was built by John Romkey which switches on and off the toaster automatically without human interaction. Father of IoT Kevin Ashton, the Executive Director of Auto-ID Labs at MIT first introduced the term IoT in the presentation designed for Procter & Gamble in the year 1999 (Shimanuki, 1999). He believed that Radio Frequency Identification (RFID) is the base of the internet and the devices that are connected via the internet can be tracked and managed from the computer. A constellation of 27 satellites created by the United States of America provides a highly stable communication system for IoT (Asenjo et al., 2014; Bravo et al., 2014; Carlsson et al., 2016; Ray, 2019). Internet Protocol address (IP address) assigns a label to every device that is connected to the Internet for communication, the introduction of IPv6 changed the course of the address allocation by assigning a 128-bit IP address to every device connected to the internet without any limit. In 2000's advancement in the IoT helped in developing smart devices like Smart Fridge (LG), Smart Homes, Google self-driving cars, Google Home, Amazon Echo, smart wearable technology, etc., (Breivold & Sandström, 2015; Kumar et al., 2020; Wang et al., 2015)

Figure 1. Development of Internet of Things- Timeline.

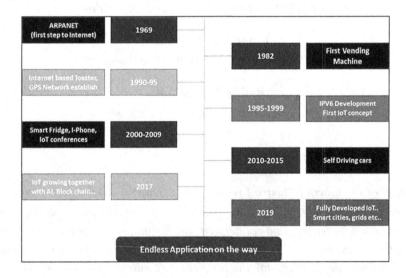

The IoT is mainly concentrated in the sectors Consumer, Industrial, Commercial, and Infrastructure varying from Smart Wearables to Smart Homes. Automation helps to fulfill the tasks with minimum human intervention. The third industrial revolution a Digital revolution began with introduction sensors, computers, robots to the industries in the 1970s (King & Mamdani, 1977). The industries are partially automated and are controlled using memory programmable controls and computers without human intervention. The current Industrial revolution Industry 4.0 replaced the third industrial revolution with the advancement in communication and connecting components of machinery, people, sensors, etc., via a wireless network (Chen et al., 2010; O'flaherty, 2005; Stenerson, 2002). Cyber-Physical production systems evolved in Industry 4.0 which enabled controlling and monitoring the production process by computer-based algorithms that helps the machines to learn from the experiences. Advanced technologies Cloud Computing, Industrial IoT, Predictive modeling, etc., overcame the challenges faced due to storage and networking. The data is exchanged between the components of machinery is collected and are applied to computer-based algorithms that use predictive modeling to track and predict the production and helped to increase the production (Bose, 2009; Waller & Fawcett, 2013)

The fourth industrial revolution, Industry 4.0 brought technological changes to production by connecting via a wireless network that connects people and machine components leading to Cyber-physical production systems that convert the factories as Smart factories. Smart Grids, Smart cities, digitally-enabled schools; Drones in military applications are some of the examples. The machines in the smart factories are integrated with technologies like artificial intelligence, Cloud Computing, and helps the machines These technologies removed the barriers of storage and communication, the data is exchanged between different components connected and helps in the prediction of the production. These algorithms use predictive modeling techniques that are designed by the statistic models helps to predict the data (Groover, 2016). For example, Vehicle insurance companies employed sage-based insurance solutions where predictive models utilize telemetry-based data to build a model of predictive risk for claim likelihood. By using the GPS and accelerometer readings one's black box predicts the risk factors. Advanced driving behavior can be predicted by considering crash records, road history, and user profiles (Jelali, 2012).

LITERATURE REVIEW

Internet of things processing with classical computing systems for the production, now dealing with megatrends like mobile computing, cloud computing, data analytics, big data is shown the Internet of things in the wide spectrum in industry requirements. Cloud-based services are greatly helping in the increment of the production with resource sharing, connectivity, time to time optimization features. Data analytics and predictive algorithms are effectively working to reduce the redundancy of the service maintenance of the types of machinery, production utilities, raw materials analysis, financial transactions assessment, sensors data, whatnot, whatever data generating within the industries region, analytics and its algorithms are greatly taking part to analyze with the pre- and post- predictive methods. These methods will help in the cost management, material maintenance, and early prediction of associate problems and provide reliable solutions (Bhogaraju & Korupalli, 2020; Eswari et al., 2015; Thramboulidis et al., 2011). When it comes to security challenges of the devices on connected platforms, a lot of data-driven safety measures are taking part, cyber-physical security systems development driving the internet of things indirectly. Recent studies have some significant results about cyber-physical attack reduction with techniques over IoT devices. Based on all the studies and relevant literature, the content on this book chapter well organized and discussed the new trends, architectures, Industrial automation, Applications of IoT concerning industries, key challenges and future spectrum etc., (Lam et al., 2010; McCue, 2014; Neumann, 2007).

INTERNET OF THINGS – INDUSTRIAL AUTOMATION DEVELOPMENT

The timeline of Internet of Things development is discussed in the introduction, in this section, we discuss the IoT - Industrial Automation Development, How devices establish a communication channel, how the Data Processing Unit (DPU) works and how the Remote Control Unit (RCU) operates in the phase of industries. Here, the focus is on the difference between Embedded Systems associated with Automation and how IoT overcome the automation process.

Figure 2. Embedded Systems – Automation – IoT Representation.

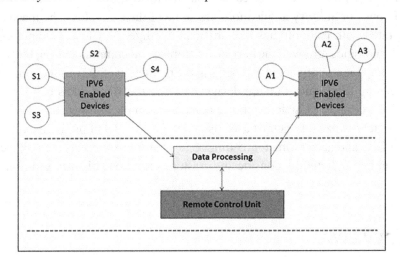

The embedded systems are an amalgam of software and hardware which helps in achieving a specific task. The embedded system does not have any networking capabilities whereas the IoT on the other hand mainly focuses on the connection of the devices via a huge network, transmitting and receiving the data is a crucial task that has more security challenges. For example, a conventional washing machine is an embedded system that is programmed such that it will wash the clothes whereas the IoT integrated washing machines not only wash clothes but can communicate via the internet and can be controlled externally by using the internet or Bluetooth connected to it. The data processing unit is a unit in the architecture that is used to process the data and includes the preprocessing techniques. The preprocessing techniques used to convert the raw data collected by different devices into a format that is understandable by the machine learning algorithms programmed in the controlling unit. The Remote Control Unit is used to control the devices connected to the internet from anywhere from the world by merely connecting with the internet, There are 2 ways of controlling the IoT connected device, either by programming the control unit or by directly programming the device itself.

Industrial IoT (IIoT) represents the smart machines and smart factories, implementation of various IoT applications, data-driven production systems, and various sensors that are integrated into the production machinery that increases the scope for efficient monitoring, production controlling of the production systems. Manufacturing companies that opt for smart factories experience a decrease in the total cost of ownership, it also increases the customer experience by smart service. For example, the drones used by Amazon for deliverables packing and transporting service efficiency and also improve customer experience. IIoT also paves a way for the flexible architecture that implements the changes very handy and efficient as it allows the implementation of the software update for hundreds of devices at the same time. IIoT enables data-driven production systems in which production on the data generated by various sensors, components of machinery. Application of Artificial Intelligent models, predictive analysis drives efficient quality assurance than the manual process and reduces the rejection rate which is one of the biggest challenges in the manufacturing industry. It helps in predicting the machine health, production rate, smart acquisition of data, etc., Cloud services provide computer monitored data-driven mechanisms with more efficient space storage, and better communication (Bedhief et al., 2019; Sengupta et al., 2020; Strauß et al., 2018).

IIoT ARCHITECTURE

As per the Industrial revolution 4.0 requirements, the IoT functional architecture deals with many wide spectral domains, each application deals with specified technology. In this concern, Fig.3 represents the functionality of IIoT w.r.t to technologies collaboration and architecture framework.

Based on applications, industrial development depends on various levels of technology adaption. When it comes to the automation process, embedded technology is the initial step of the development. The industrial region consists of sensors, actuators, and control units. Based on the industrial processing sequence, Sensors fixing at various levels happens, mainly temperature, pressure, thermal and inertial sensors are the key sensors in the industrial process, actuators act according to the sensor's outcome.

Figure 3. Functional Architecture of IIoT

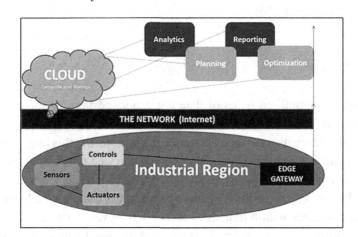

At the early stage of embedded development, according to sensors output, Actuators act to control the situations, with the support of control unit instructions. Now with IIoT techniques, Sensors sending information directly to the actuators for necessary actions, Control unit dependency reduced. But data processing happens in the Control Unit (CU), CU connected with the Cloud application for data storage and computational applications (Chehri & Jeon, 2019; Kan et al., 2018).

PREDICTIVE ANALYTICS

The predictive analytics falls under advanced analytics that takes historical data as input and forecasts the patterns and behavior of the data. Predictive analytics applies statistical, machine learning techniques to the data collected and predicts the likelihood of a particular event or its occurrence probability. There are four main steps involved in predictive analytics namely collection of the data, exploring the data, developing predictive models, installing and integrating into the hardware. Step one in the process is collecting the data. The data is collected from various sources which are termed as historical data i.e., the data from past events. The data collected is usually very noisy and unorganized .the step two in the predictive analytics cleans the untidy data i.e., removing unnecessary data, organizes the data, dividing the similar data, etc., there are various cleaning techniques practiced to clean the data efficiently. The third step consists of predictive modeling. The tidy data set now is applied to various algorithms to predict the data according to the requirement. The data applied to specific algorithms are tested and necessary changes are made to predict the data much accurately. The prediction algorithms are divided into 3 categories namely regression, tree-based, neural networks. The regression techniques are of two types depending on the linearity of the data.

The tree-based algorithms provide high accuracy and are efficient in predicting while the variables used are nonlinear. Random forest, decision trees fall under this category of algorithms. The third category is neural networks that imitate the human brain function. The neural networks help in determining the relationship between the data and are highly used in predicting the large sets of data. The neural networks are highly implemented in picture recognition, voice recognition, recommendation systems, etc., (Kan et al., 2018; Karati et al., 2018).

CLOUD COMPUTING

In this advanced world, 25 quintillion bytes of data are generated per day though the cost of storage hardware is reduced; the problem of storage and management of the huge amount of data is not resolved. Cloud computing resolves the challenges faced in managing and storing this huge amount of data. Cloud computing helps in storing and managing the data from the service provider's network. Multinational companies like Google, Microsoft, Amazon, IBM, etc., provide cloud services.

From the industrial prospect, the Cloud Computing concept is implemented to store industrial real-time data and to share resources within industrial sections. The vast data generated from the sensors, machinery, financial aspects, etc., are collected, stored, and managed with cloud computing. The computation of very large data also become handy with the introduction of cloud computing.

Cloud computing is mainly divided into 3 models namely Platform as a Service (PaaS), Infrastructure as a Service (IaaS), Software as a Service (SaaS). In the PaaS, a third party provides software and hardware resources to the user whereas SaaS is application related model that provides software applications to the end-users, the third model IaaS mainly provides the storage, server facilities to the end-user. All service providers provide resources over the internet. Major applications like Smart Vehicles (Internet

Figure 4. Cloud Computing – Framework.

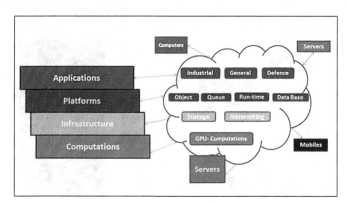

of Vehicles), Smart Industries, and Smart cities make use of Cloud resources for their own storing, maintaining, and processing of data. Cloud resources provide various resources to store and maintain the data, manipulate the data, provide the server for networking facilities. Dealing with High computational aspects, GPU based resource allocation is the main factor. Nowadays many GPU based clouds machines are useful in the simulation and analysis of high power computing applications. In addition to these functional factors, planning, and optimization factors are added advantage in the cloud resource, based on data utilization, generation, scaling, and optimization greatly utilizing the features of cloud platforms. We will discuss two major sectors of IIoT based manufacturing and its applications based on the functional aspects in the next sections (Aazam et al., 2018; Vitturi et al., 2019; Zhou et al., 2019).

DRONES MANUFACTURING INDUSTRY – IIOT APPLICATION

Unmanned aerial vehicle or Uncrewed aerial vehicle (UAV) commonly known as drones is one of the successful examples where the design and execution of the IIoT went live. Drones are used in various Military, Manufacturing, commercial and Service-based applications. Drones are aerial vehicles without pilots and are remotely controlled by humans or computers. The origins of Drones are dated around 1849 when Austria attacked Italy with hot air bombs which are attached to helium balloons. In 1918 during World War 1, the US army designed a radio-controlled aircraft named flying bomb which is unmanned. Later in 1935, a drone named Queen Bee flew unmanned with the servo-operated and radio controls in its back seat and conventional pilot seat in its front side which was rarely used. In 1946, radioactivity details are collecting after a nuclear war is collected using drones. In modern days drones are used in the US- Mexican border surveillance, transport and logistics applications of drones are introduced by Amazon to deliver the products, technological advances increased the scope of drone utilization spread to agricultural, networking, security, entertainment, photography, education, training sectors and so on (Hassanalian & Abdelkefi, 2017; Liu et al., 2015; Otto et al., 2018; Rosser Jr et al., 2018).

Drones are mainly divided into 2 types in which one is a rotor which includes single and multi-rotors such as ricopters, quadcopters, hex copters, and octocopters and the second type is fixed wings that include hybrid VTOL (Vertical takeoff and Landing) drones which do not require any runway for takeoff and Landing. The main components are propellers, brush fewer Motors, Camera, Motor mount, Land-

Figure 5. Unmanned Aerial Vehicle- Control Structure

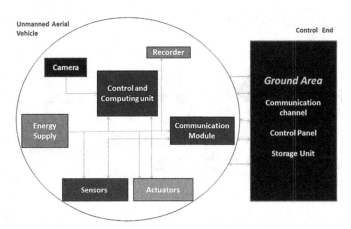

ing gear, Sensors, Electronic Speed Controllers, Battery, Antenna, Receiver, Flight Controller, Gimbal, Collision avoidance Sensors, GPS Module, etc.,

INDUSTRY SPECIFIC DRONES- APPLICATIONS AND KEY COMPONENTS

The purpose of the drones defines the design of the drones, though the basic components that are used are similar, configurations may vary according to the usage. For example, the drones used in the military served for security purposes where more control is required and also are integrated with protective systems whereas the drones used in the agriculture, industry monitoring requires more integrated sensors. The drones serve for variant applications and are shown below.

AGRICULTURE INDUSTRY – DRONES USAGE

Drones make the idea of aerial monitoring to come true a camera connected to the drone is used to monitor the growth of the plants. The camera of drones giving clear information about the crop, soil level, irrigation – water resources – water level, fungal infestations, and pest split conditions, climatic conditions, and all relevant information from the field. With this wide spectrum information, framers can freely observe the field remotely and take necessary measurements according to the current conditions. The drones also help to sprinkle the pesticides in the field with fewer human interventions and reduce the labor work. The productivity of the crop improves drastically as the problems of pests, water level, fertilizer requirements are detected on time and the proper care is taken immediately. The drones that are programmed with the advances recognition algorithms help farmers to detect the type of fertilizers or pesticides to be used by recognizing the cause of the problem from the roots. (Ahirwar et al., 2019; Daponte et al., 2019; De Rango et al., 2019; Puri et al., 2017)

Table 1. Drones and its Major Applications

Drones Applications - Domains	Usage	Processing
Agriculture Industry	Remote Monitoring of the fields,	Analytics and Analysis to find the crop level, water, soil moisture, Image processing based growth assessment, etc.
Defense Industry	Remote Monitoring of the Line of Control	Intelligence applications, Remotely Navigating opponent areas, LOC – issues. Carrying war missiles, bombers, targeting the areas, etc.
Service Industry	Servicing as a transporter for people's needs.	Drones are using as service transporters for delivering the people needs like groceries, food items, products, etc.
Transportation Industry	Future of Drones	While drones usage increased to many domains, in transportation also, there is a chance too.
Health Care Industry	Medicine transportation for Emergency cases.	Emergency medicines, organ transportation.

DEFENCE INDUSTRY – DRONES USAGE

Defense Purpose designed Drones having a high priority among all other drones. These specially designed for combat, navigation, and war-related applications and especially in defense usage drones utilized to monitor 24/7 across the line of sight. To the military, they are UAVs (Unmanned Aerial Vehicles) or RPAS (Remotely Piloted Aerial Systems). However, they are more commonly known as drones. Drones are used in situations where manned flight is considered too risky or difficult. The drones are used to monitor the ground troops which are far from the place of control. The drones help in detecting the tracks of the enemy attacks, the attacking devices attached to the drones help in attacking the enemy tracks, and helps in saving the lives of soldiers providing the video communication from tracks ahead and also paves a communication between the army bases and soldiers in the battle. They provide troops with a 24-hour "eye in the sky", seven days a week, 365 days in a year. In Defence regions, the information about the introduction of drones in the defense places is mentioned in the table given below. Drones

Figure 6. Agriculture Drone

Table 2. Drones Introduced in War Field – Country-wise list

United States	2001
Israel	2004
United Kingdom	2008
Iran	2010
U.A.E.	2011
North Korea	2012
South Africa	2013
China	2013
Spain	2015
Pakistan	2015
Nigeria	2015
Italy	2015
Iraq	2015
Georgia	2015
Egypt	2015
Ukraine	2016
Pakistan	2015
Iraq	2015
Ukraine	2016
Turkmenistan	2016
Turkey	2016
Switzerland	2016
Sweden	2016
Saudi Arabia	2016
Kazakhstan	2016
India	2016
France	2016
Azerbaijan	2016
Taiwan	2017
Poland	2017
Serbia	2018
Germany	2018
Belgium	2018
Belarus	2018

Source: https://www.newamerica.org/in-depth/world-of-drones/3-who-has-what-countries-armed-drones/

also help in identifying the enemy bases and movements of the enemy in hidden areas that are humanly inaccessible. They also help in identifying the injured soldiers and communicate with the army base to send the required troops for a medical emergency. When and which country used their first drones

Figure 7. Drones in War Field
Source: https://www.thedrive.com/the-war-zone/13284/americas-gaping-short-range-air-defense-gap-and-why-it-has-to-be-closed-immediately

in the war fields tabulated and shown in Table No.2. (Floreano & Wood, 2015; Iqbal, 2014; Martins & Küsters, 2019; Schmidt & Trenta, 2018; Sharkey, 2011)

SEVICE INDUSTRY – DRONES USAGE

The FAA - American Federal Aviation Administration has provided the rights to Giant American delivery service UPS to start drones as service agents, to carry medicals to hospitals on air. In this regard FAA granted UPS a part 135 Standard Certification that signifies, allowing the UPS to run drones in the night time and beyond the circumstances. Additionally UPS's Flight Forward Certificate to allow the company to fly an n- number of drones with remote operator's control. This increased UPS production to reach out to customer demands. Recently the usage of drones spread to various service-based companies to deliver the food items and so on (Besada et al., 2019; Koubâa et al., 2019).

HEALTH CARE INDUSTRY – IIoT APPLICATION – e -ICU CARE

Another major and course changing implementation of the IoT is in the Healthcare Industry. The diagram above demonstrates the implementation of IoT in the healthcare industry as "e-ICU Care". Critically ill patients who need to be monitored 24 hours a day highly beneficiated with e-ICU Care and can be monitored using Microphone, Video Cameras, Alarm System helps to monitor the patient anywhere in the world. A patient joined in the hospital is monitored every second precisely and any complications raised are communicated immediately to the respective duty doctors and nurses. The smart e-ICU helps in overcoming the problem of providing treatment in time for the patient with complications. The communication devices help intensivists, the care of the ICU to take action in time and avoid any further complications. he monitoring devices like Cardio meter, Ventilators, Incubators, ICU Bedside monitors, Saturation monitors, Syringe Pump, IV monitors, Dialysis machines, etc., are equipped with respective

Figure 8. Drones as service agents
Source: https://neweconomy.media/2019/10/01/ups-using-drones-to-get-medical-supplies-to-hospitals/

monitors to detect any sudden fluctuations in the heartbeat or oxygen levels or fluid level in the IV or any monitoring systems in the ICU triggers the alarm systems installed and communication is made with the intensivists and the caretakers to take immediate action. Smart monitoring can be implemented to the current e-ICU monitoring by allowing the monitoring systems to communicate with each other via the internet, the controlling system will be programmed with the predictive modeling algorithms (Amukele, 2019, 2020; Graboyes & Skorup, 2020).

The algorithm is first given inputs of the collected data ie., the data of various treated patients with different complications who joined the e-ICU, and the algorithms are programmed such that they will learn from further experiences. For example, a patient joined due to the multiple organ failure, the patient admitted has multiple sensors connected to monitor his vitals, if the patient has symptoms of a heart attack, cardiac monitoring devices triggers the alarm systems that communicated the patient's condition to the intensivists and also helps in control the oxygen level, the temperature of the ventilation unit, detect the movements of the patients to detect any sins of seizures, etc., . The implementation of smart monitoring also decreases the risk of critical patients admitted to the hospital and helps saving many

Figure 9. e-ICU Representation.

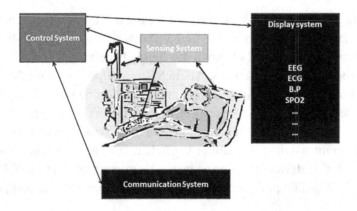

lives. The smart health care can be improved by the installation of robots for specific tasks during the absence of the intensivists like giving CPR, holding the legs and hands of the patients suffering from the seizures, detecting the number of IV fluids, etc.,

The above Figure. No 9, describes the implementation of the smart monitoring system that consists of the sensing, controlling, communication and display systems. The sensing system consists of various monitoring units that are integrating with the sensors to sense the condition of the patient in the e-ICU. The controlling system is programmed with algorithms that can predict and learn from past experiences. All the devices are connected via the internet helps for intercommunication of the devices for example: if the vitals of a patient fluctuate, then according to the programmed reasons the medications are injected in appropriate proportions and Bluetooth or any LAN devices connected to the patients and intensivists also helps in building the communication when in an emergency. The display units are connected via the internet in various places that are required and help in monitoring the patient's vitals.

CHALLENGES: INTERNET OF THINGS IMPLEMENTATION

Internet of Things (IoT) facing all key challenges from the Security issues for real-time implementation concerned.

- Device testing and Updating - Internet Issues.
- Authentication Problems
- Cyber-Physical Attacks
- IoT Malware and Ransomware
- Data security issues, cryptocurrency.
- Artificial Intelligence and Automation etc.

INDUSTRIAL IoT CHALLENGES

- Industrial maintenance – State-of-the-art technological adaption,
- Data maintenance
- Cyber Security
- Predictive Maintenance
- Aging workforce – lack of knowledge in recent advancements.

FUTURE SCOPE: IOT

Internet of Things and Predictive AI models will drive industries towards better shape by providing effective and efficient monitoring, management and control situations and scenarios. All industries will be beneficial moves towards the technological updating versions with industrial 4.0 and so on…

CONCLUSION

The application of IoT, predictive analytics, and AI helps the industries improve production quality and quantity, reduces the quality issues, detecting the issues on time, and improves communication. IIoT helps in connecting the machinery in the industry via the internet and the installation of various sensors helps in sensing various factors in the industry. The alarm system gets triggered by the actuators which take the outcome of the sensors and the communication is done with the controlling unit in case of emergency. Cloud computing overcomes the problem of storage, the collected sensors data is stored and the data is used as input to various predictive analytics models to predict future patterns and events which in turn helps in taking the right decisions at the right time. The overall efficiency of the industry will increase with the practice of IIoT. Various industries like agriculture, manufacturing, military, healthcare, etc., which practiced overcame many challenges faced due to the traditional methods. Though there are challenges while practicing IIoT, the results will maximize production and profits. Industries should adopt the latest trends and technologies to achieve higher goals and to overcome existing problems.

REFERENCES

Aazam, M., Zeadally, S., & Harras, K. A. (2018). Deploying fog computing in industrial internet of things and industry 4.0. *IEEE Transactions on Industrial Informatics*, *14*(10), 4674–4682. doi:10.1109/TII.2018.2855198

Ahirwar, S., Swarnkar, R., Bhukya, S., & Namwade, G. (2019). Application of drone in agriculture. *International Journal of Current Microbiology and Applied Sciences*, *8*(01), 2500–2505. doi:10.20546/ijcmas.2019.801.264

Amukele, T. (2019). Current state of drones in healthcare: Challenges and Opportunities. *The Journal of Applied Laboratory Medicine*, *4*(2), 296–298. doi:10.1373/jalm.2019.030106 PMID:31639681

Amukele, T. (2020). The economics of medical drones. *The Lancet. Global Health*, *8*(1), e22. doi:10.1016/S2214-109X(19)30494-2 PMID:31839132

Asenjo, J. L., Strohmenger, J., Nawalaniec, S. T., Hegrat, B. H., Harkulich, J. A., Korpela, J. L., Wright, J. R., Hessmer, R., Dyck, J., Hill, E. A., & ... (2014). *Industrial device and system attestation in a cloud platform*. Google Patents.

Bedhief, I., Foschini, L., Bellavista, P., Kassar, M., & Aguili, T. (2019). Toward Self-Adaptive Software Defined Fog Networking Architecture for IIoT and Industry 4.0. *2019 IEEE 24th International Workshop on Computer Aided Modeling and Design of Communication Links and Networks (CAMAD)*, 1–5.

Besada, J. A., Bernardos, A. M., Bergesio, L., Vaquero, D., Campaña, I., & Casar, J. R. (2019). Drones-as-a-service: A management architecture to provide mission planning, resource brokerage and operation support for fleets of drones. *2019 IEEE International Conference on Pervasive Computing and Communications Workshops (PerCom Workshops)*, 931–936. 10.1109/PERCOMW.2019.8730838

Bhogaraju, S. D., & Korupalli, V. R. K. (2020). Design of Smart Roads-A Vision on Indian Smart Infrastructure Development. *2020 International Conference on COMmunication Systems & NETworkS (COMSNETS)*, 773–778. 10.1109/COMSNETS48256.2020.9027404

Bose, R. (2009). Advanced analytics: Opportunities and challenges. *Industrial Management & Data Systems*, *109*(2), 155–172. doi:10.1108/02635570910930073

Bravo, C. E., Saputelli, L., Rivas, F., Pérez, A. G., Nickolaou, M., Zangl, G., De Guzmán, N., Mohaghegh, S. D., & Nunez, G. (2014). State of the art of artificial intelligence and predictive analytics in the E&P industry: A technology survey. *SPE Journal*, *19*(04), 547–563. doi:10.2118/150314-PA

Breivold, H. P., & Sandström, K. (2015). Internet of things for industrial automation—challenges and technical solutions. *2015 IEEE International Conference on Data Science and Data Intensive Systems*, 532–539. 10.1109/DSDIS.2015.11

Carlsson, C., Heikkilä, M., & Mezei, J. (2016). Fuzzy entropy used for predictive analytics. In *Fuzzy Logic in Its 50th Year* (pp. 187–209). Springer. doi:10.1007/978-3-319-31093-0_9

Chehri, A., & Jeon, G. (2019). The industrial internet of things: examining how the IIoT will improve the predictive maintenance. In *Innovation in Medicine and Healthcare Systems, and Multimedia* (pp. 517–527). Springer. doi:10.1007/978-981-13-8566-7_47

Chen, J., Cao, X., Cheng, P., Xiao, Y., & Sun, Y. (2010). Distributed collaborative control for industrial automation with wireless sensor and actuator networks. *IEEE Transactions on Industrial Electronics*, *57*(12), 4219–4230. doi:10.1109/TIE.2010.2043038

Daponte, P., De Vito, L., Glielmo, L., Iannelli, L., Liuzza, D., Picariello, F., & Silano, G. (2019). A review on the use of drones for precision agriculture. *IOP Conference Series: Earth and Environmental Science*, *275*(1), 12022. 10.1088/1755-1315/275/1/012022

De Rango, F., Potrino, G., Tropea, M., Santamaria, A. F., & Fazio, P. (2019). Scalable and ligthway bio-inspired coordination protocol for FANET in precision agriculture applications. *Computers & Electrical Engineering*, *74*, 305–318. doi:10.1016/j.compeleceng.2019.01.018

Eswari, T., Sampath, P., Lavanya, S., & ... (2015). Predictive methodology for diabetic data analysis in big data. *Procedia Computer Science*, *50*, 203–208. doi:10.1016/j.procs.2015.04.069

Floreano, D., & Wood, R. J. (2015). Science, technology and the future of small autonomous drones. *Nature*, *521*(7553), 460–466. doi:10.1038/nature14542 PMID:26017445

Graboyes, R. F., & Skorup, B. (2020). *Medical Drones in the United States and a Survey of Technical and Policy Challenges*. Mercatus Center Policy Brief. doi:10.2139srn.3565463

Groover, M. P. (2016). *Automation, production systems, and computer-integrated manufacturing*. Pearson Education India.

Hassanalian, M., & Abdelkefi, A. (2017). Classifications, applications, and design challenges of drones: A review. *Progress in Aerospace Sciences*, *91*, 99–131. doi:10.1016/j.paerosci.2017.04.003

Iqbal, K. (2014). Drones under UN scrutiny. *Defence Journal*, *17*(6), 68.

Jelali, M. (2012). *Control performance management in industrial automation: assessment, diagnosis and improvement of control loop performance*. Springer Science & Business Media.

Kan, C., Yang, H., & Kumara, S. (2018). Parallel computing and network analytics for fast Industrial Internet-of-Things (IIoT) machine information processing and condition monitoring. *Journal of Manufacturing Systems, 46*, 282–293. doi:10.1016/j.jmsy.2018.01.010

Karati, A., Islam, S. K. H., & Karuppiah, M. (2018). Provably secure and lightweight certificateless signature scheme for IIoT environments. *IEEE Transactions on Industrial Informatics, 14*(8), 3701–3711. doi:10.1109/TII.2018.2794991

King, P. J., & Mamdani, E. H. (1977). The application of fuzzy control systems to industrial processes. *Automatica, 13*(3), 235–242. doi:10.1016/0005-1098(77)90050-4

Koubâa, A., Qureshi, B., Sriti, M.-F., Allouch, A., Javed, Y., Alajlan, M., Cheikhrouhou, O., Khalgui, M., & Tovar, E. (2019). Dronemap planner: A service-oriented cloud-based management system for the internet-of-drones. *Ad Hoc Networks, 86*, 46–62. doi:10.1016/j.adhoc.2018.09.013

Kumar, K. V. R., Kumar, K. D., Poluru, R. K., Basha, S. M., & Reddy, M. P. K. (2020). Internet of things and fog computing applications in intelligent transportation systems. In *Architecture and Security Issues in Fog Computing Applications* (pp. 131–150). IGI Global. doi:10.4018/978-1-7998-0194-8.ch008

Lam, H. Y. K., Kim, P. M., Mok, J., Tonikian, R., Sidhu, S. S., Turk, B. E., Snyder, M., & Gerstein, M. B. (2010). MOTIPS: Automated motif analysis for predicting targets of modular protein domains. *BMC Bioinformatics, 11*(1), 243. doi:10.1186/1471-2105-11-243 PMID:20459839

Liu, Z., Li, Z., Liu, B., Fu, X., Raptis, I., & Ren, K. (2015). Rise of mini-drones: Applications and issues. *Proceedings of the 2015 Workshop on Privacy-Aware Mobile Computing*, 7–12. 10.1145/2757302.2757303

Martins, B. O., & Küsters, C. (2019). Hidden security: EU public research funds and the development of European drones. *Journal of Common Market Studies, 57*(2), 278–297. doi:10.1111/jcms.12787

McCue, C. (2014). *Data mining and predictive analysis: Intelligence gathering and crime analysis*. Butterworth-Heinemann.

Neumann, P. (2007). Communication in industrial automation—What is going on? *Control Engineering Practice, 15*(11), 1332–1347. doi:10.1016/j.conengprac.2006.10.004

O'flaherty, K. W. (2005). *Building predictive models within interactive business analysis processes*. Google Patents.

Otto, A., Agatz, N., Campbell, J., Golden, B., & Pesch, E. (2018). Optimization approaches for civil applications of unmanned aerial vehicles (UAVs) or aerial drones: A survey. *Networks, 72*(4), 411–458. doi:10.1002/net.21818

Puri, V., Nayyar, A., & Raja, L. (2017). Agriculture drones: A modern breakthrough in precision agriculture. *Journal of Statistics and Management Systems, 20*(4), 507–518. doi:10.1080/09720510.2017.1395171

Ray, P. D. (2019). *Pervasive, domain and situational-aware, adaptive, automated, and coordinated big data analysis, contextual learning and predictive control of business and operational risks and security*. Google Patents.

Rosser, J. C. Jr, Vignesh, V., Terwilliger, B. A., & Parker, B. C. (2018). Surgical and medical applications of drones: A comprehensive review. *JSLS: Journal of the Society of Laparoendoscopic Surgeons*, *22*(3), e2018.00018. doi:10.4293/JSLS.2018.00018 PMID:30356360

Schmidt, D. R., & Trenta, L. (2018). Changes in the law of self-defence? Drones, imminence, and international norm dynamics. *Journal on the Use of Force and International Law*, *5*(2), 201–245. doi:10.1080/20531702.2018.1496706

Sengupta, J., Ruj, S., & Das Bit, S. (2020). A Comprehensive survey on attacks, security issues and blockchain solutions for IoT and IIoT. *Journal of Network and Computer Applications*, *149*, 102481. doi:10.1016/j.jnca.2019.102481

Sharkey, N. (2011). Automating warfare: Lessons learned from the drones. *Journal of Library and Information Science*, *21*(2), 140. doi:10.5778/JLIS.2011.21.Sharkey.1

Shimanuki, Y. (1999). OLE for process control (OPC) for new industrial automation systems. *IEEE SMC'99 Conference Proceedings. 1999 IEEE International Conference on Systems, Man, and Cybernetics (Cat. No. 99CH37028)*, *6*, 1048–1050.

Stenerson, J. (2002). *Industrail Automation and Process Control*. Prentice Hall Professional Technical Reference.

Strauß, P., Schmitz, M., Wöstmann, R., & Deuse, J. (2018). Enabling of Predictive Maintenance in the Brownfield through Low-Cost Sensors, an IIoT-Architecture and Machine Learning. *2018 IEEE International Conference on Big Data (Big Data)*, 1474–1483. 10.1109/BigData.2018.8622076

Thramboulidis, K., & Frey, G. (2011). Towards a model-driven IEC 61131-based development process in industrial automation. *Journal of Software Engineering and Applications*, *4*(04), 217–226. doi:10.4236/jsea.2011.44024

Vitturi, S., Zunino, C., & Sauter, T. (2019). Industrial communication systems and their future challenges: Next-generation Ethernet, IIoT, and 5G. *Proceedings of the IEEE*, *107*(6), 944–961. doi:10.1109/JPROC.2019.2913443

Waller, M. A., & Fawcett, S. E. (2013). Data science, predictive analytics, and big data: A revolution that will transform supply chain design and management. *Journal of Business Logistics*, *34*(2), 77–84. doi:10.1111/jbl.12010

Wang, T., Gao, H., & Qiu, J. (2015). A combined adaptive neural network and nonlinear model predictive control for multirate networked industrial process control. *IEEE Transactions on Neural Networks and Learning Systems*, *27*(2), 416–425. doi:10.1109/TNNLS.2015.2411671 PMID:25898246

Zhou, L., Guo, H., & Deng, G. (2019). A fog computing based approach to DDoS mitigation in IIoT systems. *Computers & Security*, *85*, 51–62. doi:10.1016/j.cose.2019.04.017

Chapter 4
Internet of Things and Robotic Applications in the Industrial Automation Process

Seeja G.
VIT Chennai, India

Obulakonda Reddy R.
Institute of Aeronautical Engineering, India

Korupalli V. Rajesh Kumar
ⓘ https://orcid.org/0000-0002-7989-1824
VIT University, India

S. S. L. C. H. Mounika
Jawaharlal Nehru Technological University, Kakinada, India

Reddy Madhavi K.
Sree Vidyanikethan Engineering College, India

ABSTRACT

The recent industrial scenarios project its advancements and developments with the intervention of integrated technologies including internet of things (IoT), robotics, and artificial intelligence (AI) technologies. Industrial 4.0 revolutions have broken the barriers of all restricted industrial boundaries with the act of those interdisciplinary concepts and have taken a keen part in industrial development. Incorporation of these advancements considerably helps in improving product efficiency and in reducing the production cost. Based on categories of production, industrial automation processes may vary. In this regard, robots are playing a vital role to automate the production process at various levels of industrial operations. The combination of IoT, robotics, and AI technologies enhances the industrial productivity towards getting the success rate. This chapter focuses on how robotic technology with IoT and AI methods enhances the limitations of various industrial applications.

DOI: 10.4018/978-1-7998-3375-8.ch004

INTRODUCTION

The present technology is a mature technology where all the faults and problems are solved instantly even before its occurrence. In early days to resolve an issue, it depends on many factors, based on the technology, knowledge, skills, resource and time. Advancement bought all the working techniques into a single channel framework based on needs. Coming to Robotics and its development, past two, three decades back robots, exactly when robots are at the development stage, simple logic-based robots come as first version bots. Those robots are used to carry small things or track following robots etc. When technologies changes occurred, development of electronics, semiconductors, VLSI technology, embedded systems, and also computer designs, computer vision, simulation tools, machine learning and artificial intelligence all together enhanced the shape of Robotics as an interesting revolutionary concept. With all these integrated technologies, robots become much powerful than humans in some case studies, even in the industrial sectors, robots are replacing and working without human intervention. Internet of Things development is one of the causes of robots taking space in the growth of industrial productivity.

Now a day, the strength of robots significantly increased with materials, methods, controls and automation processes, within the fraction of time, robots can do the task neatly and cleanly. FANUC is a leading manufacturer of industrial robots; they design a robot beneficially and impressively to meet industry standards, mainly CNC machines having a high resolution of 1nm Cartesian resolution and 10-5 degrees of angular resolution. Some robots are designed in such a way that some features work dedicatedly like payload inertia, weight identification, collision detection, compliance control. Softwares are designed in such a way that it supports the networking and coordination control of two arms. Force feedback can be provided by using assembly tasks. For structured-lighting systems grasping and vision-guided fixing became a commonplace. By using the vision, robots will generally estimate and navigate the location of the spare parts for getting the exact trajectory operation type. The final cost of the end-effector tools remains a drastic fraction of the total cost of the work load. Typically the end-effector cost is often around 25% of the price of the industrial robot. Further, the value of the robot is generally about 40% of the total cost of the entire work load. The nature of the robotics has changed compared to the earlier developed robots. The nature of the robotics has changed compared to the previous robots. Robots are eliminating the mechanized transfer devices now as they are working together cooperatively but previously single robot work synchronously by handling the materials like conveyors. Robots can be seen closer to human workers, and human-robot is most likely to happen in the upcoming future. In unstructured and 3D environments, industrial robots cannot sense, control, and make decisions which are required to operate. For assembling a robot, the cost may vary based on the robot functionality, and that remains as a challenge. Till now, we are still lack of the fundamental theory and algorithms for the unstructured environments and the industrial robots. The service robotics is expecting profits in the recent advances in the mobility, algorithm advances so that will enable the robot to identify and localize the 2D map of the world and unknown 2D map of the world. Some robotics like vacuum cleaning robots which uses simple algorithms to map all the control inputs commands and cover the area by using 2D by avoiding obstacles. Security robots will be using all the sensory information to get their position as two-dimensional images that are then sent back to the source or the human operator. Even robotics is used to provide logistics services by transporting the materials. Remote-controlled robots are used to detect bombs and hazardous or unpleasant environments. The challenges for industrial robotics include all the problems in the service and personal robotics. Critical growth to the industry is by dexterous manipulation and integration of force and sense of the vision in support of the manipulation. Mobility of the robot is the key challenge as it needs to be moved when the action is required. The current generation robots are operated in both two dimensional and the indoor environment when there is a requirement. Service robots must have the capability of mobility, so it is mandatory to carry a power source with that. These service robots can be operated close to human users. While using the robots, safety is the primary constraint as it is operated under power. The industry needs to overcome significant challenges in human-robot interfaces. Figure 1 shows the background of robotics and its development ((M. Bahrin et al., 2016), (S. Jeschke et al., 2017), (H. P. Breivold et al., 2015), (H. Seraji et al., 1998)).

Figure 1. Robotics Background its Development Phase.

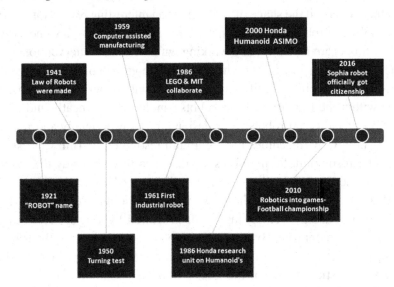

LITERATURE SURVEY

In first thing about the functionality and name "Robot" came into existence in the year 1921, after that next major stone in the robotics development phase happened in the year 1941, i.e. the law of Robotics – that is the rules and regulation passed to develop and maintain the Robotics. In enhancing the robotics power, first time turning test was conducted in 1950, later 1959 the first computer-assisted manufacturing unit started, next within a couple of years in 1961 first industrial robot developed and came into existence. 1986 the big move happened, i.e. LEGO and MIT collaboratively started research on Robotics, in the same year, Honda began to research unit to develop the Humanoid Robot, which to act like a human. After this move, a lot of developments happen, concerning technology, sensors, electronics, computers etc. Using all the advanced things in 2000, Honda presented the fully functional Humanoid Robot to the world and named it as ASIMO. After this phase, the shape of development extended like a superpower to reach high goals. In the year of 2010, robotics came into sports and football world cup tournament was also conducted for robots as players representing the countries like Japan, Israel etc. The significant achievement in the phase of Robotics – Humanoid Robots, in the year 2016-17, Sophia Robot officially got citizenship from the Saudi Arabia govt. The technologies have driven Robotics as a supernatural power. The Internet of Things (IoT) and Robotics are exponentially driving industrial automation concerning technical aspects like growth rate, supporting, information exchange, services, sensing, monitoring, navigating and productivity. With this enlargement of the industrial revolution, the combination of IoT and Robotics (IoRT) producing the better results in n numerous industrial sectors and this regard IoRT, claiming and actively functioning over the industrial automation. Behind the scene, robotics keenly works with lots of mechanics and sensors, Artificial Intelligence (AI) driven methods – predictive models, highly sophisticated embedded communication devices and networking technologies are taking part in the development of IoRT systems. This chapter mainly focuses on the background and development of Robots; IoT enabled Robots, Industrial automation with IoRT and applications of IoRT discussed ((E. Cerruto et al., 1995), (A. Pye, 2014),(M. Rüßmann et al., 2015), (K. Ashton, 2009), (L.E. Parker et al., 1998),(K. Thramboulidis et al., 2011)).

INTERNATIONAL VISION ON ROBOTICS – ANALYSIS

Industrial robots are mostly developed and manufactured in Europe and Japan due to high support and investments from the government and as well private sectors, but acceptance is found all over the world. When compared to world usage, around 40% industrial robots functioning in Japan, but the significant count is approx. 60% in 1990 and early days. This shows that industrial robots are spread widely in the world in all the leading regions in Asia. The usage and demand is increased in Japan and Europe, but there is some lag in the US. Both Korea and Taiwan are having the growth rates of above average. Compared to industrial robots, service robots are uniformly divided all over the world some small service robotics companies in the U.S. Namely Mobile Robotics, iRobot Corporation, and Evolution Robotics. In Europe, commercial products include some rehabilitation robots on pool cleaners, window cleaners, wheelchairs, tennis ball collectors and lawnmowers. Personal robots are developed by both japan and Korean for both entertainment and domestic assistants. Humanoid robots are noticed in both Japan and Korea, which are related to the internet in local companies. In this case, the US is lagging with no virtual products.

QUALITATIVE OBSERVATIONS

The difference between research and development programs in robots across continents is like the level of coordination between the government and industry. In both Japan and Korea, some concerted effort has been made to understand to develop and implement a national agenda and big picture. In Japan, one of the seven areas of emphasis includes creating industrial robotics. Robotics is being listed as one of the next generation growth engine in Korea. Humanoid project in japan is one of the examples of the national project, which involves government and industry. In Europe, with the goal of developing robotics industry, there are many projects across the continent which brings together the efforts of synergistic efforts. In Europe, the major new research initiative is driven by the joint academia/industry program. European commission for funding from 2007-2013 at the level of $100 million which was recently approved by the European Commission. There are no projects in the US for developing robots in the US. Japan, Korea and European countries have many strong professional associations for developing robotics compared to the US. Some of the national networks are present in Japan and European. Japan, Sweden and Italy have some big companies in robotics whereas the US has some small companies and start-ups. Mostly robotics is present in Europe and Asia. Human-robot interaction is significant, that leads to a lot of attention. In the US, the fundamental driver for robotics mostly comes from the Department of Defence and military programs interests. In Europe, Japan and Korea, drivers are primarily social and economic factors. Robotics is critical in the industry; in an ageing society, robots are identified as a crucial role.

INDUSTRIAL ROBOTS FOR MANUFACTURING SECTOR

Rapid changes in advanced technology and its development is enhancing the limitations of the Robotics and Automation process of the industrial product manufacturing sector. Industrial robots are being used to execute high demanded tasks, which are high precision and accurately more top quality tasks. The main feature or the ability of an industrial robot, is continuously working without taking a break compare to human capabilities, which increases the massive production output. Moreover, Robots can also work in high altitudes, dangerous conditions, extreme weather conditions and any harmful environments; this is the main motive to use robots in all applications, including industrial needs. In addition to this, industrial manufacturers encouraging the replacement of robots in place of human to reduce the salaries, Robots are like a one-time investment and working with a lot of man-hours comparing with humans and also inefficient and accurate which directly increasing the industrial profits ((H. Golnabi et al.,2007))

IoT ENABLED ROBOTICS – INDUSTRIAL AUTOMATION

As the case studies discuss various industries automation process using IoT enabled Robots or IoRT systems. Robots are the machines that are designed to do specific tasks automatically, which are programmed and are controlled by the external control device. The robot is programmed by the computer and is capable of doing recurring and complex tasks automatically with less human intervention. The robots are being used in different industries, from manufacturing to service-based, from agriculture to health care. Advanced technologies like artificial intelligence, cloud computing, fog computing, predictive analysis helped in achieving smart robots (the robots that can learn from past experiences and can predict the data using statistical methods). ((M. Wu et al., 2010), (F. Bonomi et al., 2012), (M.Wollschlaeger et al., 2017))

AUTOMOBILE INDUSTRY

Let's discuss the smart robots briefly with an example; one of the major industries that use robots in the car manufacturing sector. Robots are designed such that they can align and assemble parts of the engine and car exteriors, paint cars, testing, etc., and avoids a human intervention the dangerous tasks involved like welding, painting, installation. The robots are programmed such that they do the tasks mentioned above automatically and are controlled using a control unit. Smart robots change the traditional way of designing and controlling robots. The intelligent robots are designed such that they can communicate with each other via the internet and predict the data. One can control the entire industry from a place far away from the manufacturing unit via the internet. The robots are programmed with advanced machine learning algorithms such that they are used effectively in detecting the problems, predicting the machinery health, identifying defective parts of the cars, misalignments, etc., Smart robots are integrated with various sensors in order to monitor the temperature, pressure, gas leakage, speed of the manufacturing units, etc., the sensors integrated monitors and triggers the alarm system in case of any emergency beforehand and helps in avoiding devastating dangers. The communication system installed not only helps in controlling the robots but also helps in communication between the manufacturing units. For example, if there araised a problem in one part of the manufacturing the further manufacturing process can be automatically stopped and the robots adjacent to the damaged robots can help to examine the process and shuts down the robot to avoid further damage. The smart robots also help in detecting the damages or misalignments in the care assembly by using the cameras connected and picture recognition algorithms and can divide the defective units and increases efficiency. The data of the manufacturing is collected and is stored using the cloud technology that helps in tracking the manufacturing process and helps the manufacturer to identify the number of raw materials, number of defective units produced, ways of improvement, and improves the quality of the manufactured units and also helps in reducing the overall manufacturing cost.

The cybersecurity paves a way to securely control a large number of manufacturing units, smarts units and other machines connected and regulated in the same network. The smart robots keep track of the no of units produced; no units returned and predicted the number of units to be produced according to the market demand by using predictive analysis. Smart robots are also used in service-based applications by various industries. Amazon uses robots in the warehouse to effectively manage the deliverables. The robots are installed such that they are capable of driving the deliverables according to the product type, segregate the deliverables according to the locations, they are also used in food delivery. The robots are programmed to reach a specific location via GPS connected and are secured from the stealing of food ((C. Balaguer et al., 2008),(K. Kamei et al., K 2018), (K.V.R. Kumar, 2020))

Figure 2. Car manufacturing Robots
Source: http://roboticsandautomationnews.com/2019/10/15/robotics-poised-to-increase-productivity-and-reshore-manufac-turing-says-new-report/26295/

Additionally, while discussing the benefit for robot automation was apparent as car body assembly has become the predominant robot application. The car body assembly is very critical to handle as handling and positioning the metal sheets, transport of the body frames and spot welding is hazardous as physically demanding to the worker, as it is difficult to realize on fixed automation lines given the desired variety of car body configurations to be assembled on production line. In the stamping section, all the required metal sheets are cut into small pieces and ready for pressing into body panels. The robots will load the panels onto the tray, which helps themselves in fixing the panels for one robot to another spot welded. After that inspection happens to investigate whether the panels are correctly fixed or not after that robot will be moved to the paint section. For every robot, there will be some individual assembly units that will differ based upon the functionality such as axles, door, motors and transmissions, which will be pre-mounted in separate areas. Car body will be delivered to the body assemble unit to check whether right time and place on the assembly line. The climax part of the assembling process is that the engine should be mounted on the car. Today in industrial robots which can take the workload of a category of 100-300kg, which is the result of stemming from this application (J. Yun et al., 2018).

Figure 3. Automotive manufacturing - Robots
Source: https://kafkadesk.org/2018/11/09/slovakia-leads-central-and-eastern-europe-in-industrial-robotics/

HEALTH CARE RELATED ROBOTICS

Robots are involved in applications like medical surgery. The robot will have some characteristics like it can be interactive with the human and take some suggestions in the operating room. The main aim of the surgical robots is to help or provide the surgeon with some new set of versatile tools that may help in the treatment of the particular patient but not to replace the surgeon. These surgical robots are very cooperative that works with the surgeon by giving some ideas. For remote surgery operations, these surgical robots are being used. In this surgical robot, there are two types one among that is surgeon extender which will be directly operated or controlled by the surgeons available in the operation theatre. These surgeon extenders are used to do some extracurricular activities like superhuman activities which cannot be performed by the humans like the elimination of hand tremor or ability to perform skillful operations inside the patient's body so that the surgeon extender helps in reducing the operation time also. The second one is auxiliary surgical support these generally usually work with the surgeon side by side these control interface like joysticks, voice control, etc. One important thing is that the surgeon should be more attentive when surgical robots are in the operation theatre.

Figure 4. Robotics as Surgeons
Source: https://www.mayoclinic.org/tests-procedures/minimally-invasive-surgery/ovc-20256864

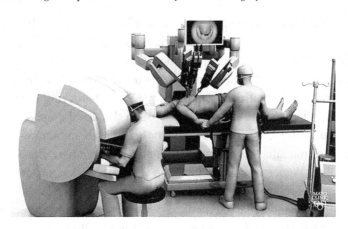

In the year 2000, the Food and Drug Administration has given rights and clearance to "da Vinci Surgical System" to perform Robotic surgery. Later this technique has become popular and spread across the global hospitals, mainly in the United States and Europe for the treatment of a wide range of applications as well as surgeries. (Z. Pang et al., 2018) This Clinical Robotic system consists of an inbuilt camera arm and mechanical arms with surgical instruments. The clinician or surgeon controls the arms from a control panel connected to a computer near the operating table within the chamber. To monitor the process, the inbuilt camera transmits the video to the high-definition, magnified, 3-D video monitor. Figure 3 shows the Robotics doing the operation at Miyo Clinic – model (G. Melich, et al., 2018).

ADVANTAGES OVER TRADITIONAL METHODS

The Robotic system helps the surgeons and clinicians in many ways; it enhances the precision, flexibility and controls the functionality during the operation and them to a better vision from the internal organisms compared to traditional methods. Based on advanced technologies, Robotic surgery claiming 99% accurate, even surgeons can perform complex operations with a high success rate with the help of Robotics.

Additionally, Invasive based surgery procedures have increased with Robotics. The key benefits of invasive based surgery:

- Fewer complications, such as surgical area infections,
- Less pain and blood loss,
- Quicker recovery,
- Smaller, less noticeable scars.

IoT ENABLED EMBEDDED MEDICAL DEVICES MANUFACTURING

Medical or Healthcare related research or product development industry rapidly increasing its outcome, day by day smart health technique is developing. IoT enabled Medical devices helps patients to connect to their doctors remotely for their treatment or diagnosis and also reducing the burden to the doctors or clinicians with advanced patients monitoring systems using Internet-enabled medical devices. Not only monitoring of the patient's conditions sometimes helps in identifying the diseases for the people who not able to reach the clinics, from distant locations. In the development of IoT enabled medical devices, wireless communications, VLSI - Nanotechnology-based miniature-sized devices, sensors, network providers and other manufacturing materials are taking keen part. Based on smart cities, smart vehicles, smart things, all the devices are going in a loop in the network connection. Due to this connection, healthcare also became a reliable and feasible source for IoT enable Healthcare. Many healthcare industries started investing in the IoT enabled medical devices manufacturing sector. In an aspect, in the manufacturing industry, Robotics based unit manufacturing is essential, to provide better results. (D.S. Elson et al., 2018) Will discuss who are the leading manufacturers of the medical robots, what are the functionalities in the Table 1.

Table 1. Global Medical-Robotics Manufacturers

Manufacturer	Country	Specialization
Intuitive Surgical	United States	This is the global leader in the field- Intuitive Surgical. Also famous for its " Da Vinci Surgical System" which uses the 3D HD vision system to function
Hansen Medical	Silicon Valley in California	They are the developer of medical robots called Magellan robotics system and Sensei X robotics to carry cardiac and vascular surgeries.
Medrobotics	Massachusetts	They manufactured "Flex Robotics System" to aid surgeons in their operations.
Verb Surgical	United States	Develops the digital surgery platform with an amalgamation of AI, Big data analytics, Visualization and instrumentation and with the collaboration of Google, working to bring forward the digital surgeries.
Microbot Medical	United States	They work on micro-robotics medical technologies, and also specialize in researching, designing and developing new technologies for surgeries.
Titan Medical	Toronto, Canada	It is developing the SPORT Surgical systems to cover areas that have not been touched like abdominal, gynecological and urologic.
CyberKnife System	United States	It is a fully robotic radiation delivery system. It works on treating cancerous and non-cancerous tumours. It is designed to treat a tumor by sending radiation at the right point and hence maintains accuracy.

Source: https://www.analyticsinsight.net/top-7-healthcare-robotics-companies/

ROBOTICS IN ROBOTS MANUFACTURING

"Robots making robot" is one of the smart factories whose initiative is taken and launched by Geek+. This initiative results in robot arms will be manufacture mobile robots which is nothing but Geek+ is a supplier of robots and uses AI technologies for warehouses. This Geek+ robot uses AI algorithms and many other automated solutions to manufacture the robots. They are used by the Nanjing factory. This factory produces and makes all the company robots. The process is very flexible for the production as robots are manufactured by robots which is the essential part and the demand is increased in such a way that customization of the products are limited. The parts are getting shorter as the robot manufacture is increasing day by day. As the robots are manufactured by the robot this supply chain is increased day by day as this is the best process to get the robot manufacture and flexible production and help the companies to get the smart robot in a concise time. By using this smart solution for manufacturing the robots, this has become a blueprint for intelligent manufacturing and flexible production used by the Nanjing facility.

Worldwide, many industrial solutions and a solution to meet production this is fulfilled by Geek+ as it can adapt and implement template of smart factory and customize the robots. Data knowledge which develops the smart logic solutions and experience is gained by Geek+ for warehousing and environment manufacturing as it has established 200+ projects across the world. Geek+ has taken the initiative to develop from retail and apparel to pharmaceutical companies. Geek+ has built an eco-system by using international technology partners to develop a smart warehouses and smart factories which use AI technology, robot arms to develop a mobile robot, IoT, management system like logistics and production, advanced robotics and big data analysis. This smart solution helped many companies or industries to upgrade their operations to an intelligent and agile supply chain. ((G. Riva et al., 2019), (S. R. Fletcher et al., 2019))

Figure 5. Robotics Manufacturing Robots
Source: https://roboticsandautomationnews.com/2019/10/18/geek-launches-smart-factory-with-robots-making-robots/26352/

ROBOTS MAKING ROBOTS

By using Geek+ technique, Nanjing factory production has exceeded 10,000 robots in a single-shift annual production with an output of almost manual production capacity of doubled tradition. Robots operate together under a logistics management system like production. For automated production, that includes robot arms, smart camera, automatic tightening machine and PLC. In Geek+, it introduces a moving system where the robots use QR code navigation to carry their shelves foe their inventory management; this process is followed by P800 robots. The robot which also uses QR code navigation and SLAM navigation too for allowing flexible distribution in automated production station to load, store and unload WIP's this process is followed by M1000 robots. It also introduces Forklift system, to access the entire pallet without ground navigation by using SLAM navigation this process is done by unnamed forklifts.

AI technologies are used by robots which include:

Vision Technology

The industrial vision system is equipped in the robot arm station as this will help in manufacturing mobile robot. The process of positioning and assembling function is required for the production of a mobile robot to achieve the more flexible automated process robot arms can identify the process of repairing the product and remaining auxiliary materials and current work in progress.

Scheduling

Algorithms are scheduled in such a way that robot can figure and identify the best path to communicate and move around the mobile robot.

Business Related Issues

An intelligent operating system for the innovative and production area is being configured in such a way that it can understand and implement all the required business process like bottleneck process optimization, shelf heat management and work in process management. Once the process of manufacturing mobile robot is completed those will automatically navigate them to the calibration area to get the necessary parameter settings. The final testing of the mobile robot will be done after the basic parameters received and they will navigate to finished product area for the final product inspection. They will be navigated to the packaging area and will be ready to ship. Geek+ smart factory management system is powered by AI to operate all the smart factories, which has the production logistics management system with a new integrated system developed. The production area is connected with the stock area, and the management unifies different robotic solutions. All the aspects are provided with the proper facility from the inventory process to the production line, for a flexible and efficient system integrating logistics and production is implemented. This replaces the traditional conveyor belt system with a new model for the automated mobile robots, namely island production mode. This island production mode is entirely flexible and scalable, which can be easily duplicated. The single production line can be replaced with multiple products which can be produced in one production line. By providing an effortless solution to the bottleneck process, the whole process can be adjusted in one system. The production line is adjusted based upon the business requirements and can be implemented stepwise without planning it. These flexible production model and new intelligent model offers the best model for the costly and rigid conveyor belts ((F. Schmitt et al., 2018), (F. Lima et al., 2019))

CHALLENGES – FACING BY ROBOTICS MANUFACTURES

Industrial manufacturers facing some key challenges, those are classified into major four categories.

Employee Skillset and Training

For the newly implemented robotic systems from the available employees, a new level of expertise is required. Every associate will be trained on how to operate in new environments, but the other employees will be hired based upon the skillset, robotic experience, certifications and education.

Safety Measures

Safety measures are introduced in the industrial robots with some strict regulations, stiff penalties and robot safety in the surroundings. Manufacturers have to ensure a safe environment for the workers before the robots are installed for ensuring compliance.

Budgeting for the Cost of the System

For industrial robots, there will be significant investment associated, though the price is steadily dropping. To compensate for the initial investment during ROI period production volumes and sales levels need to remain steady.

Managing Product Workflow

When a robot is installed there will be many considerations for determining the product workflow. Calculations must be ensured carefully for the maximum productivity by not producing existing systems by the orientation and speed part of the presentation.

FUTURE SCOPE

In every second, there are many advances in technology, rapidly in all fields. In the global automation industry, there is a quick development in technology and tremendous growth in every area. In future, by using all the automation techniques, there will be a continuous growth which is anticipated and predicted. In the automation industry, robotic process automation is one of the revolutionary processes; in the upcoming year's staff implementation and higher potential in terms of utilization is expected to increase. This automated robotic process focuses on industrial automation and business-oriented, which are more likely to be handled by humans. Future robotics are oriented and modelled by the automated tools which are very useful that will replace human jobs. Using RPA, these operations can be automated easily.

CONCLUSION

In this book chapter significantly discussed the Robotics and its era, development phase, time-line and later discussed briefly robotics in the industrial sector, and how robots are helping the manufacturers. Coming to technologies collaboration – Internet of Things with Robotics, AI and IoT with Robotics and how these technologies are enhancing the revolutions in the industrial automation process. Based on Industrial automation process, discussed how the Automobile industry is working, Medical sector automation and how Robotics is helping in the robots manufacturing industry. Additionally discussed the key challenges facing by the robotics manufacturing industry, the future scope of the industrial automation with Robotics discussed briefly and showed the future directions.

REFERENCES

Ashton, K. (2009). That 'internet of things' thing. *RFID Journal, 22*(7), 97-114.

Bahrin, M. A. K., Othman, M. F., Azli, N. N., & Talib, M. F. (2016). Industry 4.0: A review of industrial automation and robotics. *Jurnal Teknologi, 78*(6-13), 137-143.

Balaguer, C., & Abderrahim, M. (2008). Trends in robotics and automation in construction. In *Robotics and Automation in Construction*. IntechOpen. doi:10.5772/5865

Bonomi, F., Milito, R., Zhu, J., & Addepalli, S. (2012, August). Fog computing and its role in the internet of things. In *Proceedings of the first edition of the MCC workshop on Mobile cloud computing* (pp. 13-16). ACM. 10.1145/2342509.2342513

Breivold, H. P., & Sandström, K. (2015, December). Internet of things for industrial automation—challenges and technical solutions. In *2015 IEEE International Conference on Data Science and Data Intensive Systems* (pp. 532-539). IEEE. 10.1109/DSDIS.2015.11

Cerruto, E., Consoli, A., Raciti, A., & Testa, A. (1995). A robust adaptive controller for PM motor drives in robotic applications. *IEEE Transactions on Power Electronics, 10*(1), 62–71. doi:10.1109/63.368459

Elson, D. S., Cleary, K., Dupont, P., Merrifield, R., & Riviere, C. (2018). Medical Robotics. *Annals of Biomedical Engineering, 46*(10), 1433–1436. doi:10.100710439-018-02127-7 PMID:30209705

Fletcher, S. R., Johnson, T. L., & Larreina, J. (2019). Putting people and robots together in manufacturing: are we ready? In *Robotics and Well-Being* (pp. 135–147). Springer. doi:10.1007/978-3-030-12524-0_12

Golnabi, H., & Asadpour, A. (2007). Design and application of industrial machine vision systems. *Robotics and Computer-integrated Manufacturing, 23*(6), 630–637. doi:10.1016/j.rcim.2007.02.005

Jeschke, S., Brecher, C., Meisen, T., Özdemir, D., & Eschert, T. (2017). Industrial internet of things and cyber manufacturing systems. In *Industrial Internet of Things* (pp. 3–19). Springer. doi:10.1007/978-3-319-42559-7_1

Kamei, K., & Arai, T. (2018). Optimization for Line of Cars Manufacturing Plant using Constrained Genetic Algorithm. Journal of Robotics. *Networking and Artificial Life*, *5*(2), 131–134. doi:10.2991/jrnal.2018.5.2.13

Kumar, K. V. R., Kumar, K. D., Poluru, R. K., Basha, S. M., & Reddy, M. P. K. (2020). Internet of Things and Fog Computing Applications in Intelligent Transportation Systems. In *Architecture and Security Issues in Fog Computing Applications* (pp. 131–150). IGI Global. doi:10.4018/978-1-7998-0194-8.ch008

Lima, F., de Carvalho, C. N., Acardi, M. B., dos Santos, E. G., de Miranda, G. B., Maia, R. F., & Massote, A. A. (2019). Digital Manufacturing Tools in the Simulation of Collaborative Robots: Towards Industry 4.0. *Brazilian Journal of Operations & Production Management*, *16*(2), 261–280. doi:10.14488/BJOPM.2019.v16.n2.a8

Melich, G., Pai, A., Shoela, R., Kochar, K., Patel, S., Park, J., Prasad, L., & Marecik, S. (2018). Rectal Dissection Simulator for da Vinci Surgery: Details of Simulator Manufacturing With Evidence of Construct, Face, and Content Validity. *Diseases of the Colon and Rectum*, *61*(4), 514–519. doi:10.1097/DCR.0000000000001044 PMID:29521834

Pang, Z., Yang, G., Khedri, R., & Zhang, Y. T. (2018). Introduction to the special section: Convergence of automation technology, biomedical engineering, and health informatics toward the healthcare 4.0. *IEEE Reviews in Biomedical Engineering*, *11*, 249–259. doi:10.1109/RBME.2018.2848518

Parker, L. E., & Draper, J. V. (1998). Robotics applications in maintenance and repair. Handbook of Industrial Robotics, 1378.

Pye, A. (2014). The internet of things: Connecting the unconnected. *Engineering & Technology*, *9*(11), 64–64. doi:10.1049/et.2014.1109

Riva, G., & Riva, E. (2019). SARAFun: Interactive Robots Meet Manufacturing Industry. *Cyberpsychology, Behavior, and Social Networking*, *22*(4), 295–296. doi:10.1089/cyber.2019.29148.ceu PMID:30958039

Rüßmann, M., Lorenz, M., Gerbert, P., Waldner, M., Justus, J., Engel, P., & Harnisch, M. (2015). Industry 4.0: The future of productivity and growth in manufacturing industries. *Boston Consulting Group*, *9*(1), 54-89.

Schmitt, F., Piccin, O., Barbé, L., & Bayle, B. (2018). Soft robots manufacturing: A review. *Frontiers in Robotics and AI*, *5*, 84. doi:10.3389/frobt.2018.00084

Seraji, H. (1998). A new class of nonlinear PID controllers with robotic applications. *Journal of Robotic Systems*, *15*(3), 161–181. doi:10.1002/(SICI)1097-4563(199803)15:3<161::AID-ROB4>3.0.CO;2-O

Thramboulidis, K., & Frey, G. (2011). Towards a model-driven IEC 61131-based development process in industrial automation. *Journal of Software Engineering and Applications*, *4*(04), 217–226. doi:10.4236/jsea.2011.44024

Wollschlaeger, M., Sauter, T., & Jasperneite, J. (2017). The future of industrial communication: Automation networks in the era of the internet of things and industry 4.0. *IEEE Industrial Electronics Magazine*, *11*(1), 17–27. doi:10.1109/MIE.2017.2649104

Wu, M., Lu, T. J., Ling, F. Y., Sun, J., & Du, H. Y. (2010, August). Research on the architecture of Internet of Things. In *2010 3rd International Conference on Advanced Computer Theory and Engineering (ICACTE)* (Vol. 5, pp. V5-484). IEEE.

Yun, J., Jeong, E., Lee, Y., & Kim, K. (2018). The effect of open innovation on technology value and technology transfer: A comparative analysis of the automotive, robotics, and aviation industries of Korea. *Sustainability*, *10*(7), 2459. doi:10.3390u10072459

Chapter 5
Fog Computing in Industrial Internet of Things

Maniyil Supriya Menon
K. L. University, India

Rajarajeswari Pothuraju
K. L. University, India

ABSTRACT

Fog computing, often projected as an extension to cloud, renders its design to deal with challenges of traditional cloud-based IoT. Fog enlightens its features of low latency, real-time interaction, location awareness, mobility support, geo-distribution (smart city), etc. over cloud. Fog by nature does not work on cloud instead on a network edge for facilitating higher speeds. Fog pulls down the risk of security attacks. Industrial sector is revolutionized by ever changing technical advancements and IoT, which is a young discipline embraced by industry thereby bringing in IIoT. Fog computing is viable to Industrial processes. IIoT is well supported by the middleware fog computing as industrial process requires most of the task performed locally and securely at end points with minimum delay. Fog, deployed for industrial processes and entities which are part of internet, is gaining importance in recent times being titles as fog for IIoT. Additionally, as industrial big data is often ill structured, it can be polished before sending it to cloud resulting in an enhanced computing.

INTRODUCTION

Fog Computing, is an embodiment that uplifts the edge gap of resource availability by placing operations and resources at the edge of a network. This is a decentralized architecture providing computation at a level ahead of cloud computing which succeeded in bringing down the services to an intermediate level in the hierarchical structure at the tip, generally network components, and yet holding data up with the cloud. It immensely brings the essence of cloud intimate to the network where the data originates and survives operated, promising proponents that reduce the bandwidth requirement. With all the above-mentioned

DOI: 10.4018/978-1-7998-3375-8.ch005

characteristics, Fog Computing offers leverage inefficiencies like faster processing and lowered rate of resource consumption at a reducing expensive factor.

Besides offering a path full of enhancements, fog computing is characterized by Geographical distribution, mobility factor, interfacing real-time applications, Environmental heterogeneity, and synchronizing interoperability (Pouryousefzade & Akbarzadeh, 2019). Fog computing, spelled as fog networking strives to build control, management, and format on the backbone i.e. Internet.

CHARACTERISTICS OF FOG COMPUTING

- **Geographical Distribution**: Fog offers uplifting uninterrupted deployment services for providing QoS to mobile components and motionless Edge devices. The node is geographically spread over various environmental at different phases.
- **Mobility factor:** Communication between mobile devices is made possible by this mobility factor using SDN protocols, which separates the identity of the host from the identity of location with a distributed indexing system.
- **Interfacing Real-time Applications**: Interacting with real-time applications and devices is an urging requirement for Fog (Peter, 2015). These may include supervising requirements like critical processes with sensors or fog devices, real-time exchange of information for a traffic monitoring system (Vasey, 2018). By default, fog applications are capable of handling real-time processing providing QOS instead of batch processing structurally.
- **Environmental Heterogeneity:** Structurally defined, Fog computing is virtualized offering storage, computation, and services of network with the main cloud and devised, components at the termination. FOG heterogeneity servers maintain hierarchical blocks at distributed locations.
- **Synchronizing Interoperability**: The usage of interoperability between fog devices, support and guarantee services over large range like streaming data and real-time processing supporting the analysis of data and predictive decision making.
- **Huge Wireless Access:** Here wireless access protocols and gateways are examples of node proximity in FOG to end-user.

ROLE OF FOG

- Real-time big data industrial mining intending an increased performance.
- Being able to gather data from various types of sensors parallel.
- Speed computation of collected data for generating commands for actuators, robots with agreeing on latency.
- Use of translation protocols and mapping for sensors and robots that are incompatible.
- Handling the power management system.

Figure 1. Layers of Fog Computing

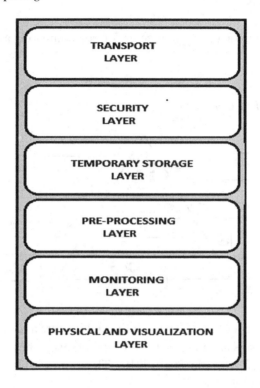

LAYERS OF FOG COMPUTING

The physical and visualization layer sitting at the bottom in Figure 1 includes nodes like physical modes, virtual modes, and sensor networks. Sensor devices which distributed over a geographical area to gathered sensed information is served to the upper layers via components like gateway after advanced filtering and processing.

Monitoring Layer receiving information keeps monitoring what task is accomplished at what time by which mode and what it requests further etc. This means that this layer completely takes over the responsibility of keeping track of resource utilization, the operational processing and availability of fog and sensor modes for further future aspects.

The Preprocessing layer justifying its definition of preprocessing the gathered data by analyzing them using extraction tools and pruning the undesired to come up with relevant meaningful information.

Temporary Storage Layer generally buffers information on temporary storage which also supports data distribution, replication, and de-duplication. After the information is loaded into the cloud, the temporary storage no longer retains it. It acts as a bridge connecting preprocessed data to be served on the cloud.

The security layer performs its responsibility of providing secured encryption and decryption on information from the lower layer. It also considers the scope of integrity to avoid unwanted disruption of data.

The transport layer transports secure reliable pre-processed data to the cloud for further sophisticated computations. These computations include processing advanced services requested with low power consumption feature supporting partial data upload to cloud by with processing carried out by Gateway connecting IoT and cloud.

ARCHITECTURE OF FOG

Figure 2 gives a detailed description of the architecture of Fog computing and explains each component as follows

Figure 2. The architecture of Fog Computing

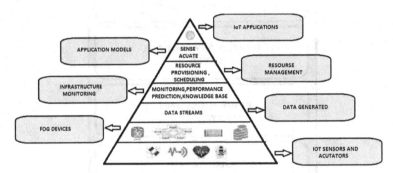

- **Sensors**: These IoT sensors occupy the base of the architecture and are spread globally. It collects information from the sensors and radiates these observations to the layers above through gateways for advanced processing (Chalapathi, et al., 2019). These immutable values gathered from sensors are considered as IoT data streams.
- **Iot Actuators**: These actuators play a vital role in handling the variation in the sensor device environments. They also occupy the bottommost area of the architecture is responsible for controlling the system.
- **Fog Device:** This is the one that is capable of hosting the modules of the application and termed fog devices. Any entry with this capability is considered as Fog device in the architecture design. These devices in which interface sensors to the internet are generally gateways. The upper layer of the architecture is occupied by service for Fog and IoT. They are discussed below.
- **Component Monitoring:** This component tracks the utilization of resources, availability of actuators, sensors, networking elements, and devices of Fog. The resources performance is also monitored and respective actions are proposed by defining its status. This status information is forwarded to other services as demanded.
- **Management of Resources***:* This is a critical part of the FOG architecture containing modules that coherently handle resources in a way that QoS criteria are fulfilled and resource of application utilization is maximized. Finally, placement and schedule modules keep monitoring the phase of resources available to pick the best fit resources for anchoring an application component.
- **Handling Power:** One of the unsolved problems foreseen by the IoT environment is the resource utilization of nodes in IoT, which focuses on energy consumption (Kychkin, et al., 2019). Power management is complex to handle as FOG devices, unlike cloud deal with devices having variable power consumption factors.

LITERATURE REVIEW

Several Authors contributed their ideas for bringing in FOG into IoT. Doghman, et al., 2016 contributed a detailed justification of Fog concepts, approaches, and practices in the IoT world that motivated towards Fog with paving interest in its applicability. Aazam, et al., 2018 shared information regarding support in Industrial sector, large corporations that gave insight into research challenges which helped to get a deep knowledge in IIoT. Ai, et al., 2018 presented a tutorial of various types of edge computing techniques like mobile, cloud and Fog with a detailed comparison factors that increased the scope for IoT applications. Caiza, et al., 2020 came up with enhanced review about security features, architecture energy consumption of Fog at Industrial level. Puliafito, et al., 2017 discussed about mobility support in Fog and the challenges to be handled. They are close to the adverse effecs discussed in the chapter. Chalapathi, et al., 2019 elaborated the influence of edge and Fog computing scenarios in IIoT environment. Harish, et al., 2019 gave a profound review on Fog computing devices, its role in IoT and applications which helped in understanding of IIoT components. Stojmenovic, et al., 2014 discussed regarding the security concerns like attacks comprising integrity. This is an issue of concern that was just mentioned by the Authors. Jalasri, et al., 2018 presented how fog computing with IoT benefits in providing security benefits. Jason Anderson, 2018 contributed the way Organizations need to understand requirement of Fog and Edge depending on the environments. These contributions successfully shared a different knowledge in Fog computing perspective. Karmakar, et al., 2019 presented an evaluation of present scenario of IIoT and focused on Research options. Kychkin, et al., 2019 proposed architecture for energy management for enhanced results. Loupos, et al., 2019 presented an article on security in IIoT. Luiz, et al., 2018 reviewed on IoT, Fog and cloud computing in application perspective. Vasey, 2018 presented detailed information on Open Fog. Mittal, et al., 2018 provided a combination of Edge and Cloud computing with their applications. Peter, 2015 extended the review on Fog with their detailed application in Real world. Pouryousefzadeh, et al., 2019 narrated literature review and analytical research on with a IoT with a cultural heritage perspective. Mahmud, et al., 2017 analyzed the Fog challenges, and presented a Fog Taxonomy and proposed research gaps. Saroa, et al., 2018 contributed the Fog applications with their implementation in smart cities. Schleicher, et al., 2019 proposed a model on peer- to-peer Fog computing and extended comparative studies on Fog and cloud computing. Yi, et al., 2016 reviewed representative application scenarios, issues in design and challenges on a brief edge. Shi, et al., 2019 discussed about Edge computing issues like storage, supercomputing and cloud services entailing their shortcomings. Shrouf, et al., 2014 presented a peer review with reference architecture for IoT based smart factories. Wang, et al., 2019 contributed a detailed comparison of various channel models related to IIoT. Wollschlaeger, et al., 2017 put light on role of IoT, CPS (Cyber Physical System)in 5G networks. Xu, et al., 2018 presented a survey on cloud, edge, and Hybrid computing platforms with Machine Learning in IIoT inspiring evolutionary directions for research. Yu, al., 2019 performed a detailed survey on security issues of IoT and IIoT with related Security challenges. Zhang, et al., 2017 contributed to numerous issues of IIoT and their related applications. The contributions of fore mentioned Authors help in deep analysis of Fog in IoT and IIoT in comparison Edge and Cloud with enhanced scope of recommendations pertaining to IIoT components and overwhelming Challenges.

FOG VS CLOUD VS EDGE COMPUTING

Organizations in recent times are spreading their wings towards the cloud, Fog, and Edge computing paradigms (Harish, et al., 2019). Organizations are benefitted with various computing and data storage devices and IIOT. Fog, Edge, Cloud (Ai, Peng, & Zhang, 2018) sound similar but a closer view brings in the differences, and are separate layers of discussion in IoT.

Cloud Computing

A Technology that aims at vast storage, high computational power, maximum resource utilization, and high-performance networking abilities is now a de facto standard for several industries. Fog and Edge are extensions comprising a distributed network. The basic feature of a cloud-based system is that they allow information to be retrieved from different places in the world. Encapsulated hardware captures data from IIOT devices and moves it to the fog layer (Puliafito, Mingozzi, & Anastasi, 2017). Pertinent information passed to the cloud in distinct geographical areas. Thus cloud is taking advantage of IIOT devices by getting data from other layers. Organizations are attaining improved results by combing cloud with on-site fog or edge machines, and therefore shifting in the directions of fog or Edge environment for enhanced usage of IoT devices.

Fog Computing

They target in pulling down processing and intelligence nearer to data construction. The basic difference between Fog and Edge which sounds similar lie in the placement of intelligence and computing power (Mahmud, Kotagiri, & Buyya, 2017). Getting inside Fog gives a place for intelligence at LAN i.e. transfers data from endpoints to Gateway, later to source for processing. Edge places intelligence and processing at embedded automated controllers. Fog uses gateways and edge devices with LAN for offering processing abilities.

Edge Computing

The characteristic of IoT bought about virtually indefinite endpoints, which is challenging the consolidation of data and processing at a single data center demanding for edge computing. Edge computing performs computation closer to the edge of the network. This can be thought of as an extension of legacy peer to peer, Edge and distributed and remote cloud services. The processor of the edge computing device is more secure and consumes low power comparatively.

IIOT uses a combination of Fog, Edge, and cloud technologies (Bittencourta, Immicha, & Sakellari-oub, 2018). But comparing edge computing offers benefits over traditional architectures i.e. performing computation at network edges brings down the network traffic. Security is also offered through encryption closer to the network. Few factors of fog and cloud are shown in the Table 1.

In Table 1 several characteristic features of cloud and Fog are considered for comparison in several aspects.

Table 1. Comparative Features of Cloud and Fog

Characteristic	Cloud	Fog
Architecture	Centralized	Distributed
Communication with devices	From Distance	From edge
Data processing	Away from Source	Nearer to Source
Computing capabilities	Higher	Lower
No. Of nodes	Few	Many
Analysis	Long Term	Short-term
Latency	High	Low
Connectivity	Internet	Several Protocol Standards
Security	Lower	Higher

FOG COMPUTING IN IoT

Due to the growth of IoT, the use of the cloud for connected devices overridden by new fog technologies paved the path to FOG computing in IoT (Jalasri & Lakshmanan, 2018). The IoT offers few challenges whose solutions lie with fog (Anderson, 2018).

Challenges

1. The internet of things is expected to put huge pressure on the current internet and data center infrastructure.
2. Centralizing Cloud data operations at a single site.
3. Data analyzed on devices close to IoT is reaching 40 percent, which begs for an urgent alternative approach.

The IoT even though exerts intense pressure on the current internet, is now adapting to architectures accordingly by adjusting to the requirement like bandwidth, deployment factors, usage, and operation on data-centered sources (Schleicher, Graffi, & Rabaya, 2019). This projects a problem of serving multiple requests from various remote sites by a single centralized repository as the network pops up cons like limitations in available bandwidth, synchronization constraints (Doghman, et al., 2016), Normalization issues, and further. Further approaches like distributed computing need to be put in place to avoid computational overhead at devices close to IoT.

FOG COMPUTING IN IIoT

IIoT: Industrial Internet of Things

The industrial sector is evolving with rapid advancements in technology ranging from industrial process automation to autonomous industrial processes. IOT, which emerged few years back, has been embedded by industry giving rise to IIOT (Karmakar, et al., 2019). As per the reports CISCO and OVUM, 2017 Industrial Environment is ranked first in utilization of IOT devices. This environment includes production, smart homes, grid computation, and transportation.

A large amount of discrete data is being generated due to smart devices and the Industrial process constraints operations to be executed locally about the delay, privacy issues thereby transferring organized data over the internet to web services and cloud (Stojmenovic, & Wen, 2014). For acquiring the above, support of middleware becomes mandatory between cloud/web services and industrial environment. At this point, fog acts as a potential bridge for supporting several industrial scenarios.

IIOT concentrates on business, majorly private institutions, and some academic institutions and deals with main transfer and managing mission-critical data, and responses depending on machine to machine conversations. IIOT succeeded in encapsulating cyber-physical systems (cps), Maximum automation, Business robots, smart factory, so on. IIOT still needs to be realized as research challenges are being raised against by society of researches (Shrouf, Ordieres, & Miragliotta, 2014). The challenges addressed are like stabilization, scalability, usability, privacy, etc. The main sectors of industry requiring the utilization of IIOT are health care, education, transportation, retail, and production (Xu, et al., 2018).

Figure 3. Industrial IITt Data Processing Stack Layer

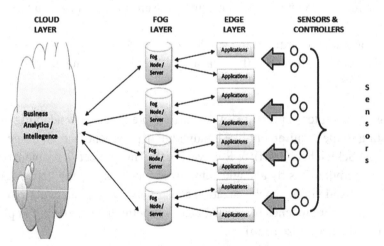

In figure 3 the cloud layer encompasses salient features like Big data processing where large volumes of data clusters are analyzed, computed for extracting patterns that help in wise decision making considering several disciplines of data formats. This layer also has the essential capabilities of handling custom rules as algorithms that support the exchange of information between the user interface and database. All the processing handled at this layer bases on large velocities of data residing on storages termed Data warehouses. These warehouses are constructed by collecting data from several heterogeneous sources during data warehousing.

Contemporarily the Fog Layer lying between the cloud layer and edge layer is aiming at maintaining local networks, performing advanced analysis, and resulting in data reduction which is one of the requirements. It also orients in handling control responses and Data visualization pattern support with a scope of standardization.

At the ground level i.e. the Edge layer deals with the user applications by collecting data from the sensors and controllers of IoT devices. It also deals with processing real-time huge volumes of data, handling Industrial gateways that implement IIoT plans, and collaborate to rise with a future support environment. It also adds up to on-premises data representation, embedded systems, and data storage even at micro-levels.

The processing capabilities vary as we move from Cloud layer to Edge Layer i.e. the response time and speed of processing are slower at the cloud when compared with the edge layer which directly receives data from different IoT devices. Edge Layer is a close neighbor to sensors and connectors hence exhibit better response time than the cloud layer. The fog layer also puts its share in compensating these gaps by adopting high-speed servers thereby bringing down the speed and response issues.

IIOT: CHALLENGES IN ADOPTING

- **Interoperability**: This is spotted as the largest challenge by IoT experts. The manufacturing enterprises are pouring in new machines and protocols which are yet to be interconnected and at times do not support interoperability. Hence bringing about an association between legacy industrial systems and promises tough interoperability within is a stunning challenge.
- **Analyzing Data And Transmission**: The amount of information generated by edge devices or sensors in a computational scenario is vast in size, which is tough to store & compute (Wang, et al., 2019) .Hence advanced analytics and machine learning processes cannot be supported and transmission of sensitive information over the web for analytics is one more milestone to overcome
- **Security**: As the technology of manufactory processes and production processes is rapidly becoming smarter, hence for being connected to the cloud directly, increasing the probability of being vulnerable to threats and attacks (Loupos, et al., 2019) . This is one more raising challenge for IIOT as and more the features increase, the risk entailing them becomes sharper.
- **IT and OT Hand on Hand**: An unseen question of doubt is how would IT & OT converge and go hand on hand. This doubt rose as IT is mature with a well-authorized policy system but OT is a younger trend still upcoming and not default a networking technology. This bridging gap is throwing challenges to align IT systems with manufacturing processes.

OVERCOMING IIOT CHALLENGES

- **IIoT Gateway:** Also interchangeable IoT gateway provides provision for security establishing a connection between existing device infrastructure and industrial infrastructure, practically speaking IoT gateway could successfully connect Distributed control system or SCADA straightly with the cloud using well-accepted protocols like OPC. This surrenders a solution to interoperability problems and M2M interaction.

- **Trusted Terminal, Network, And Perception Modules:** Modules of industry adopt trusted perception module(TPM), Trusted Terminal Module(TTM) and Trusted Network attacks . Additionally, data center security measures like web application firewalls, secure Gateway (Jalasri & Lakshmanan, 2018) delivery controller ensure the security of data using various encryption techniques when data continuous to more between network stable at a node.
- **Edge Computing:** This allows only relevant information to be moved further for analytics edge computing supports connection of gateways with diversifying functionalities forming a cluster, which taken to distributed edge computing (Shanhe, Li, & Qun, 2016). These permit data processing at the edge and near the source before sending It to the cloud for further evaluation. Individual clusters forming fog node, when combined indulge distributed fog computing benefitting fast transfer of data and real-time analysis with faster response (Shi, Pallis, & Xu, 2019).
- **Clusters of IoT Gateway:** The convergence problem of IT and OT is driver off buy the IoT gateway ensuring integration of IT systems like ERP, CRM with OT like MES, SCAPA. So, communication of cloud and data storage is also feasible.

Enabling IIOT is a positive move to pick manufacturing companies ahead by complete the manufacturing process Implementing IOT Gateway .clusters with fog &edge computing abilities fight back challenges and improve the overall performance of industries.

IIoT COMPONENTS

Basic requirements or building blocks of IIOT are considered here.

WSN's and WSAN's –Localizing

Without the existence of WSN's and WSAN's IIOT is not considered complete. The industrial environment contains numerous varying sensors. To handle and interface with these heterogeneous sensors and retrieve data, there is a need for a sophisticated well-equipped gateway. This gateway is a component of Fog architecture (Aazam, Zeadally, & Harras, 2018). Fog filters retrieve data and forwards only structured data over the internet (Yu & Guo, 2019). Consumption of energy also plays a vital role in a smart industrial environment. For proving such efficient consumption of energy many other alternate sources like solar or thermal, need to be adopted which demand enriched intelligence and reduced time responses. Therefore an efficient and enriched fog is inevitable in Fog.

WSAN's perform enumerable actions depending on data sensed that need to be more quality focusing and reliable data. In an IIOT, handling WSAN's over network is using remote applications about quick responses and security issues (Loupos, et al., 2019). In contrast to sensors, actors have more resources in their reach with huge data transfer and increased battery life (Caiza, et al., 2020). In some cases, WSAN's largely delay sensitivity. Fog is efficient in taking quicker local actions and attains a discrete interface of four different actors and sensors with applications handled remotely.

CPS-Monitoring and Operations

CPS permits the connections and networking between legacy systems and modern devices of the cyber world. A cyber-physical system is a part of IoT where devices and machines are bridged directly, or via an internet application. CPS agrees for communication within products, devices, and humans similarly. When coming to an industrial system CPS takes care of automation, monitoring, controlling, maintaining, and assisting inexpensively (Wollschlaeger, Sauter, & Jasperneite, 2017). However, a CPS handles weather fluctuations and the needs of machinery being handled. The control unit of CPS is efficient at handling devices, sensors, actuators, and machines with an interface directly with the real-world for gathering data and processing.

The main considering factor about CPS is the exchange of data between entities mentioned, as the processing is performed centrically. CPS is blessed to give its maximum as the advancements in areas of artificial intelligence and machine learning are prevailing as an add-on for manufacturing with intelligence, dynamism, scalability, and flexibility. Few awesome applications of CPS are robotic surgery, automating car manufacturing, smart, and so on.

Analysis of Industrial Big Data

Automating industrial environment generates a vast amount of magnitude data, which needs to be analyzed before utilization. WSN'S and WSAN'S, robots populate the system with streaming data which need to be tailored using data analysis techniques. An IIoT could not promise a better performance until collected data from various sources undergo analysis improving predictive analysis, tolerating faults, avoiding failures, lowering expensive factors, monitoring, and diagnosis, etc.

CONCLUSION

Fog computing bridging the gap between resource and network, clearly self defines its characteristics promoting its usage and benefits into various areas (Zhang, Jonsson, & Li, 2017). The IoT which is potentially targeting an increase of information now grows to conquer and transform companies and organizations in the world around were discussed in detail. The challenges of IoT are strongly considered with a proposed opinion of solutions. IIoT is of consideration which is very well discussed in this chapter taking the features of IoT for enhancing Industrial operations and peripherals with advanced features.

FUTURE SCOPE

Fog computing with its space in IoT being a new technology needs to be furnished in several areas like communication aspects, deployment issues, rendering its benefits in parallel computations ruling over, security levels concentrating even on end-user privacy as it can be a peephole into user systems. IIoT which relies on IoT should spread its strength despite the above challenging factors in evolving technology phases with a scope of identifying research gaps.

REFERENCES

Aazam, M., Zeadally, S., & Mr. Harras, K. A. (2018). Deploying Fog Computing in Industrial Internet of Things and Industry 4.0. *IEEE Transactions on Industrial Informatics*, *14*(10), 1–1. doi:10.1109/TII.2018.2855198

Ai, Y., Peng, M., & Zhang, K. (2018). Edge computing technologies for the Internet of Things: A primer. *Digital Communications and Networks*, *4*(2), 77–86. doi:10.1016/j.dcan.2017.07.001

Al-Doghman, F., Chaczko, Z., Ajayan, A. R., & Klempous, R. (2016). A review on Fog Computing technology. *IEEE International Conference on Systems, Man, and Cybernetics (SMC)*. DOI:10.1109/SMC.2016.7844455

Andersen, J. (2018). *Why the Future of IIoT Needs Both Edge and Fog Computing*. Rtinsights.

Bittencourta, L., Immicha, R., & Sakellarioub, R. (2018). *The Internet of Things, Fog and Cloud Continuum: Integration and Challenges*. arXiv:1809.09972v1

Caiza, G., Saeteros, M., Oñate, W., & Garcia, M. V. (2020). Fog computing at an industrial level, architecture, latency, energy, and security: A review. *Heliyon*, *6*(4), e03706. doi:10.1016/j.heliyon.2020.e03706 PMID:32300668

Harish, Nagaraju, Harish, & Shaik. (2019). A Review on Fog Computing and its Applications. *International Journal of Innovative Technology and Exploring Engineering, 8*(6C2).

Ivan Stojmenovic, S. I. T., & Wen, S. (2014)The Fog Computing Paradigm: Scenarios and Security Issues. *Proceedings of the 2014 Federated Conference on Computer Science and Information Systems*. 10.15439/2014F503

Jalasri, M., & Lakshmanan, D. L. (2018). A Survey: Integration of IoT and Fog Computing. *Second International Conference on Green Computing and Internet of Things (ICGCIoT)*. 10.1109/ICG-CIoT.2018.8753010

Karmakar, A., Dey, N., Baral, T., Chowdhury, M., & Rehan, M. (2019). Industrial Internet of Things: A Review. *International Conference on Opto-Electronics and Applied Optics (Optronix)*. 10.1109/OPTRONIX.2019.8862436

Kychkin, A., Deryabin, A., Neganova, E., & Markvirer, V. (2019). IoT-Based Energy Management Assistant Architecture Design. *IEEE 21st Conference on Business Informatics (CBI)*. DOI:10.1109/cbi.2019.00067

Loupos, K., Caglayan, B., Papageorgiou, A., Starynkevitch, B., Vedrine, F., Skoufis, C., ... Boulougouris, G. (2019). Cognition Enabled IoT Platform for Industrial IoT Safety, Security, and Privacy — The CHARIOT Project. *IEEE 24th International Workshop on Computer-Aided Modeling and Design of Communication Links and Networks (CAMAD)*. DOI:10.1109/camad.2019.8858488

Mahmud, Kotagiri, & Buyya. (2017). Fog Computing: A Taxonomy, Survey and Future Directions, Internet of Everything. Internet of Things. In *Technology, Communications, and Computing*. Springer. Doi:10.1007/978-981-10-5861-5_5

Mittal, S., Negi, N., & Chauhan, R. (2017). Integration of edge computing with cloud computing. *International Conference on Emerging Trends in Computing and Communication Technologies (ICETCCT).* DOI:10.1109/icetcct.2017.8280340

Peter, N. (2015). Fog Computing and It's Real-Time Applications. *International Journal of Emerging Technology and Advanced Engineering, 5*(6).

Pouryousefzadeh, S., & Akbarzadeh, R. (2019). Internet of Things (IoT) systems in future Cultural Heritage. *International Conference on Internet of Things and Applications (IoT).* DOI:10.1109/iicita.2019.8808838

Saroa, M. K., & Aron, R. (2018). Fog Computing and Its Role in the Development of Smart Applications. *2018 IEEE Intl Conf on Parallel & Distributed Processing with Applications, Ubiquitous Computing & Communications, Big Data & Cloud Computing, Social Computing & Networking, Sustainable Computing & Communications, (ISPA/IUCC/BDCloud/SocialCom/SustainCom).* DOI:10.1109/bdcloud.2018.00166

Schleicher, E., Graffi, K., & Rabaya, A. (2019). Fog Computing with P2P: Enhancing Fog Computing Bandwidth for IoT Scenarios. *International Conference on the Internet of Things (iThings) and IEEE Green Computing and Communications (GreenCom) and IEEE Cyber, Physical and Social Computing (CPSC), and IEEE Smart Data (SmartData).* DOI: com/cpscom/smartdata.2019.0003610.1109/ithings/green

Shi, W., Pallis, G., & Xu, Z. (2019). Edge Computing. *Proceedings of the IEEE, 107*(8), 1474–1481. doi:10.1109/JPROC.2019.2928287

Shrouf, F., Ordieres, J., & Miragliotta, G. (2014). Smart factories in Industry 4.0: A review of the concept and of energy management approached in production based on the Internet of Things paradigm. *2014 IEEE International Conference on Industrial Engineering and Engineering Management.* 10.1109/IEEM.2014.7058728

Wang, W., Capitaneanu, S. L., Marinca, D., & Lohan, E.-S. (2019). Comparative Analysis of Channel Models for Industrial IoT Wireless Communication. *IEEE Access: Practical Innovations, Open Solutions, 7*, 91627–91640. doi:10.1109/ACCESS.2019.2927217

Wollschlaeger, M., Sauter, T., & Jasperneite, J. (2017). The Future of Industrial Communication: Automation Networks in the Era of the Internet of Things and Industry 4.0. *IEEE Industrial Electronics Magazine, 11*(1), 17–27. doi:10.1109/MIE.2017.2649104

Xu, H., Yu, W., Griffith, D., & Golmie, N. (2018). A Survey on Industrial Internet of Things: A Cyber-Physical Systems Perspective. *IEEE Access: Practical Innovations, Open Solutions*, 1–1. doi:10.1109/ACCESS.2018.2889501

Yu, X., & Guo, H. (2019). A Survey on IIoT Security. *IEEE VTS Asia Pacific Wireless Communications Symposium (APWCS).* 10.1109/VTS-APWCS.2019.8851679

Zhang, Y., Jonsson, M., & Li, M. (2017). Guest Editorial Special Issue on Industrial IoT Systems and Applications. *IEEE Systems Journal, 11*(3), 1337–1339. doi:10.1109/JSYST.2017.2702940

KEY TERMS AND DEFINITIONS

Actuators: A component of machine intended to control a system or mechanism.

Cloud Computing: A computing environment that promises services like storage, databases, software over Internet for improved performance.

Edge Computing: A computing environment that fills gap between data storage and computation location.

Fog Computing: Architecture supporting edge devices for computation and connecting networks.

Fog IIoT Stack: A fog-based technology stack for a complete IoT solution.

Gateway: A device for connecting two different networks over internet.

IIoT: IoT extended for Industrial sectors.

Sensors: A device that detects a physical property indicates and responds accordingly.

Chapter 6
Voice–Controlled Biped Walking Robot for Industrial Applications

B. Pavitra

Anurag Group of Institutions, India

D. Narendar Singh

Anurag Group of Institutions, India

Mohamamd Farukh Hashmi

National Institute of Technology, Warangal, India

ABSTRACT

Mechanical IoT is proceeding to extend from assembling and keen homes to retail, nourishment bundling, social insurance, and other use cases. For the present makers, modern IoT is quickly turning into an unquestionable requirement on creation lines for reasons of adequacy and hazard relief. Checking sensors must coordinate with generation line gear, including robots. Information streams and checking/announcing modules can give basic way data, trigger reaction conventions, and influence different exercises being followed on different modern systems. Modern IoT can expand perceivability into the states of robots in a production line, ranch, stockroom, or emergency clinic. Data about parts and gear can consistently observe to augment profitability and to decrease line personal time. Voice assistant robot takes the commands of the operator or mentor sets the functionality accordingly which is stored in the database and then execute the given task.

DOI: 10.4018/978-1-7998-3375-8.ch006

INTRODUCTION

An enormous proportion of examination has been done with biped walking robots from 1970. During this time biped walking robots have been changed by imaginative improvement into biped humanoid robots. Additionally, the biped humanoid robot has got a one of operator evaluation subjects in the cautious robot research society. Regularly researchers perceive the humanoid robot industry to be the twenty-first century industry pioneer and we will in the end enter a period with one robot in each home. The strong focus on biped human robots starts from a high standing prerequisite for humanoid robots. Furthermore, a similar appearance human robot is engaging for synchronization in a human robot society. Regardless, while it isn't hard to develop a human-like biped robot stage, the verification of stable biped robot walking tends to a far reaching test. This is a brief result of a nonappearance of perception on how individuals walk dependably. Moreover, biped walking is a precarious reformist improvement of a specific assistance stage.

Earlier stage biped walking around robots included static walk around a slow walking speed. The improvement time was in excess of 10 s for every turn of events and the value control structure was performed utilizing COG (Centre Of Gravity). Thusly the extensive motivation driving Centre Of Gravity on to the ground reliably fall inside the supporting polygon that is made by two feet. During the static walk, the robot able to stop the walking improvement at any time without tumbling down. The hindrance of static walking is that the improvement is exorbitantly moderate and wide for moving the Centre Of Gravity. Masters likewise began to focus in on profound walking around biped robots. It is enthusiastic walking around a speed of under 1 s for every turn of events. In case the dynamic switch can be kept up, dynamic walking is smoother and even more astounding regardless, while using little body upgrades. In light of everything, if the inertial forces created utilizing the re-establishing of the robot body are not sensibly controlled, a biped robot enough tumbles down. Also, during dynamic walking, a biped robot may tumbles down from aggravations and can't stop the walking headway all of a sudden.

Hence, the possibility of Zero Moment Point was agreeable from the earliest starting point with control inertial forces. In the anticipated single assistance stage, the Zero Moment Point is commensurate to the COP Centre of Pressure on the sole. The potential gain of the Zero Moment Point is that it is the detect the motivation behind mix of gravity is extended onto the ground in the static state and a point where the unbending inertial force made out of the gravitational force and inertial power of mass encounters the ground in the dynamic state. If the Zero Moment Point cautiously exists inside the supporting polygon made by the feet, the robot never tumbles down. Most of assessment packs have used the Zero Moment Point as a smaller relentless measureable quality with standard of dynamic walking of biped robots. To till end, the robot is controlled so much that the Zero Moment Point is kept up inside the supporting polygon.

LITERATURE OVERVIEW

Laws of Biped Robot

- Zeroth Law: "A Robot must not harm mankind in spite of the fact that in real life, permit mankind come to hurt".

- First Law: - "A Robot must not hurt a human body despite the fact that in real life; permit one to come to hurt".
- Second Law: - "A Robot should consistently comply with individual, except if it is in strife with a higher request law".
- Third Law: - "A Robot must shield itself from hurt except if that it is struggle with a higher request law". (Zaier, 2012)

Background

Right when people believe AI (AI), the critical picture that springs up in their psyches is that of a robot skimming around and offering mechanical responses. There are various kinds of AI yet humanoid robots are one among the foremost notable structures. they need been depicted during a few Hollywood movies and just in case you're an aficionado of fantasy, you'll have run over several humanoids. Maybe the soonest kind of humanoids was made in 1495 by Leonardo Leonardo . it had been a defensive layer suit and it could play out tons of human limits, as an example, sitting, standing and walking. It even moved similarly as a licensed human was inside it. From the beginning, the critical purpose of AI for humanoids was for ask about purposes. They were getting used for analyze on the foremost capable strategy to enhance prosthetics for people . By and by, humanoids are being made for a few of purposes that aren't compelled to research . Progressed humanoids are made to end differing human tasks and have different positions within the work territory. A segment of the positions they might include are the activity of a private right, collaborator, front work territory official, then forth. The route toward envisioning a humanoid is exceptionally confusing and an enormous amount of labor and examination is put into the system. Most events, designers and draftsmen face a few of troubles. First-grade sensors and actuators are huge and a minor blunder could achieve glitching. Humanoids move, talk and convey out exercises through explicit features, as an example, sensors and actuators. Individuals expect that humanoid robots are going to be robots that are basically like people. That is, they need a head, middle, arms and legs. Nonetheless, this is not generally things as certain humanoids don't totally take after people. Some are designed consistent with just a few particular human parts, as an example, the head . Humanoids are typically either Androids or Gynoids. An Android may be a humanoid robot intended to require after a male human while gynoids appear as if female people. Humanoids run through specific highlights. they need sensors that guide them in detecting their surroundings. Some have cameras that empower them to ascertain unmistakably. Engines put at vital focuses are what direct them in moving and making motions. These engines are generally alluded to as actuators. tons of labor, assets and examination are set into making these humanoid robots. The physical body is assumed of and dissected first to urge a wise image of what is going to be imitated. By then, one must choose the task or reason the humanoid is being made for. Humanoid robots are made for a few of purposes. Some are made cautiously for preliminary or examination purposes. Others are made for incitement purposes. a few of humanoids are made to try to to unequivocal endeavors, for instance, the tasks of a private partner using AI, helping at old homes, then forth. The resulting stage specialists and trailblazers got to take before a completely helpful humanoid is readied is making instruments like physical body parts and testing them. By then, they need to encounter the writing which is one among the foremost basic stages in making a humanoid. Coding is that the stage whereby these creators program the bearings and codes that might engage the humanoid to end its abilities and offer reactions when represented an invitation . Doesn't sound so irksome, right. Regardless, it's incautious to simply accept that creating a humanoid is as straightforward as making a kite or a slingshot in your yard. Disregarding the way that humanoid robots are becoming conspicuous, trend-setters face a couple of troubles in making totally valuable and sensible ones.A segment of these challenges include:

- **Actuators:** These are the motors that help moving and making signals. The physical body is dynamic. you'll without a doubt get a stone, heave it over the road, activate various occasions and do the three stage move. of these can happen in around ten to fifteen seconds. to form a humanoid robot, you would like strong, profitable actuators which will duplicate these exercises deftly and inside a comparable slot or maybe less. The actuators should be gainful enough to expire a good extent of exercises. Sensors: These are what help the humanoids to spot their condition. Humanoids need all the human distinguishes: contact, smell, sight, hearing and equality to figure properly. The gathering sensor is critical for the humanoid to listen to rules, decipher them and do them. The touch sensor shields it from getting things and causing self-hurt. The humanoid needs a sensor to switch improvement and correspondingly needs warmth and anguish sensors to understand when it faces hurt or is being hurt. Facial sensors moreover need to be flawless for the humanoid to point out up, and these sensors should have the choice to expire a good extent of verbalizations. Guaranteeing that these sensors are open and capable may be a hard task.
- **AI-based Interaction**: the extent at which humanoid robots can accompany individuals is restricted This where AI is important . It can help decipher bearings, questions, declarations and should even have the selection to offer cunning, wry answers and obtain unpredictable, dubious human ramblings.

Configuration Process for Biped Robot

The last objective of the advancement of biped robots is to repeat the capacities of a person in a specialized framework. Despite the fact that few biped robots as of now exist and noteworthy exertion is placed into this exploration field, we are still a long way from arriving at this objective. Biped robots are unpredictable frameworks which are described by high useful and spatial mix. The plan of such frameworks is a test for architects which can't yet be agreeably explained and which is regularly a long and iterative cycle. Biped robots are a genuine model for complex and exceptionally coordinated frameworks with spatial and utilitarian interconnections among segments and gathering gatherings. They are multi-body frameworks in which mechanical, electronic, and data innovative segments are incorporated into a little plan space and intended to interface with one another.

Demand for Effective Plan of Biped Robots

Mechanical robots are being utilized in many assembling plants everywhere on the world. This Product class has arrived at a significant level of development and an expansive assortment of robots for extraordinary Applications is accessible from various makers. Despite the fact that both sort of robots, modern and humanoid, control objects and similar kinds of parts, for example symphonies drive gears, can be found in the two sorts, the objective frameworks contrast fundamentally. Modern robots work in detached conditions carefully isolated from people. They play out a set number of unmistakably characterized tedious errands. These machines and the apparatuses they use are frequently intended for a specific reason. High precision, high payload, high speeds and firmness are regular improvement objectives. Biped robots cooperate in a mutual space with people. They are planned as general partners and ought to have the option to learn new abilities and to apply them to new, beforehand obscure errands. Human like kinematics permits the robot to act in a domain initially intended for people and to utilize similar devices as people along these lines. Human appearance, conduct and movements which are recognizable to the client from collaboration with peers make biped robots more unsurprising and increment their acknowledgment. Wellbeing for the client is a basic prerequisite. Because of these noteworthy contrasts, a great part of the advancement information and item information from modern robots can't be applied to biped robots. The multi-modular connection between a biped robot and its condition, the human clients and in the end different humanoids can't completely be recreated in its whole unpredictability. To research this cognizance, real biped robots and investigations are required. Right now just toy robots and a couple of exploration stages are financially accessible, regularly at significant expense. Most biped robots are planned and worked by the unique concentration or objectives of a specific exploration venture and a lot more will be worked before develop and normalized robots will be accessible in bigger numbers at lower prizes. A couple of biped robots have been created by organizations, yet very little is thought about their plan cycle and only from time to time is there any data accessible that can be utilized for expanding the time and cost proficiency in the advancement of new improved humanoid robots. Planning a biped is a long and iterative cycle as there are different collaborations between for example mechanical parts and the control framework.

PROPOSED SYSTEM

The underneath figure. shows the square outline of the 6-DOF biped advance sorting out part: In this framework, a biped robot is orchestrated and tried the level and hazardous surfaces. To change the ground contact and for convincing advancement on unsafe surface, we utilize principal philosophy by covering the foot of the robot by any hard materials. despite it, the android application is used to investigate the way; the biped way can be changed solicitation to evade influence inside block. In this structure, a Pulse width modulation signal is given as a commitment to the mechanized pins of Arduino UNO. Considering the on time and slow time of year of Pulse width modulation signal the essential for degree to the development of the biped is resolved for each degree of chance. In like way the servo motors turn and therefore there is a development in the biped. In any case, there is an obstacle for Arduino UNO. Unquestionably the current from all I/O pins is 200milli amps and 5v yield pin for instance 400-900(Ma). Regardless, current essentials for high power metal servos to develop a versatile robot are between 500ma-900ma per servo and besides require higher working voltage. To work the servos without hurting the Arduino board we need another board called as PWM based servo driver shield. It is a board that can be halted to Arduino to improve the value. This shield drives high power servos with high current essentials. Outside power source is expected to drive the servos. It is driven by pcA9685 controller. It has ability to drive 16 servos at the same time. This board is interfaced to Arduino using I2C show. It works with 12 cycle objective. Working voltage is 2.4v to 5v and working repeat is 25 Hz to 1526 Hz.

The Figure 2 flowchart shows how the robot functions. At the point when we give power flexibly to Arduino board and Servo engine driver, it turns ON. We utilize an android application in versatile to control the robot. We have to combine up the portable Bluetooth with the Bluetooth module HC-05. Orders given from the application will be communicated through Bluetooth. Arduino IDE peruses the orders given by the client and it confirms with that string which is available in the program. In the event that the string matches, the fitting activity is performed. On the off chance that the order doesn't coordinate with the string, it doesn't play out any activity. Also, if there is no order from the client, it doesn't play out any activity. After an order got and played out any activity, it sits tight for next order.

HARDWARE REQUIREMENT

Arduino Uno

The Arduino Uno is a microcontroller device dependent on the ATmega328 Archicteture. These are generally utilized super arrangement of AVR microcontrollers created by Atmel. It has 14 advanced information/yield pins (of which 6 can be utilized as pulse width modulation yields), 6 simple data sources, a 16 MHz fired resonator, a USB association, a force jack, an ICSP header, and a reset button. It contains all that expected to help the microcontroller; just interface it to a PC with a USB link or force it with an AC-to-DC connector or battery to begin. The Uno contrasts from all previous sheets in that it doesn't utilize the FTDI USB-to-chronic driver chip. Rather, it includes the Atmega16U2 (Atmega8U2 up to form R2) customized as a USB-to-chronic converter.

The underneath figure 3 shows the Arduino pinout.

Servo Motor

Servo engine is a turning actuator that takes into account exact control of rakish or direct position. It utilizes shut circle servomechanism which is blunder detecting negative criticism to address its position. It comprised of numerous parts, for example, gears, engine, potentiometer, control circuit and so on.

Working Principle

From the underneath figure 4 microcontroller conveys contribution to control beat. The beat width to voltage converter convert input signal into voltage it is one info given to comparator and another information is given from position sensor to detect point of arm and impart identical voltage sign. In view of blunder between the voltages of two data sources the mistake I imparts sign to the engine to pivot a specific way. Apparatus box is utilized to venture down the rpm and increment force at yield. The beneath table 1 portrays the point position to pivot the engine.

Channel Servo Driver

From the figure 5 Driving servo engines with the Arduino Servo library is entirely simple, yet every one expends a valuable pin - also some Arduino preparing power. The Adafruit 16-Channel 12-cycle PWM/Servo Driver will drive up to 16 servos over I2C with just 2 pins. The on-board PWM regulator will drive each of the 16 channels at the same time with no extra Arduino handling overhead. Additionally, you can bind up to 62 of them to control up to 992 servos - all with similar 2 pins. The Adafruit PWM/Servo Driver is the ideal answer for any undertaking that requires a great deal of servos.

Bluetooth Module HC-05

HC-05 Bluetooth Module is a simple to utilize Bluetooth SPP (Serial Port Protocol) module, intended for straightforward remote sequential association arrangement. Its correspondence is through sequential correspondence which makes a simple method to interface with regulator or PC. HC-05 Bluetooth module gives exchanging mode among ace and slave mode which implies it ready to utilize neither accepting nor sending information. The figure 6 gives Bluetooth module and Table 2 gives the pin depiction.

RESULTS AND DISCUSSIONS

Fabrication

Let's start with the basic assembly of Biped Robot from the figure 7. Some basic tools like screw drivers, pliers, small spanners, soldering iron, wire cutter, nipper, stripper are required to complete the assembly. The figure 7 gives the required components for assembling biped robot.

The following screws, nuts, bolts and other parts are like these:

- M4 12 Screws with M4 Screw and Nut
- 3 by 6 Servo Screw - Servo Screw, head screw, big button, CSK screw, bearing
- Miniature Ball Radial Bearing - Bearing
- 3 mm Nylock Nut - Lock nut
- 3 mm Nut – Nut and required components for building Biped robots.

CONCLUSION AND FUTURE WORK

The structure execution portrays step organizing of biped humanoid robot in different scene conditions. In the past cases the step orchestrating was arranged interestingly for level surfaces without contemplating the obstacles. The speed of robot was very well balanced in level surfaces yet in skewed and subtle surfaces it was moderate. Here the biped humanoid robot is planned for developments in different manners to be explicit forward, the two different ways turn. The servo motors are set in six joints of the biped robot. Since the various developments of a biped robot are compelled by six servo motors it has six degree of chance. These motors are compelled by the Pulse width modulation signal is delivered from the Arduino development board. Taking into account the ON time and OFF period of the beat width balanced sign the various plots for the development of the biped robot is being resolved. Dependent upon these focuses the development of the biped robot occurs in different scene conditions like level, precarious and skewed surfaces. For controlling the crushing in hazardous surfaces the foot of the biped robot is secured with various harsh materials. Despite this ultrasonic sensors are used to discover the deterrents in arranged courses. Considering the squares present the biped robot takes substitute ways.

In future, developed humanoid robots accept a crucial capacity for different domain applications. From this time forward the above biped humanoid model can be connected with develop a humanoid robot which modifies particular biological conditions. Such a humanoid robots are commonly used for saving lives in defence and military, present day help and various applications.

REFERENCES

Brandao, M., Hashimoto, K., Santos-Victor, J., & Takanishi, A. (2014, November). Gait planning for biped locomotion on slippery terrain. In *2014 IEEE-RAS International Conference on Humanoid Robots* (pp. 303-308). IEEE. 10.1109/HUMANOIDS.2014.7041376

Buchli, J., Theodorou, E., Stulp, F., & Schaal, S. (2011). Variable impedance control a reinforcement learning approach. *Robotics Science and Systems: Online Proceedings, VI*, 153–160.

Buschmann, T., Favot, V., Schwienbacher, M., Ewald, A., & Ulbrich, H. (2013). Dynamics and Control of the Biped Robot Lola. In H. Gattringer & J. Gerstmayr (Eds.), *Multibody System Dynamics, Robotics and Control*. Springer. doi:10.1007/978-3-7091-1289-2_10

Dai, H., Valenzuela, A., & Tedrake, R. (2014). Whole-body motion planning with centroidal dynamics and full kinematics. In *2014 IEEE-RAS International Conference on Humanoid Robots* (pp. 295-302). IEEE. 10.1109/HUMANOIDS.2014.7041375

Deits, R., & Tedrake, R. (2014, November). Footstep planning on uneven terrain with mixed-integer convex optimization. In *2014 IEEE-RAS international conference on humanoid robots* (pp. 279-286). IEEE.

Fabisch, A., Petzoldt, C., Otto, M., & Kirchner, F. (2019). *A Survey of Behavior Learning Applications in Robotics--State of the Art and Perspectives*. arXiv preprint arXiv:1906.01868

Huang, W., Kim, J., & Atkeson, C. G. (2013, May). Energy-based optimal step planning for humanoids. In *2013 IEEE International Conference on Robotics and Automation* (pp. 3124-3129). IEEE. 10.1109/ICRA.2013.6631011

Jo, H. S., & Mir-Nasiri, N. (2013). Development of minimalist bipedal walking robot with flexible ankle and split-mass balancing systems. *International Journal of Automation and Computing, 10*(5), 425–437. doi:10.100711633-013-0739-4

Kaneko, K., Kanehiro, F., Morisawa, M., Akachi, K., Miyamori, G., Hayashi, A., & Kanehira, N. (2011, September). Humanoid robot hrp-4-humanoid robotics platform with lightweight and slim body. In *2011 IEEE/RSJ International Conference on Intelligent Robots and Systems* (pp. 4400-4407). IEEE. 10.1109/IROS.2011.6094465

Kasina, H., Bahubalendruni, M. R., & Botcha, R. (2017). Robots in medicine: Past, present and future. *International Journal of Manufacturing, Materials, and Mechanical Engineering, 7*(4), 44–64. doi:10.4018/IJMMME.2017100104

Khusainov, R., Shimchik, I., Afanasyev, I., & Magid, E. (2015, July). Toward a human-like locomotion: modelling dynamically stable locomotion of an anthropomorphic robot in simulink environment. In *2015 12th International Conference on Informatics in Control, Automation and Robotics (ICINCO)* (Vol. 2, pp. 141-148). IEEE. 10.5220/0005576001410148

Kim, J. Y., Park, I. W., & Oh, J. H. (2007). Walking control algorithm of biped humanoid robot on uneven and inclined floor. *Journal of Intelligent & Robotic Systems, 48*(4), 457–484. doi:10.100710846-006-9107-8

Kumaran, B. S., & Kirubakaran, S. J. (2018). Implementation of 6-DOF Biped Footstep Planning Under Different Terrain Conditions. *Dimension, 18*(13).

Lin, C. Y., Tseng, C. K., Teng, W. C., Lee, W. C., Kuo, C. H., Gu, H. Y., ... Fahn, C. S. (2009, June). The realization of robot theater: Humanoid robots and theatric performance. In *2009 International Conference on Advanced Robotics* (pp. 1-6). IEEE.

Maiorino, A., & Muscolo, G. G. (2020). Biped robots with compliant joints for walking and running performance growing. *Frontiers of Mechanical Engineering, 6*, 11. doi:10.3389/fmech.2020.00011

Manikanthan, S. V., & Padmapriya, T. (2017). Relay Based Architecture For Energy Perceptive For Mobile Adhoc Networks. *Advances and Applications in Mathematical Sciences, 17*(1), 165–179.

Nikolić, M., Branislav, B., & Raković, M. (2014). Walking on slippery surfaces: Generalized task-prioritization framework approach. In *Advances on Theory and Practice of Robots and Manipulators* (pp. 189–196). Springer. doi:10.1007/978-3-319-07058-2_22

Ott, C., Baumgärtner, C., Mayr, J., Fuchs, M., Burger, R., Lee, D., . . . Hirzinger, G. (2010, December). Development of a biped robot with torque controlled joints. In *2010 10th IEEE-RAS International Conference on Humanoid Robots* (pp. 167-173). IEEE. 10.1109/ICHR.2010.5686340

Rubio, F., Valero, F., & Llopis-Albert, C. (2019). A review of mobile robots: Concepts, methods, theoretical framework, and applications. *International Journal of Advanced Robotic Systems, 16*(2), 1729881419839596. doi:10.1177/1729881419839596

Silva, M. F., & Machado, J. T. (2012). A literature review on the optimization of legged robots. *Journal of Vibration and Control, 18*(12), 1753–1767. doi:10.1177/1077546311403180

Suliman, W., Albitar, C., & Hassan, L. (2020). Optimization of Central Pattern Generator-Based Torque-Stiffness-Controlled Dynamic Bipedal Walking. *Journal of Robotics, 2020*, 2020. doi:10.1155/2020/1947061

Yokoyama, K., Handa, H., Isozumi, T., Fukase, Y., Kaneko, K., Kanehiro, F., . . . Hirukawa, H. (2003, September). Cooperative works by a human and a humanoid robot. In *2003 IEEE International Conference on Robotics and Automation* (Cat. No. 03CH37422) (Vol. 3, pp. 2985-2991). IEEE. 10.1109/ROBOT.2003.1242049

Zaier, R. (Ed.). (2012). *The Future of Humanoid Robots: Research and Applications.* BoD–Books on Demand. doi:10.5772/1407

Zhou, X., Guan, Y., Zhu, H., Wu, W., Chen, X., Zhang, H., & Fu, Y. (2014). Bibot-u6: A novel 6-dof biped active walking robot-modeling, planning and control. *International Journal of Humanoid Robotics, 11*(02), 1450014. doi:10.1142/S0219843614500145

APPENDIX: FIGURES AND TABLES

Figure 1. Block diagram of Biped walking robot

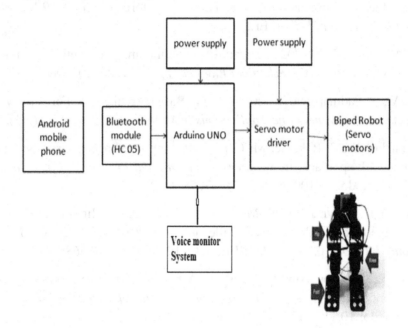

Figure 2. Flow chart

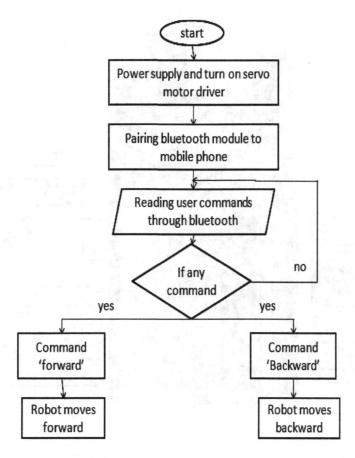

Figure 3. Arduino Uno Pinout

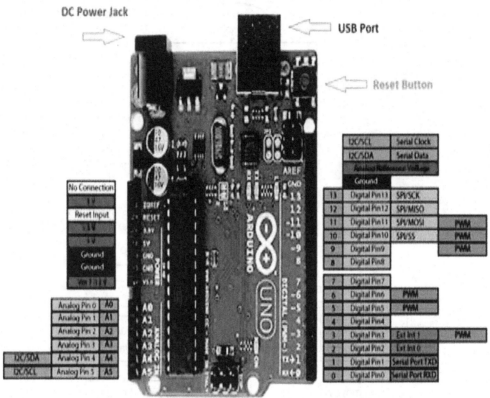

Figure 4. Working Principal of servomotor

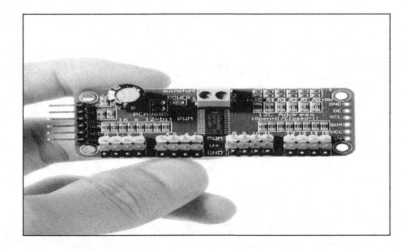

Figure 5. Servo driver
Reference [https://diyi0t.com/shift-register-tutorial-for-arduino-and-esp8266/]

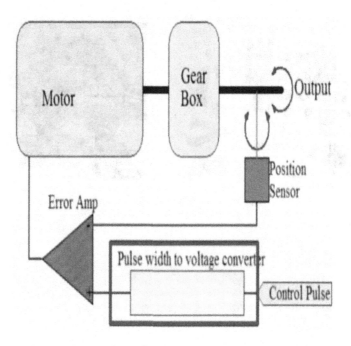

Figure 6. Bluetooth Module
Reference: [https://www.skyfilabs.com/online-courses/biped-walking-robot-project?v2]

Figure 7. Components required for assembling biped robot
Reference: [https://www.skyfilabs.com/online-courses/biped-walking-robot-project?v2]

Table 1. Angle position to rotate the motor

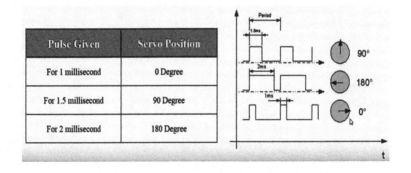

Pulse Given	Servo Position
For 1 millisecond	0 Degree
For 1.5 millisecond	90 Degree
For 2 millisecond	180 Degree

Table 2. Pin description of HC-05 Module

Pin	Description	Function
VCC	+5V	Connect to +5V
GND	Ground	Connect to Ground
TXD	UART_TXD, Bluetooth serial signal sending PIN	Connect with the MCU's (Microcontroller and etc) RXD PIN.
RXD	UART_RXD, Bluetooth serial signal receiving PIN	Connect with the MCU's (Microcontroller and etc) TXD PIN.
KEY	Mode switch input	If it is input low level or connect to the air, the module is at paired or communication mode. If it's input high level, the module will enter to AT mode.

Chapter 7
Realizing a Ultra–Low Latency M–CORD Model for Real–Time Traffic Settings in Smart Cities

Kathiravan Srinivasan
https://orcid.org/0000-0002-9352-0237
Vellore Institute of Technology, Vellore, India

Aswani Kumar Cherukuri
Vellore Institute of Technology, Vellore, India

Senthil Kumaran S.
https://orcid.org/0000-0001-9994-9424
Vellore Institute of Technology, Vellore, India

Tapan Kumar Das
Vellore Institute of Technology, Vellore, India

ABSTRACT

At present, the need for an ultra-high speed and efficient communication through mobile and wireless devices is gaining significant popularity. The users are expecting their network to offer real-time streaming without much latency. In turn, this will result in a considerable rise in network bandwidth utilization. The live streaming has to reach the end users mobile devices after traveling through the base station nodes, core network, routers, switches, and other equipment. Further, this will lead to a scenario of content latency and thereby causing the rejection of the mobile devices users' request due to congestion of the network and mobile service providers' core network witnessing an extreme load. In order to overcome such problems in the contemporary 5G mobile networks, an architectural framework is essential, which offers instantaneous, ultra-low latency, high-bandwidth access to applications that are available at the network edge and also making the task processing in close proximity with the mobile device user.

DOI: 10.4018/978-1-7998-3375-8.ch007

INTRODUCTION

The present-day consumer electronics gadgets like smartphones, intelligent mobile devices, and smart televisions and so on, have extraordinary demand for real-time streaming of videos (Guo, Liu, & Zhang, 2018). The multi-access edge computing (MEC) provides the solution to such problems by moving the computation and data processing of services and applications nearer to the mobile gadget user. Rather than processing the data in the cloud, the analyses, processing and storing of data is done at the edge of the network. Moreover, the computing is done with the deployment of petite edge servers at the network edge instead of the centralized servers in the conventional architecture. The data related to applications and services are analyzed, processed and stored closer to the user equipment (UE), thereby minimizing the network traffic and improving the overall efficiency and performance. Further, MEC offers high-scalability, high-flexibility, and high-availability besides leading to the growth of novice applications and tailored network services provided to the consumers. Also, this specific technology offers an extraordinary quality of experience (QoE) and extraordinary quality of service (QoS). Due to the latest progress in the MEC, it has become easy for consumer Electronic devices and gadgets for implementing and using several real-time applications in virtual reality, augmented reality, mixed reality, Internet-of-Things (IoT) and other emerging technological fields in comparison with conventional technologies.

In the 5G communication technologies, it can be witnessed that the virtual platform is built using the MEC architecture by amalgamating the essential concepts of the software-defined networking (SDN) with the network function virtualization (NFV) (Golestan, Mahmoudi-Nejad, & Moradi, 2019). Also, the point to be noted is that the 5G communication networks can be easily programmed and the virtualization can be done with ease leading to superior evolution of technologies. Besides, the features like automation, maximum throughput, ultra-low latency is possible with the MEC enabled 5G edge computing networks. Furthermore, MEC architecture is highly-scalable due to the decentralization of the network core, which aids to address the demands of real-time and high-speed applications and services (Sabella, Vaillant, Kuure, Rauschenbach, & Giust, 2016).

The NFV provides a novel direction for devising, implementing and handling the network traffic and services. The multi-access edge computing technology makes use of the NFV for virtualizing the network functions and executing them at the network edge. Further, the networking needs of the NFV and the MEC are more or less analogous. Hence the NFV's infrastructure services and management shall be re-used for designing a significant proportion of the MEC's architecture. Subsequently, it might lead to the scenario, where, in a separate virtual platform the network service providers/ operators will be hosting applications and services involving the NFV and the multi-access edge computing technology (Srinivasan & Agrawal, 2018). This paradigm might lead to the possibility of achieving minimal capital expenditure (CAPEX) and operational expenditure (OPEX) for the network service providers (M-CORD, 2020)

Recently, it could be observed that the total amount of smart gadgets and consumer electronic devices connected to the internet have increased several manifolds. The business organization IHS Market's investigation establishes the fact that the amount of IoT gadgets and devices around the globe will rise from 31 billion in 2018 to 125 billion in 2030 (Markit, 2017). Also, several IoT gadgets, devices and applications might demand instantaneous, high throughput, high-speed data transmission with ultra-low latency (Srinivasan, Agrawal, Cherukuri, & Pounjeba, 2018). Hence, the MEC technology along with the NFV paradigm would enable a large number of connected IoT gadgets and devices with superior Quality of Experience.

The present-day cellular networks have to encounter the significant issue of the phenomenal and extraordinary amount of data generated by smart and intelligent gadgets and devices. The applications of M-CORD enabled 5G Networks is suitable for scenarios where the consumers deprived of conventional cable or DTH services subscription, use the over the top streaming and media services for the video, audio, movie and television data through the internet (Abbas, Khan, Mahmood, Rivera, & Song, 2018) (Saha, Tsukamoto, Nanba, Nishimura, & Yamazaki, 2018) (Trindade, et al., 2017) (Visvizi & Lytras, 2018) (Boyes, Hallaq, Cunningham, & Watson, 2018) (Merino, Bediaga, Iglesias, & Munoa, 2019) (Popescu, Dragana, Stoican, Ichim, & Stamatescu, 2018) (Langmann & Stiller, 2019).

Applications of M-CORD Enabled 5G Networks

- Video-on-demand services such as Amazon Prime, Netflix, and so on (Zhang & Hassanein, 2010) (Niu, Xu, Li, & Zhao, 2012).
- Smart-cities and Internet-of-things based applications (Rajab & Cinkelr, 2018)
- Mixed, Virtual and Augmented reality applications (Aggarwal & Singhal, 2019)
- Live streaming of High-definition videos (Nallappan, Guerboukha, Nerguizian, & Skorobogatiy, 2018).
- Self-driven autonomous vehicles (Daily, Medasani, Behringer, & Trivedi, 2017).
- Instantaneous object detection and tracking, real-time intruder detection, and surveillance (Reddy, Hari Priya, & Neelima, 2015) (Amaresh, Rao, & Hallikar, 2014).
- Real-time traffic control and management (Meghana, Kumari, & Pushphavathi, 2017).
- Industrial Internet of things enabled smart manufacturing technologies (Liao, Loures, & Deschamps, 2018)

Overlook Into Multimedia Traffic Analysis and Initiation of Traffic Offloading

In recent years, the progress and advancements in mobile communications have been prodigious. Moreover, this is primarily due to the increase in the demand for smartphones and tablets, combined with the augmented popularity of social networks and multimedia streaming. Also, it is expected that IoT technology has facilitated the network traffic to grow at a faster pace in the upcoming years. MEC reduces traffic bottlenecks towards the core networks, and the IoT demands high support from the underlying network. An efficient network can help in sustaining the services across the devices with prolonged connectivity. Additionally, the low power consumption would help the devices last for months with a single charge.

Instantaneous applications in virtual reality, augmented reality, mixed reality, autonomous cars, smart traffic control systems, face recognition and tracking, interactive gaming and so on, can significantly benefit from the arrival of state of the art MEC paradigm. This technology aims to address computation offloading and latency reduction, and by enabling the data analysis, processing and storage at the edge of the network, which might also positively impact network capacity constraints. The fast and high-speed computation and easy access provided by the MEC are attracting considerable attention. Furthermore, this has opened a scope for even more automated artificial intelligence (AI) services with high computation that were very difficult to implement using the traditional approaches. However, resource shortage or software breakdown can lead to failure in the MEC architecture affecting the quality of experience (QoE). Thus, the ever-increasing traffic requires extensive research to correctly understand the traffic and develop methods to optimally utilize the network resources and deploy them in a commercial network setting.

Significance of Traffic Analysis

Data collection is an arduous process due to the humongous size of the Internet. Further, organizing and analyzing such a massive quantity of data is an even more challenging task. For the analysis of multimedia traffic to be efficient, every type of traffic needs to be classified by the level of details required to perform the analysis. Traffic patterns may consist of voice traffic, video traffic, data traffic, and so on. Traffic classification is an essential first step as many different applications coexist while simultaneously transmitting different types of multimedia traffic. Analyzing traffic behavior will have a significant implication while the two-way traffic is being investigated during the client-server communication paradigm. Moreover, this makes traffic analyses an even more arduous task. Taking the measurements on the network's edge solves this problem as here both the outgoing and incoming flows of traffic are coupled together. The peering contract among the providers disjoints the forward and backward traffic in the core network. Furthermore, this provides us with a massive advantage while performing traffic analysis in the MEC architecture as MEC brings all the processing to the edge of the network architecture.

In general, various traffic analysis tools can be utilized, such that it can reconstruct the traffic characteristics at all levels, starting right from the packet level in the bottom to application level in the top along with a proper collection of the traffic data. Tools like OpenAirInterface, Open vSwitch, OpenStack and Wireshark can be used to perform the analysis. They will offer relevant information about the statistical data and the required performance indices about the real-time multimedia traffic. These facts will allow us to measure the data on the edge node to satisfy the client-server communication paradigm. Open vSwitch aids to deal with the hardware virtualization that helps us to deploy the M-CORD architecture in the traffic offloading framework along with OpenStack, which helps to compute the data in the cloud and also addresses the infrastructure-as-a-service (IaaS) needs. OpenAirInterface provides an ecosystem for the core EPC, and Wireshark further supports as a packet analyzer and a network troubleshooter. The measuring setup for traffic analysis is shown in Figure 1. A variety of statistical data comprising of both IP and TCP statistics can be efficiently collected using the tools for analyzing the traffic accurately.

Figure 1. Measuring setup for performing traffic analysis in the Multi-access Edge Computing (MEC) architecture.

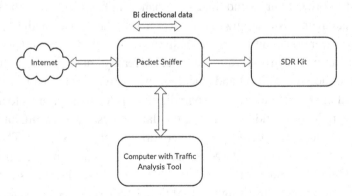

Need for Quality of Service and Vitality of Its Parameters

The transmission characteristics between the two-data link service (DLS) users can be described through the QoS parameters. These characteristics are observable by the DLS users but are solely dependent on the DLS provider. The main aim is to create setups to improve the QoS parameters. Different applications produce different types of traffic having different sets of requirements. There might be applications that can tolerate some amount of traffic delay in the network whereas there might be others with high sensitivity towards any delay and a very high penalty for deadline miss. For example, small delays in text chat might be allowed, but any delay in the network for traffic generated from vehicle anti-collision system will cause a problem. Some applications can tolerate traffic loss whereas others cannot like packet drops in video streaming can be ignored, but the same cannot be said for text chat. These different types of requirements are expressed using the QoS parameters such as latency, loss, bandwidth, jitter, and throughput and so on. The QoS parameters of the network can be measured to successfully create a MEC architecture that allows an efficient flow of traffic while providing a high Quality of Service.

Traffic Offloading

The problem of high latency and excessive load on the network is best solved using traffic offloading. In 5G communication networks MEC aids in planning to offload the real-time network traffic near the UE, which is closer to the network edge. Furthermore, this allows the applications and services to be implemented near the user equipment/subscriber, which minimizes the end-to-end network delay. Moreover, minimization of end-to-end network delay can be one of the most critical requirements for the 5G architecture. The delay at every node in the instantaneous traffic offloading process of the core network can be categorized into four types namely the processing delay, queuing delay, transmission delay, and the propagation delay. Since now the offloading is being performed nearer to the UE, the request will travel a shorter distance thus meaning fewer nodes in the route leading to a shorter delay.

Various other services are provided by the MEC like pre-processing of large chunks of data, performing compute-intensive applications at the edge allowing optimization of mobile resources and various other context-aware services and pre-processing large amounts of data enabling the reduction of network load by sending only the results to the core network instead of the entire data that is collected from all the sources as shown in Figure 2.

Separation of the control and user plane, as demonstrated in Figure 3, provides an efficient way for the coexistence of various radio technologies like high capacity small cells and the traditional control network architecture which helps us offload the traffic in MEC architecture. Splitting the control plane and user (C/U) plane of the core network functions enables the operators to optimize real-time performance and manage the loads for various use cases. Efficient management of real-time resources is provided by the macro cells in the network domain which perform as the network control plane in the C/U split.

The logical/physical decoupled C/U plane is integrated with an enormous amount of the heterogeneous small cells. The excessive data requirements of smart gadgets can be managed by deploying the very dense small cells with low power capabilities, thereby unlocking the potential for high spatial reuse. The capacity for spatially localized transmission is increased by the small cell network allowing a superior and effective spectrum utilization along with numerous simultaneously connected gadgets. This scenario is achieved by bringing the small access points near the smart gadgets. Also, this enhances mobility robustness. Moreover, the failure rate of the handover can be minimized by perpetuating the

Figure 2. MEC architecture and its utilization.

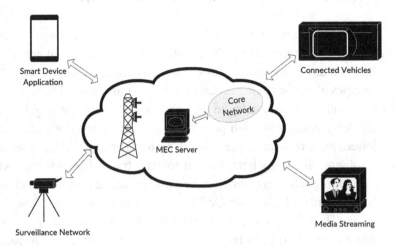

connection of the macro-cell C-plane. It can be witnessed that the network core's signaling overhead can be minimized through the C/U split.

The SDN is deployed for establishing the logical separation amidst the C-Plane and U-Plane. Further, this split assists the SDN in the formation of novice network functions, thereby isolating the complexity by providing a programmable hardware framework. In the MEC architecture, the functionalities like network management, network monitoring, and policy installation are offered by the SDN with the essential elasticity, thereby acting as a catalyst for a better architecture.

In the current scenario of the Mobile Cloud Computing (MCC) architecture (see Figure 4), at the remote-cloud data-centers, the massive amount of data processing can be performed at a high rate and high reliability. However, increased use of MCC architecture leads to the high amount of propagation delay as a result of the considerable distance amidst the remote-cloud data-center and the end user. Further, this results in very high latency making MCC inadequate for the emerging wide-ranged latency-critical

Figure 3. Architecture for Control (C) plane and User (U) Plane Separation

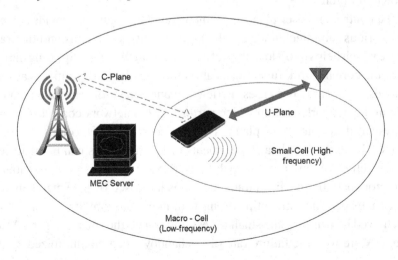

real-time mobile applications. The data exchange between the remote cloud and the UE induces enormous amounts of data flow, thus bringing down the backhaul network. Hence, it is essential that state of the art technology of MEC supplements cloud computing, as demonstrated in Figure 5, by pushing the traffic, computing and the network functions to the edge of the network.

Figure 4. Mobile Cloud Computing Architecture

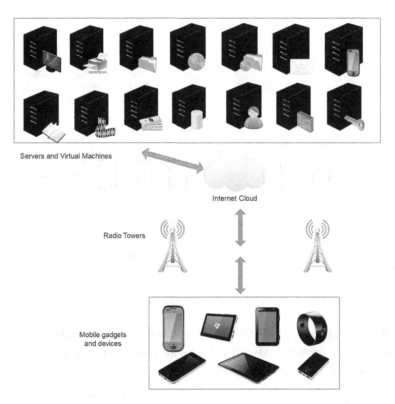

Recent advancements in the cloud architecture, SDN and the network function virtualization (NFV) allow implementation of MEC on virtualized platforms. Computing services to multiple mobile devices are provided by a single edge device using multiple virtual machines (VMs) in the NFV. In the future, the cloud storage would come to the rescue of many developers considering the amount of data that necessities to be stored. The challenges of effective resource utilization have to be dealt with separately. In the MEC milieu, the offloading mechanism is implemented on various virtualization servers at numerous localities near the network edge. These small cell virtual servers will be placed at the radio access network in the MEC environment (see Figure 6).

Also, this will further reduce the distance between the UE and the virtual servers. The virtualized servers will perform related services such as NFV and SDN apart from the MEC offloading services. Moreover, this would lead to the efficient use of the resources and reduce the deployment cost. At the network edge, developing a virtual architecture based on MEC will solve the problem of latency and improve the mean opinion score. However, designing a robust virtual architecture for the local internet breakout is a state of the art process and an arduous task. Thus, we deploy the existing Mobile- Central

Figure 5. MCC architecture incorporated with MEC

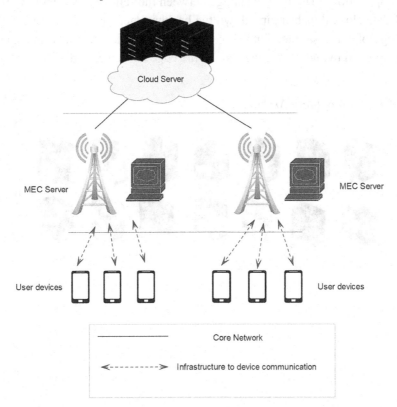

Office Re-architected as a Datacentre (M-CORD) framework [5] as the architecture for the virtualized small cell servers to solve the problem. Also, it will provide efficient mechanisms for data dissemination that supports better QoS and Quality of experience.

M-CORD Framework

The connectivity for wireless User Equipment (UEs) to Packet Data Networks (PDNs) can be accomplished through the Mobile- Central Office Re-architected as a Datacentre framework [5]. The PDNs are Service Provider specific networks, such as Voice over Long-Term Evolution (VoLTE) networks, and public networks, or such as the Internet. Connectivity wise, at a high-level M-CORD, uses two networks

Figure 6. Traffic is offloading at the edge in the MEC environment.

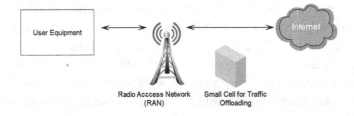

namely an Evolved packet core (EPC) network and a radio access network (RAN). The RAN comprises of some base stations (eNBs in LTE) that provide wireless connectivity to UEs while they are in motion.

M-CORD infrastructure is built on the principles of using fewer commodity building blocks. It uses a combination of open source software and white boxes as its basic building blocks. Further, this allows the M-CORD architecture to be very economical and reduces the infrastructure setup cost of the data-centres. It is built by combining the basics of SDN, cloud technologies and NFV as shown in Figure 7. The virtualization of Radio Access Network (RAN) and disaggregation are the critical features of the architecture. It provides an ideal platform for further renovation and research work in the 5G networks. The next-generation core network deploys the SDN and NFV paradigms along with the capabilities of cloud computing. This scenario indeed assists the network service providers in minimizing the CAPEX and OPEX costs. Further, this paradigm also supports high-scalability, high-flexibility and quick development and implementation of novel network applications and network services and would allow for the future development of Unmanned Aerial Vehicle (UAV) based 5G networks that are cost intensive. These UAVs would function like a human-controlled search unit or might assemble as an autonomous aerial swarm having the ability to group as a 5G enabled aerial communication network.

Figure 7. CORD framework as a combination of SDN, NFV and cloud architecture.

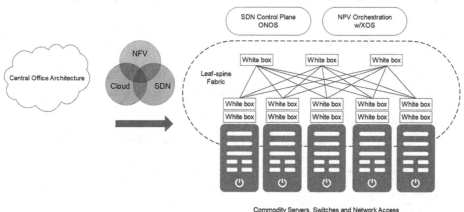

M-CORD aims to provide enhanced resource utilization by providing real-time resource management, monitoring the framework and exploiting the use of multiple Radio Access Technologies (RATs). It can be used to provide customized service composition and differentiate between QoE based upon different service requirements to provide better-customized service and highest quality of experience to the customers. Further, this model provides a virtualized and disaggregated RAN and EPC that uses a commodity hardware and open source software resulting in a low cost and an efficient deployment. In the RAN, both the non-disaggregated eNB and the RU, are linked to the M-CORD POD through a physical association to one of the leaf switches. The traffic generated by the UEs first traverses wirelessly to the eNBs, which are connected to the M-CORD fabric. M-CORD is a robust edge cloud solution: all the core functionalities can be pushed to the edge by the service providers by using M-CORD. They distribute CORD across their edge and central clouds, thereby extending core services across multiple clouds.

M-CORD Architecture for Ultra-low Latency and Real-time Traffic Offloading

A unified platform that offers extensive service delivery can be formed by combining the open-source projects such as OpenStack, Docker, Everything-as-a-Service (XaaS) Operating System - XOS and Open Network Operation System (ONOS) along with the CORD architecture. The ONOS possesses the capability in organizing the leaf-spine topology through the ability to control the underlying white box switch fabric. The XOS is responsible for accomplishing functions such as assembling and composition services. The remote-cloud data-center management and software containers services interconnection can be achieved utilizing the OpenStack and Docker. Figure 8 illustrates the M-CORD architecture that amalgamates the disaggregated/ virtualized RAN; disaggregated/ virtualized EPC and the mobile edge services like edge caching, Self-Organizing Networks (SON) and so on. The RAN with the support of radio access technologies, aids connecting the UE/ smart gadget with the network core. The tailored agile cloud functionalities and services along with the effective network slicing and deep observability are provided by the unified platform designed with the support of the CORD architecture. Further, this paradigm offers great robustness and high-flexibility by offering functionalities like programmable data plane and dynamic radio resource optimization.

Figure 8. M-CORD Architecture for ultra-low latency and real-time traffic offloading

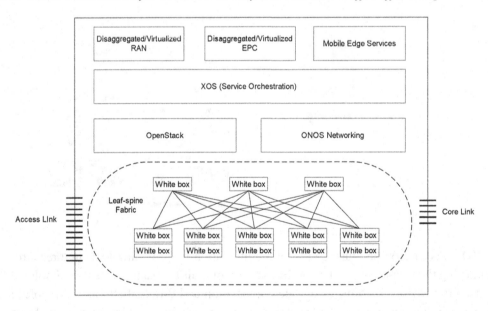

The high-performance offered physical hardware can be replaced through virtualization. This scenario can be achieved by making use of off-the-shelf hardware. The virtualization of the RAN becomes a challenging job due to the system's instantaneous retort to the radio-frequency signals. The baseband unit (BBU) and a remote radio unit (RRU) combine to form a wireless base station. Further, the RRU is placed on the top or base of the tower, whereas the BBU will be typically located near the UE. Each and every BBU get shifted to a centralized locality having space sharing capabilities with other BBUs, in a centralized RAN architecture. An emerging notion in the 5G communication networks ambiance provides the opportunity for diverse partitioning approaches for the baseband that is more appropriate for small cells. Moreover, this might help in minimizing the maintenance expenses and setting up the cost. The virtualized RAN holds a vital role in the development of the present network architecture that will result in a superior framework for 5G communication networks. Moreover, this network setting possesses the capability to handle and progress towards high-demand applications in an agile ambiance.

CONCLUSION

This article offers a conceptual insight into the M-CORD architecture for ultra-low latency and real-time traffic offloading applications in the 5G communication networks environment. It can be observed that this architecture performs all the computing tasks and data processing near the network edge. Subsequently, this will result in the establishment of robust models for instantaneous high-speed traffic offloading in the future. We have described the M-CORD framework, and also we have established a unified virtual platform that integrates the NFV, SDN, remote-cloud data-center and, open source software along with the 5G- enabled M-CORD architecture. Moreover, the M-CORD architecture based MEC platform provides the computation at the network edge, thereby minimizing the network core's burden and also creating an environment for the latency-sensitive applications and services to work effortlessly and precisely. Furthermore, the M-CORD virtual paradigm offers an ambiance for collaborative development in addition to providing an open source solution and thereby minimizing the overall CAPEX and OPEX. There are still some open issues to be addressed such as build issues and downtime problem. Several business organizations working in 5G communication technology are developing real-time, high-scalability, high-flexibility, high-availability, high-speed, ultra-low latency applications and services based on this M-CORD framework for 5G enabled smart city.

REFERENCES

Abbas, M. T., Khan, T. A., Mahmood, A., Rivera, J., & Song, W.-C. (2018). Introducing network slice management inside M-CORD-based-5G framework. In *NOMS 2018 - 2018 IEEE/IFIP Network Operations and Management Symposium* (pp. 1-2). Taipei: IEEE. doi:10.1109/NOMS.2018.8406113

Aggarwal, R., & Singhal, A. (2019). *Augmented Reality and its effect on our life*. IEEE. doi:10.1109/CONFLUENCE.2019.8776989

Amaresh, H., Rao, Y., & Hallikar, R. (2014). Real-Time Intruder Detection System Using Sound Localization and Background Subtraction. *2014 Texas Instruments India Educators' Conference (TIIEC)*, 131-137. doi:10.1109/TIIEC.2014.030

Boyes, H., Hallaq, B., Cunningham, J., & Watson, T. (2018). 10). The industrial internet of things (IIoT): An analysis framework. *Computers in Industry*, *101*, 1–12. doi:10.1016/j.compind.2018.04.015

Daily, M., Medasani, S., Behringer, R., & Trivedi, M. (2017). Self-Driving Cars. *Computer*, *50*(12), 18–23. doi:10.1109/MC.2017.4451204

Golestan, S., Mahmoudi-Nejad, A., & Moradi, H. (2019). A Framework for Easier Designs: Augmented Intelligence in Serious Games for Cognitive Development. *IEEE Consumer Electronics Magazine, 8*(1), 19–24. doi:10.1109/MCE.2018.2867970

Guo, H., Liu, J., & Zhang, J. (2018). Computation Offloading for Multi-Access Mobile Edge Computing in Ultra-Dense Networks. *IEEE Communications Magazine, 56*(8), 14–19. doi:10.1109/MCOM.2018.1701069

Langmann, R., & Stiller, M. (2019). The PLC as a Smart Service in Industry 4.0 Production Systems. *Applied Sciences (Basel, Switzerland), 9*(18), 1–20. doi:10.3390/app9183815

Liao, Y., Loures, E., & Deschamps, F. (2018). Industrial Internet of Things: A Systematic Literature Review and Insights. *IEEE Internet of Things Journal, 5*(6), 4515–4525. doi:10.1109/JIOT.2018.2834151

M-CORD. (2020, July 30). http://opencord.org/

MarkitI. (2017). https://news.ihsmarkit.com/prviewer/release_only/slug/number-connected-iot-devices-will-surge-125-billion-2030-ihs-markit-says

Meghana, B., Kumari, S., & Pushphavathi, T. (2017). Comprehensive traffic management system: Real-time traffic data analysis using RFID. In *2017 International conference of Electronics, Communication and Aerospace Technology (ICECA)* (pp. 168-171). Coimbatore: IEEE. 10.1109/ICECA.2017.8212787

Merino, R., Bediaga, I., Iglesias, A., & Munoa, J. (2019). Hybrid Edge–Cloud-Based Smart System for Chatter Suppression in Train Wheel Repair. *Applied Sciences (Basel, Switzerland), 9*(20), 1–18. doi:10.3390/app9204283

Nallappan, K., Guerboukha, H., Nerguizian, C., & Skorobogatiy, M. (2018). Live Streaming of Uncompressed HD and 4K Videos Using Terahertz Wireless Links. *IEEE Access: Practical Innovations, Open Solutions, 6*, 58030–58042. doi:10.1109/ACCESS.2018.2873986

Niu, D., Xu, H., Li, B., & Zhao, S. (2012). *Quality-assured cloud bandwidth auto-scaling for video-on-demand applications. In 2012 Proceedings IEEE INFOCOM.* IEEE. doi:10.1109/INFCOM.2012.6195785

Popescu, D., Dragana, C., Stoican, F., Ichim, L., & Stamatescu, G. (2018). A Collaborative UAV-WSN Network for Monitoring Large Areas. *Sensors (Basel), 18*(12), 1–25. doi:10.339018124202 PMID:30513655

Rajab, H., & Cinkelr, T. (2018). IoT based Smart Cities. In *2018 International Symposium on Networks, Computers and Communications (ISNCC)* (pp. 1-4). Rome: IEEE. doi:10.1109/ISNCC.2018.8530997

Reddy, K., Hari Priya, K., & Neelima, N. (2015). Object Detection and Tracking — A Survey. In *2015 International Conference on Computational Intelligence and Communication Networks (CICN)* (pp. 418-421). Jabalpur: IEEE. doi:10.1109/CICN.2015.317

Sabella, D., Vaillant, A., Kuure, P., Rauschenbach, U., & Giust, F. (2016). Mobile-Edge Computing Architecture: The role of MEC in the Internet of Things. *IEEE Consumer Electronics Magazine, 5*(4), 84–91. doi:10.1109/MCE.2016.2590118

Saha, R. K., Tsukamoto, Y., Nanba, S., Nishimura, K., & Yamazaki, K. (2018). *Novel M-CORD Based Multi-Functional Split Enabled Virtualized Cloud RAN Testbed with Ideal Fronthaul.* IEEE. doi:10.1109/GLOCOMW.2018.8644390

Srinivasan, K., & Agrawal, N. K. (2018). A study on M-CORD based architecture in traffic offloading for 5G-enabled multiaccess edge computing networks. In *2018 IEEE International Conference on Applied System Invention (ICASI)* (pp. 303-307). Chiba: IEEE. 10.1109/ICASI.2018.8394593

Srinivasan, K., Agrawal, N. K., Cherukuri, A. K., & Pounjeba, J. (2018). An M-CORD Architecture for Multi-Access Edge Computing: A Review. In *2018 IEEE International Conference on Consumer Electronics-Taiwan (ICCE-TW)* (pp. 1-2). Taichung: IEEE. doi:10.1109/ICCE-China.2018.8448950

Trindade, E. P., Hinnig, M., Costa, E., Marques, J., Bastos, R., & Yigitcanlar, T. (2017). Sustainable development of smart cities: A systematic review of the literature. *Journal of Open Innovation*, *3*(1), 1–14. doi:10.118640852-017-0063-2

Visvizi, A., & Lytras, M. (2018). It's Not a Fad: Smart Cities and Smart Villages Research in European and Global Contexts. *Sustainability*, *10*(8), 1–10. doi:10.3390u10082727

Zhang, X., & Hassanein, H. (2010). Video on-demand streaming on the Internet — A survey. In *2010 25th Biennial Symposium on Communications* (pp. 88-91). Kingston: IEEE. doi:10.1109/BSC.2010.5472998

Chapter 8
Traffic Analysis Using IoT for Improving Secured Communication

Lakshman Narayana Vejendla
Vignan's Nirula Institute of Technology and Science for Women, India

Alapati Naresh
Vignan's Nirula Institute of Technology and Science for Women, India

Peda Gopi Arepalli
Vignan's Nirula Institute of Technology and Science for Women, India

ABSTRACT

Internet of things can be simply referred to as internet of entirety, which is the network of things enclosed with software, sensors, electronics that allows them to gather and transmit the data. Because of the various and progressively malevolent assaults on PC systems and frameworks, current security apparatuses are frequently insufficient to determine the issues identified with unlawful clients, unwavering quality, and to give vigorous system security. Late research has demonstrated that in spite of the fact that system security has built up, a significant worry about an expansion in illicit interruptions is as yet happening. Addressing security on every occasion or in every place is a really important and sensitive matter for many users, businesses, governments, and enterprises. In this research work, the authors propose a secret IoT architecture for routing in a network. It aims to locate the malicious users in an IoT routing protocols. The proposed mechanism is compared with the state-of-the-art work and compared results show the proposed work performs well.

DOI: 10.4018/978-1-7998-3375-8.ch008

INTRODUCTION

The Internet of Things may simply be referred to as the Internet of the Whole, which is a network of objects enclosed with software, sensors, electronics, which allows them to capture and distribute data. Smart Homes and Cities, Connected Cars, Healthcare, Smart Farming, Industrial Internet, Manufacturing, Smart Retail are some of IoT's applications (AbuMansour, 2018). IoT has many advantages that it provides more reliable communication and is very efficient and saves time and money, increases business opportunities, increases productivity, improves quality of life. Not only advantages but also some disadvantages in IoT are less privacy and less security, compatibility and technology over-reliance (Aman, 2017). Security is the key issue and challenge in IoT. Some of the security challenges in IoT are Authentication, Access Control, Data Confidentiality, Trust, Secure Mediaware, Privacy. It is very important that data transmission between IoT devices is very secure (Ardissono, 2002). Communication is possible through routing protocols and data should be secured during routing.

Figure 1. Basic IOT architecture

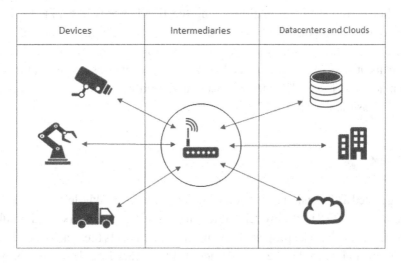

Routing is a key factor in IoT that helps to communicate between devices and also to transmit data (Baker, 2017). Running a good routing protocol will boost the efficiency of Low Power and Lossy networks, known in short as LLN (Carlier, 2016). To assess the efficiency of a protocol, we should consider factors such as energy consumption, overhead control, throughput, packet transmission ratio, latency. Routing is the key component in IoT 's full IPV6 network. The routing protocols should make the IoT a reality(Castiglione, 2013).

Throughout this study, I discuss the security of routing protocols throughout IOT primarily in the network layer and the thorough explanation of the attacks on these routing protocols and some of their counter measures and the perfomance evaluation of these routing protocols when an attack occurs. Let's see some of the IOT routing protocols proposed by IETF which stands for Internet Engineering Task Force, RPL is intended for LLN's which represents Low Power and Loss Network (LLN's) routing protocol, and also a source routing protocol, an significant aspect about RPL is the specific routing result for LLN's (Chen, 2018), another protocol is IPV6 over Low Power Wireless Personal Area Network title.

AODV named Ad-hoc On Demand Distance Vector Routing Protocol, which accounts for a single, loop-free, on-demand route (Chiu, 2005). In AODV, it selects the best path and avoids the others, even though several paths have been found by the source. XMPP's XML-based and IM-based protocol (Instant Messaging) stands for Extensible Messaging and Presence Protocol, another MQTT protocol known as IOT message protocol, is a machine-to - machine, IOT connectivity protocol, also an extremely light-weight publisher-subscriber message model known for remote connections. As a result, there are several routing protocols in IOT when entering the IOT protocol stack that consist of layers such as Physical layer, Data link layer, Network layer, Transport layer and Application layer that are five in number, there different layers consist of different routing protocols and each layer undergoes different types of attacks (Conti, 2017).

Here we focus mainly on the network layer, which is the internet layer in the TCP-IP model, and when it comes to the OSI reference model, it is level 3 layer (Jones, 2004). The network layer is also part of the IOT reference architecture networking layer, such as PHY, MAC layers. The main task of this layer is to handle and route data packets. At this layer, the datagram from the transport layer is enclosed to the data packets allocated to their destinations using the IP address (Li, 2017). This survey discusses the network layer routing protocols in IOT and the various types of attacks that occur on these routing protocols and the effect of these attacks on these routing protocols and the performance evaluation of these protocols during attacks (Maheshwari, 2018). The research work section-2 addresses the literature survey 3rd section addresses the safe routing process and section-4 describes the experimental assessment and section-5 ends the research study.

RELATED WORK

Allcock (2018) suggested a "Intrusion Detection Program to Detect Sinkhole Attack on the RPL Protocol on the Internet of Things." IOT is mainly related to wireless sensing networks and is subject to security issues, such as sinkhole attacks. The proposed IDS mechanism detects these attacks on RPL and prompts the leaf nodes (sensor nodes) to the the importance of the packet loss. Here, the proposed algorithm measures the Intrusion Ratio for the detection of malicious nodes in the network.

Anderson (2004) suggested "Analysis of Mechanisms for the Detection of Sinkhole Attacks on RPLs" In this research work, major security issues were focused around the network layer and all approaches were examined and considered, and their uses and drawbacks and resource usage were identified. In the long run, a brief connection was made, which demonstrates the historical organization of methods for detecting attacks such as sinkhole, followed by the current efficient technology.

Ardissono (2002) has written "An Intrusion Detection Program for Selective Transmission of Attacks in IPv6-based Mobile WSNs." This research focuses primarily on detecting selective forwarding intruders, the proposed IDS-based solution is to detect selective forwarding attack, even removing modified node, which is proposed to improve high performance within the mobile network at the expense of overhead control due to the loss of Hello packets.

Castiglione (2013) suggested "Implementation of a Wormhole Attack Against the RPL Network: Challenges and Results." Framed an attack in response to IEEE 802.15.4 WSAN by creating a wormhole. The proposed attack was applied to a genuine RPL topology. The analyzes suggested that the proposed attack could be convincing to undergo a particular attack, such as a DoS. In the long run, we ended up investigating the possibility of conceivable counter-measures.

Chiu (2005) have suggested "Performance Evaluation of the RPL Protocol under Mobile Sybil Attacks." Here, a trust based IDS (T-IDS) solution was proposed in order to reduce the mobility of sybil attacks in the RPL. When the RPL is undergoing sybM, it is observed that the overhead control and energy utilization have been increased and the packet delivery ratio has been reduced. The proposed T-IDS tackles issues that occur when the RPL undergoes mobility-based attacks.

Gochhayat (2019) suggested the "Securing RPL Routing Protocol for Blackhole Attacks Using a Trust-Based Mechanism" A new trustworthy routing protocol providing feedback-aware trust-dependent protection mechanism for IoT systems has been introduced in this research work. This framework sets out the importance (trust) of the nodes that depend on the broad sending actions of the neighboring hubs within the network. As shown in the tests, trust-based value will be subject to knowledge feedback between the nodes as well as to the assessment of trust.

Jones (2004) suggested "Rank Attack using the RPL Goal Function for LLNs." They present a position attack for the routing protocol in LLN systems that alters the objective function (OF) alongside the value of the rank, the proposed rank attacks are progressively distracting in nature, and the attacker node can insist, without using much stretch power, on its adjacent nodes to track their information through the attacker node. However, the results showed that the proposed attack could decrease 30 percent-57% of the data delivery proportion depending on the situation of the attacker node inside the network.

Li (2017) suggested "RIAIDRPL: Rank Increased Attack (RIA) Detection Algorithm for Avoiding Loop in the DODAG RPL." This research work proposes an algorithm named RIAIDRPL to classify the increased attack rank (RIA) that generates a loop in the RPL network. It is also stimulated by various limitations such as PDR, packet delay, attacker identification ratio, attacker node identification rate. The efficiency of the proposed algorithm will be checked with other types of routing attacks on the basis of different network parameters.

Oh (2017) suggested "The Effect of Rank Attack on Network Topology of the Low-Power and Loss Network Routing Protocol." This survey analyzed the numerous possible hazards of attack that are being revised to minimize the execution of the Routing Protocol foe LLN. The influence of the attacks can be seen by imitating the assault on different areas inside the network. Outcomes that have been discovered would have a significant impact on the execution of the program, especially if they are actualized in the sending of the burden zone or in the case of specific attackers. Distinctive types of attacks have been investigated in order to conceal unimproved information for routing or turning favored parents.

PROPOSED WORK

Our layout requires the client to decide which router(s) to fill as the monitor(s), but it is not clear how to select the router(s) for this purpose. In this section, we propose an approach to select the area(s) of the monitor(s) astutely in order to achieve a high precision rate. The terms DIO and DODAG refers to the DODAG Information Object and the Destination Directed Acyclic Graph respectively.

Algorithm for working of router in RPL:

```
Step 1: Receive a DIO(DODAG Information Object)
Step 2: Receive DIO the 1st time
```

```
If yes then follow the steps
Add the sender to the list of parent
Calculate the rank on the basis of objective function
Forward DIO's to others in multicast
If no then follow the steps
Satisfy criteria
If no
Then discard the packets
If yes
Then process the DIO
        If rank not less than own_rank
Maintain the location in the DODAG(Destination Oriented Directed Acyclic Graph)
Go to 3rd condition in step 2
                                            If rank less than own_rank
Then improvise the location and get lesser rank
The parents with the less rank will be denied
Go to 3rd condition in step 2
```
Step 3: end

Another option is to use the measure of between's centrality, which is a measure of centrality in a graph based on shortest paths. The between's centrality of a node v is given by the expression $g(v) = \Sigma$ $s f = v f = t \, \sigma st(v) \, \sigma st$

Where σst is the total number of shortest paths from node s to node t and $\sigma st(v)$ is the number of those paths that pass through v.

Another smart choice for selecting a location for a monitor is to select a router that has the highest potential for transmitting a significant amount of traffic through it. This router is called the 'max-flow router' (Myles, 2003). In order to determine which router has the ability to move a large amount of traffic, we used a technique called 'random walking' (Torre, 2016). We have developed an algorithm that performs a large number of random walks, n walks, on a given topology. During each walk, a random router is chosen to be the source of the walk, and at each step the algorithm randomly chooses to finish the walk or to proceed and randomly move to one of the neighbours of the current router. For each router, we establish a counter that increases by one every time a walk arrives at this router. Finally, we want to put the display on the router with the highest counter value.

Our versatile architecture helps us to build another fascinating technique for choosing a router for the computer. We train the detector on each of the possible routers and estimate its output. We then select the router that achieves the highest accuracy to be the monitor.

Flow between nodes (N = (V, E))

```
1:    counters ¬ zeros (|V |)        D zeros-array
2: for i = 1, . . ., nwalks do
3:        r ¬ choose the first router randomly
4:        counters[r] + +
5:        to continue = choose if to continue the walk, with prob Pc
6:        while to continue do
```

```
7:          choose an arbitrary neighborrJ of r
8:          r ¬ rJ
9:          counters[r] + +
10:          to continue = choose if to continue the walk, with prob Pc
selecting node for route traffic (nmonitors, k, N = (V, E))
 1: M ¬ Æ
2: R ¬ V
3: for i = 1, . . ., nmonitors do
4:          if M = Æ then
5:          M ¬ {the router r Î R which achieved the highest accuracy}
6:          R ¬ R \ {r}
7:          continue
8:          BESTk ¬k% routers from R with the highest accuracy results
9:          rJ ¬ the router r ÎBESTk that is farthest from M
10:          M ¬ M È {rJ}, R ¬ R \ {rJ}
```

Here our proposed algorithm working with mainly two phases. In first phase we are going to identify the highest flow routers. Then we can distribute the traffic based on other routes. Based on selecting node for traffic diversion.

Identifying the attacker nodes (max flow nodes, traffic)

```
{
If (node)
Max traffic > threshold;
Place in a suspected list;
Evaluate the parents of those nodes;
If(node contains fake parents)
Take the id of the node and place them in a blocked list;
}
```

Experimental Results

The suggested approach is implemented in ANACONDA SPYDER, which conducts a traffic analysis for safe data communication. The proposed method is compared with the traditional methods and the results show that the proposed method has better performance than the traditional methods.

a) Throughput

The rate at which packets were successfully transmitted via a network channel is known as the network performance. Therefore, in order to determine the value for small networks, we should apply the packets received by all nodes. There are many ways to calculate performance (instantaneous or average) in a wired or wireless network using a network simulator.

Formula:

$$Throughput = \frac{sum\Big(\big(Total\,count\,of\,true\,packets\big) * \big(average\,size\,of\,the\,packet\big)\Big)}{Total\,time\,sent\,to\,deliver\,that\,amount\,of\,data}$$

b) Packet Delivery Ratio

PDR is simply defined as the ratio between the packets that were generated by the source and the packets that were received by the destination.

Formula: Algebraically, it can be defined as:

PDR= N1 ÷ N2

Where, N1 is the total sum of data packets which was received by the destination and N2 is the total sum of data packets produced by the source.

c) End to End Delay

Difference between the time at which the sender generated the packet and the receiver received the packet. The end to end delay is also known as one way delay which was being referred to time taken for the packet to transmit across the network from sender to receiver.

Formula:

End to End Delay = Sum of (Delay at sender + Delay at receiver + Delay at intermediate nodes)

The proposed method monitors every node and check for attackers based on their behaviour where as the existing method doesnot monitor every node for secure data communication. The throughput of

Figure 2. Throughput

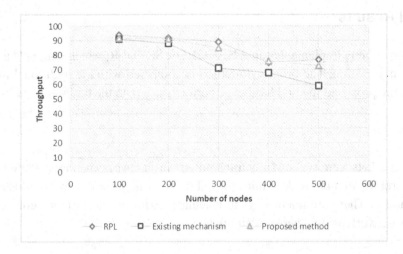

the proposed method is high when compared to the traditional methods as the malicious users are effectively identified.

Here fig-2 represents the throughput comparison between regular RPL protocol, existing secure RPL and our proposed mechanism. Here we simulate regular RPL protocol with different number of nodes varying from 100 to 500 without any attacker nodes. Existing and proposed mechanisms contains attacker nodes of 5, 10, 20, 22 and 25 attacker nodes in each case. And we observe the performance and plotted in fig-2. Here regular RPL protocol has highest throughput than compared to existing and proposed but proposed is very near to standard RPL and more dominating than Existing work.

Figure 3. E2E Delay

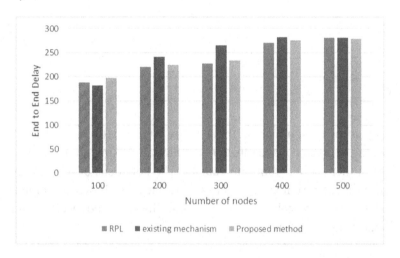

Here fig-3 represents the end to end delay comparison between regular RPL protocol, existing secure RPL and our proposed mechanism. Here we simulate regular RPL protocol with different number of nodes varying from 100 to 500 without any attacker nodes. And existing and proposed mechanisms contains attacker nodes of 5, 10, 20, 22 and 25 attacker nodes in each case. And we observe the performance and plotted in fig-2. Here regular RPL protocol has very slight delay than compared to existing and proposed but proposed is closer delay to standard RPL and more dominating than Existing work.

Here fig-4 represents the packet delivery ratio comparison between regular RPL protocol, existing secure RPL and our proposed mechanism. Here we simulate regular RPL protocol with different number of nodes varying from 100 to 500 without any attacker nodes. And existing and proposed mechanisms contains attacker nodes of 5, 10, 20, 22 and 25 attacker nodes in each case. And we observe the performance and plotted in fig-2. Here regular RPL protocol has highest delivery than compared to existing and proposed but proposed is very near to standard RPL and more dominating than Existing work.

Figure 4. Packet delivery ratio

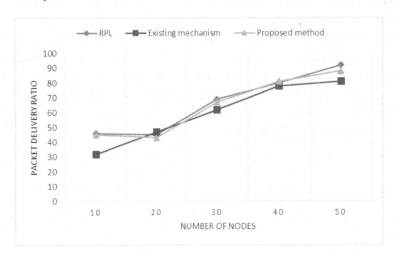

CONCLUSION

Secure communication is the most important thing in any kind of network. IOT is a very large network and a very difficult thing to do is to make safe contact. Several routing protocols are proposed for routing in IOT. Yet most of them suffer from safe contact. The key focus of this research is on safe communication between different IOT nodes, so that we use network-based surveillance mechanisms to detect malicious nodes and ensure stable communication. The proposed mechanism is performing well compared to the literature mechanisms.

REFERENCES

Abu Mansour, H. Y., & Elayyan, H. (2018). IoT theme for smart datamining-based environment to unify distributed learning management systems. In *2018 9th International Conference on Information and Communication Systems (ICICS)* (pp. 212–217). Academic Press.

Allcock, B., Bester, J., Bresnahan, J., Chervenak, A., Kesselman, C., Meder, S., Nefedova, V., Quesnel, D., Tuecke, S., & Foster, I. (2001). Secure, efficient data transport and replica management for high-performance data-intensive computing. In *Eighteenth IEEE Symposium on Mass Storage Systems and Technologies, 2001. MSS '01* (pp. 13–13). 10.1109/MSS.2001.10001

Aman, M. S., Quint, C. D., Abdelgawad, A., & Yelamarthi, K. (2017). Sensing and classifying indoor environments: An iot based portable tour guide system. In *2017 IEEE Sensors Applications Symposium (SAS)* (pp. 1–6). 10.1109/SAS.2017.7894055

Anderson, G., Burnheimer, A., Cicirello, V., Dorsey, D., Garcia, S., Kam, M., Kopena, J., Malfettone, K., Mroczkowski, A., Naik, G., Peysakhov, M., Regli, W., Shaffer, J., Sultanik, E., Tsang, K., Urbano, L., Usbeck, K., & Warren, J. (2004). Demonstration of the secure wireless agent testbed (swat). In *Proceedings of the Third International Joint Conference on Autonomous Agents and Multiagent Systems, 2004. AAMAS 2004* (pp. 1214–1215). Academic Press.

Ardissono, L., Goy, A., Petrone, G., Segnan, M., & Torasso, P. (2002). Ubiquitous user assistance in a tourist information server. In *Adaptive hypermedia and adaptive web-based systems* (pp. 14–23). Springer. doi:10.1007/3-540-47952-X_4

Baker, S. B., Xiang, W., & Atkinson, I. (2017). Internet of things for smart healthcare: Technologies, challenges, and opportunities. *IEEE Access: Practical Innovations, Open Solutions, 5*, 26521–26544. doi:10.1109/ACCESS.2017.2775180

Carlier, F., & Renault, V. (2016). Iot-a, embedded agents for smart internet of things: Application on a display wall. In *2016 IEEE/WIC/ACM International Conference on Web Intelligence Workshops (WIW)* (pp. 80–83). 10.1109/WIW.2016.034

Castiglione, A., Santis, A. D., Castiglione, A., Palmieri, F., & Fiore, U. (2013). An energy-aware framework for reliable and secure end-to-end ubiquitous data communications. In *Proceedings of the 2013 5th International Conference on Intelligent Networking and Collaborative Systems, INCOS '13* (pp. 157–165). IEEE Computer Society. 10.1109/INCoS.2013.32

Chen, D., Bovornkeeratiroj, P., Irwin, D., & Shenoy, P. (2018). Private memoirs of iot devices: Safeguarding user privacy in the IoT era. In *2018 IEEE 38th International Conference on Distributed Computing Systems (ICDCS)* (pp. 1327–1336). IEEE.

Chiu, D. K., & Leung, H. F. (2005). Towards ubiquitous tourist service coordination and integration: A multi-agent and semantic web approach. In *Proceedings of the 7th International Conference on Electronic Commerce* (pp. 574–581). ACM. 10.1145/1089551.1089656

Cicirello, V., Peysakhov, M., Anderson, G., Naik, G., Tsang, K., Regli, W., & Kam, M. (2004). Designing dependable agent systems for mobile wireless networks. *IEEE Intelligent Systems, 19*(5), 39–45. doi:10.1109/MIS.2004.41

Conti, M., Kaliyar, P., & Lal, C. (2017). Remi: A reliable and secure multicast routing protocol for IoT networks. In *Proceedings of the 12th International Conference on Availability, Reliability and Security, ARES '17* (pp. 84:1–84:8). 10.1145/3098954.3106070

Conti, M., Kaliyar, P., Rabbani, M. M., & Ranise, S. (2018). SPLIT: A secure and scalable RPL routing protocol for Internet of Things. In *14th International Conference on Wireless and Mobile Computing, Networking and Communications (WiMob), Limassol* (pp. 1–8). 10.1109/WiMOB.2018.8589115

Gochhayat, S. P., Kaliyar, P., Conti, M., Tiwari, P., Prasath, V. B. S., Gupta, D., & Khanna, A. (2019). LISA: Lightweight context-aware IoT service architecture. *Journal of Cleaner Production, 212*, 1345–1356. doi:10.1016/j.jclepro.2018.12.096

Hasan, H. (2017). Secure lightweight ECC-based protocol for multi-agent IoT systems. In *IEEE 13th International Conference on Wireless and Mobile Computing, Networking and Communications (WiMob), Rome* (pp. 1–8). IEEE.

Jones, V., & Jo, J. H. (2004). Ubiquitous learning environment: An adaptive teaching system using ubiquitous technology. In *Beyond the comfort zone: Proceedings of the 21st ASCILITE Conference* (*vol. 468*, p. 474). Academic Press.

Li, Q. Q., Prasad Gochhayat, S., Conti, M., & Liu, F. A. (2017). EnergIoT: A solution to improve network lifetime of IoT devices. *Pervasive and Mobile Computing, 42*, 124–133. doi:10.1016/j.pmcj.2017.10.005

Li, R., Asaeda, H., & Li, J. (2017). A distributed publisher-driven secure data sharing scheme for information-centric IoT. *IEEE Internet of Things Journal, 4*(3), 791–803. doi:10.1109/JIOT.2017.2666799

Liu, X., Leon-Garcia, A., & Zhu, P. (2017). A distributed software-defined multi-agent architecture for unifying IoT applications. In *8th IEEE Annual Information Technology, Electronics and Mobile Communication Conference (IEMCON)* (pp. 49–55). 10.1109/IEMCON.2017.8117142

Maheshwari, N., & Dagale, H. (2018). Secure communication and firewall architecture for IoT applications. In *2018 10th International Conference on Communication Systems Networks (COMSNETS)* (pp. 328–335). 10.1109/COMSNETS.2018.8328215

Myles, G., Friday, A., & Davies, N. (2003). Preserving privacy in environments with location-based applications. *IEEE Pervasive Computing, 2*(1), 56–64. doi:10.1109/MPRV.2003.1186726

Oh, S., & Kim, Y. (2017). Security requirements analysis for the IoT. In *2017 International Conference on Platform Technology and Service (PlatCon)* (pp. 1–6). Academic Press.

Roth, J. (2002). Context-aware web applications using the pinpoint infrastructure. In *IADIS International Conference WWW/Internet 2002* (pp. 13–15). IADIS Press.

Torre, I., Koceva, F., Sanchez, O. R., & Adorni, G. (2016). A framework for personal data protection in the IoT. In *2016 11th International Conference for Internet Technology and Secured Transactions (ICITST)* (pp. 384–391). 10.1109/ICITST.2016.7856735

Wazid, M., Das, A. K., Odelu, V., Kumar, N., Conti, M., & Jo, M. (2018). Design of secure user authenticated key management protocol for generic iot networks. *IEEE Internet of Things Journal, 5*(1), 269–282. doi:10.1109/JIOT.2017.2780232

Zhang, K., Yang, K., Liang, X., Su, Z., Shen, X., & Luo, H. H. (2015). Security and privacy for mobile healthcare networks: From a quality of protection perspective. *IEEE Wireless Communications, 22*(4), 104–112. doi:10.1109/MWC.2015.7224734

Chapter 9
Challenges to Industrial Internet of Things (IIoT) Adoption

V. Shunmughavel

Computer Science & Engineering, SSM Institute of Engineering and Technology, India

ABSTRACT

The industrial internet of things (IIoT) has made its development within a short span of time. Initially it was considered as a novel idea and currently it is a major driver in industry applications. It has created productivity and efficiency for industries worldwide. This innovative technology can become a practical reality if engineers overcome a variety of challenges. They are connectivity, cost, data integration, trust, privacy, device management, security, interoperability, collaboration, and integration. In this chapter, several facts behind the above-mentioned challenges are being explored and addressed.

INTRODUCTION

The rapid development of IIoT within a short span of time has benefited industry applications. There are certain prime challenges in the adoption of IIoT. In the absence of a secure and properly encrypted network, the adoption of IIoT could lead to brand new security challenges and vulnerabilities (Gudlur, et al., 2020). The prime concern of IIoT is to focus on connecting more and more devices together. This leads to more entry points for malware. Each of these entry points need to be adequately protected against malware and malicious hacking, as well as accidental damage and penetration due to digital damage. The sensitive personal data gathered from devices must be protected from unauthorized access. The users have to be provided with required tools that help them to define the policies for sharing their personal data with authorized persons and applications for data privacy.

Due to the connection of different systems through IIoT, a difficulty to create real cross-domain services that will allow continuous movement of devices and data arises. The problem identified is that various protocols are being utilized and there is no standardization that will lead to interoperability (Phan, & Kim, 2020).There are multiple platforms, numerous APIs, and protocols available for IIoT integration. This leads to the inability of industry workers to operate the new solutions and over all productions process effectively due to their lack of understanding. Device management becomes a big challenge

DOI: 10.4018/978-1-7998-3375-8.ch009

since the usage and number of sensors, gateways and devices will be very large and populated over large geographical areas. The key to enable the Industrial Internet of Things will be connecting all the devices over long distances using cellular and satellite technology. A constant and reliable connectivity is the entire focus of IIoT. Thus, unreliable source of connectivity causes problems in implementing IIoT. When deploying an IIoT application, streams of data move from different sources such as sensors, contextual data from mobile device information, and social network feeds and other web resources. To preserve the semantics of the data in such cases of the data integration from multiple sources is very important.

The close collaboration between individuals and industries with diverse skills and domain knowledge is required to build an IIoT solution. Expertise in two different domains – Information technology (IT) and Operations Technology (OT) is required to build IIoT infrastructure. The major concern for most industries is the huge investment towards the cost of implementing an IIoT infrastructure and obtaining Return on Investment (ROI). Adopting more efficient production processes, and achieving better productivity might also result in increased economic sustainability (Nagy, et al., 2018).In IIoT, things usually move around and are not connected to a power supply, so their smartness needs to be powered from a self-sufficient energy source. Power saving is another challenge since radio frequency identification transponders do not need their own energy source. Their functionality and communications range are very limited. Trust management in IIoT is a major issue for integrating any advanced and automated technological solutions into the manufacturing process.

BACKGROUND

Many detailed studies, regarding vital Challenges and open research issues regarding the implementation of Industry 4.0, have been carried out. Several challenges and fundamental issues in various circumstances that occur throughout the implementation of Industry 4.0 were addressed (Wang, et al., 2016; Vaidya, et al., 2018). They were: 1) decision-making and negotiation, 2) industrial wireless network (IWN) protocols, 3) big data and its analytic, 4) system modelling and analysis, 5) cyber security issues and 6) interoperability, 7) Investment issues. Some of the technology challenges concerning the implementation of Industry 4.0 involve the development of smart devices, the establishment of network environments, big data analysis and processing and digital production (Zhou, Liu, & Zhou, 2015). Currently the volume of data collected by industrial internet of things (IIoT) applications is very huge, making it a challenge to offer platforms with adequate capacity and performance. It is essential to make a detailed analysis of several IIoT platforms in the market and before making the decision of which one to adopt (Moura, et al., 2018).

The lack of a digital strategy in line with resource shortage as well as the lack of standards and poor data security is the main obstacle for the technological implementation of Industry 4.0 (Schroder, 2016). The three greatest challenges connected with implementing Industry 4.0 proposed are standardization, work organization and product availability (Kagermann, et al., 2013).There are still issues and challenges to be coped with in regard to equipment intelligent requirements, deep integration networks and knowledge-driven manufacturing (Chen, et al., 2017).Interoperability is the main open issue in Industry 4.0.Accessibility, multilingualism, security, privacy, the use of open standards, open source software and data integration are the major facts to ensure high accuracy and efficiency of processes (Lu, et al., 2017). An interoperability framework is implemented in this research in which the system components can cooperate and offer the seamless operation from the device to the backend framework. The overall framework is adopted by the EU funded project SEMIoTICS, even when devices from different vendors are utilized (Hatzivasilis, et al., 2018).

Some manufacturers and enterprises hesitate to implement Industry 4.0 due to certain barriers. These include uncertainties about return on investment (financial benefits), lack of strategies of coordinating organizational units at different locations, lack of skill set and hesitation to go through radical transformation and various security threats (Kusters, et al., 2017).A novel analysis framework for IIoT devices provides a practical classification schema for those with an interest in security-related issues surrounding IIoT (Boyes, et al., 2018).Huge advancements and commercialization in IIoT has exposed several security vulnerabilities of the IoT systems. The security attacks on IIoT systems are classified into four different categories on the basis of objects of attack (i.e. devices, networks, software or data) (Sengupta, J.,Ruj, R.& Bit, S., D., 2020). Security vulnerabilities of industrial (e.g. ItronCentron Smart Meter) IIoT devices are analyzed. Their experiments reveal that the industrial automation system is vulnerable to brute force attacks revealing the passwords. It also reveals that the smart meters can be hijacked to launch ransom ware attacks against other systems (Wurm, et al., 2016).

The significant advantages of the proposed architecture in resource utilization and energy consumption are demonstrated by implementing a sleep scheduling and wake-up protocol which balances the traffic load and enables a longer lifetime of the whole system (Wang, et al., 2016). Industrial IoT influences the modern industries to take up new data-driven strategies and manage the global competitive pressure more easily. However, the implementation of IoT increases the entire volume of the generated data transforming the industrial data into industrial Big Data. The adoption of IoT in manufacturing, considering sensory systems and mobile device and industrial Big Data is presented (Mourtzis, et al., 2016).

DEVICE MANAGEMENT

Industrial IIoT systems consist of numerous devices, such as smart phones, temperature sensors, actuators, connected in various environments. These sensors, devices, gateways are connected via communication networks to cloud services and applications. These things could be surrounded or distributed by long distances in different environments but controlled and managed centrally in the cloud, thus named cloud computing. On the other hand, a decentralized solution known as edge/fog computing is an alternative to be realized when processing is required to be carried out closer to the source of the data to improve the quality of service provided.

IIoT System Architecture

The Industrial IoT (Figure1.) system architecture is presented as layers and these layers are: sensing layer, communication layer, cloud layer, management layer, and services and applications layer (Umar, et al., 2018).

The sensing layer comprises of sensors, actuators and smart devices that collect the data from external surroundings. The communication layer focuses on moving the collected data to the cloud. The cloud layer combines the data and stores it for future usage. The stored data is used by the services and application layer. The output data from the services and application layer is presented as services, applications and features to the users. The IIoT communication protocols in the communication layer acts as an interface between sensing layer and cloud.

Figure 1. IIoT System Architecture
(Umar, et al., 2018)

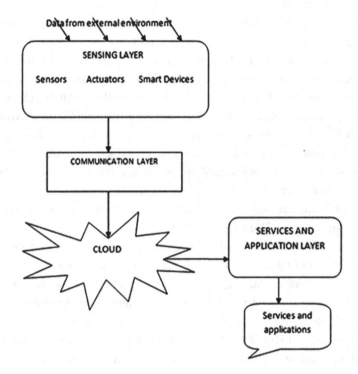

Challenges

Consequent to the accelerated evolution in IIoT, service providers encounter several challenges in satisfying the management requirements. These challenges include the following.

Connectivity of Heterogeneous IIoT Devices

This heterogeneity in connectivity is considered as an important challenge to the interoperability of protocols and solutions developed by different vendors. Therefore, a considerable concern within developing IIoT solutions handles the interaction with heterogeneous IIoT devices.

Device Management Challenges

Device management is the prime challenge in an IIoT environment. Keeping the device status and logs is a continuous process and device statistics is very much required.

Next Generation of IIoT Management Tools

The state-of-the-art methods, protocols, and applications in this new emerging area have been proposed in their research (Sethi & Sarangi, 2017). The variety of tools in the market that have the potential to play an essential role in monitoring smart things in the IIoT solutions are Xively Connected Product

Management (CPM), DevicePilot, Wind River Helix Device Cloud (HDC), QuickLink, ThingWorx Utilities, Particle, Losant Helm and DataVIIoT Device Management tool(Umar, et al., 2018).

CONNECTIVITY

Network Connectivity

The devices hosted by IIoT networks are more when compared to traditional networks. Hence it is important to improve the reliability of IIoT network. Several benefits are brought by the convergence of multiple networks to form IIoT network. If any small connectivity problem occurs, then there is chance for a failure of the entire network in IIoT environment. A novel method to enhance the reliability of IIoT networks is proposed to ensure network availability and high bandwidth (Yang,& Kim,2019). To avoid a single point of failure, high availability network design is required to keep the network in stable condition. Making use of this type of redundant network design minimizes the downtime required when troubleshooting issues on the network. With the addition of more devices to IIoT networks, it should also be ensured that these devices and applications can function without experiencing delay, without overloading the network and making it to crash. The best method to protect a network against cyber-attacks is to use the defense-in-depth security architecture, which is designed to protect individual zones and cells (Chen, & Stegner, 2018).

The next step is to ensure that users cannot change settings by accident or on purpose. The IEC 62443 standard has been viewed by many cyber security experts as the most relevant standard to secure devices on industrial networks. It consists of a series of guidelines, reports, and other relevant documentation that define procedures for implementing electronically secure IACS (Industrial Automation & Control Systems) networks.

Issues With Legacy System

Industrialists prefer to purchase equipment which they can use for decades. They do not prefer to immediately implement new technology such as IIoT. Instead,

they prefer to incorporate their traditional devices into modern solutions. It is a challenge for Industrialists is to develop innovative and more efficient methods of extracting valuable data from all of the equipment deployed across their networks.

Challenges With Facilitating Communication Between the OT and IT Worlds

People, who are working in IT (Information Technology) systems, are not aware of OT (Operational Technology). People, who are working in OT (Operational Technology) systems, are not aware of IT (Information Technology). Also, most of the OT devices are not compatible with IT devices and networking protocols. Both IT and OT engineers have failed to face this challenge. A new platform must be developed by the industrialists so that OT and IT professionals can work together. Hence they can mutually share their ideas and arrive at a clear solution.

Making Devices Smarter via an IIoT Gateway

Gateways deployed in IIoT are used to generate a data transfer between IT devices and OT devices. Due to the absence of use of universal protocols, this data transfer between IT and OT is always an issue. The development of new universal protocols and devices with smart processing capabilities is a major challenge for the Industrialists. Remote monitoring is one of the prime features of the gateways. One of the biggest challenges in IIoT environment is secure remote communication.

Simplify Data Acquisition

In IIoT, gateways must be designed in such a way that they should support multiple protocols. This simplifies the data acquisition process.

DATA INTEGRATION

Intelligent Automation to Solve the Data Integration

A semantic understanding of information across different systems is provided by Bit Stew's integration technology. Work benches and analysis methods help to apply a systematic approach. The MIx Core technology has a novel approach to data integration. The purpose of Mix Core technology is to eliminate the typical point solution technique to integration and replace it with a technique that can be dynamically updated and adjusted using Machine Intelligence. This is the basics of the MIx Core integration stack as discussed below.

Traditional methods of integration depend on definitions, contracts and interface bindings. Hence they are too inflexible to support the requirements of large-scale industrial environments. These methods cannot fit to new environmental factors, new data sources, real-time business impact scenarios and revised configurations. New methods for integration must be based on a more simple design and this is the important concept behind the Mix Core technology. In Mix Core technology the architecture and abstraction layers are designed in such a way that to allow the integration to fit into changing needs. The high-performance demands and the need for cognitive processing of information are still preserved.

Integration Approach

The MIx Core integration solution (Figure 2.) is segmented into five technology layers (Varney, 2018). They are

- System Connectors
- Byte Sequencing & Marking
- Composite Information Patterns
- Semantic Model Mapping
- Intelligence, Analytics and Knowledge

The main features of the MIx Core integration stack include the following:

- High performance access to data based on byte segment marking, byte-level referencing of any data type, and Low-level byte-sequencing support protocol/data translation

Figure 2. Mix Core Integration Stack
(Varney, 2018)

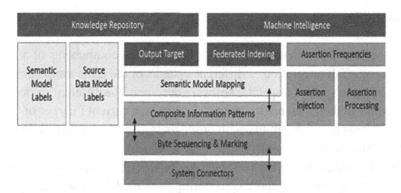

- Data interpretation is performed in higher layers. This minimizes overhead in low level processing. It also allows for several interpretation based on composite information patterns.
- These composite information patterns access the data stream. These information patterns are rapid and intelligent methods are based on syntax notation for discovery and interpretation. These patterns give the adaptation and specification for intelligent processing of data streams.
- Asynchronous Independence Assertion based processing allows rules processing, continuous testing and intelligent control. This is done by using MIx Core, using sequencing, frequency tracking, injection and scope. The assertion-processing layer concentrates on algorithms in the assessment and measurement of data and the execution flow of analytic methods. This is done for the purpose of integration and information analysis.
- Dynamic mapping is supported by semantic model mapping. Hence models and formation of Meta data for contextual understanding is accomplished.
- Code generation is supported by the machine learning and knowledge repository bound in MIx Core.

The MIx Core data integration technology has been deployed by many industrialists who have established the Industrial Internet. This consists of industries that have 10 million different sensors deployed geographically and are controlling critical operational systems. All these industrialists face a common problem of combining data from different backend systems including:

- Operational systems such as production control, historians and Supervisory Control and Data Acquisition (SCADA)
- Business systems such as SAP (Systems, Applications and products), Oracle and mobile workforce management systems
- IIoT communications management systems such as external systems, routers and gateways

- External system such as environmental data, Light Detection and Ranging (LiDAR) data and satellite images
- Open source systems such as Hadoop Distributed File System (HDFS) and Apache HiVE (Query Interface)

INTEROPERABILITY

Interoperability is the ability of a system to work with or use the components of another system. It is simple enough to accomplish integration of different systems of the same domain or between different implementations within the stack of a specific software designer. The different devices and applications are installed and operate in their own platforms and clouds in the current Internet-of-Things (IIoT) environment. However there is no compatibility with products from different brands. For example, a smart phone developed in Android cannot interact with a smart bulb without the appropriate interface provided by the same vendor. For the successful implementation of IIoT infrastructure, the following standards are required to enable the following without considering their model or manufacturer.

- Horizontal and vertical communication
- Operation
- Programming across devices and platforms

Hence the following four levels of interoperability introduced are

- Technological interoperability
- Syntactic interoperability
- Semantic interoperability
- Organizational interoperability

Technological interoperability includes the seamless operation and cooperation of heterogeneous devices that employ different communication protocols on the transmission layer. Syntactic interoperability develops clearly defined and agreed formats for data, interfaces, and encodings. Semantic interoperability resolves commonly agreed information models and ontologies for the used terms that are processed by the interfaces. Organizational interoperability is nothing but cross-domain service integration and orchestration through common semantics and programming interfaces (Zeid, 2019).

Ethernet and wireless are the common interfaces used to connect devices to the application layer while connecting devices to the IIoT. In an industrial operation, not all devices may utilize these interfaces for communicating. Most traditional industrial devices communicate through serial. As an example, a serial device server connects to a serial based device and converts the interface to Ethernet or wireless depending on the model used. The data can be extracted from the equipment and sensors through Ethernet or wireless connections using a device server. This would not have been possible before.

An additional step is required to enable connectivity to the IIoT, when a common interface is introduced in an industrial operation. This step helps to establish a common protocol or language amongst all the devices. For instance, all devices may be communicating through different protocols though they have an Ethernet interface. This can be resolved by using an Ethernet gateway to convert different protocols to a common one. Since the Ethernet gateways are very flexible, they can provide error free protocol conversion. This will suit the future needs of industrial networking.

SECURITY ISSUES IN IIoT

In Industry 4.0 the manufacturing process is connected to information and communication technologies. End user data is merged with machine data which enables machines to communicate with each other. This has made the possibility for components and machines to manage manufacturing process independently with flexibility. Hence manufacturing process becomes more efficient and saves resources. Its benefits include higher product quality, greater flexibility, shorter product launch times, new services and business models (Kamel & Hegazi, 2018). Industry 4.0 enables a new approach to manufacturing, namely Smart Manufacturing by introducing latest technologies with numerous capabilities.

Numerous security challenges are caused by a large number of factors in the Industry 4.0 environment. Hence new technologies need to be prepared to handle a wide variety of cyber security threats. The various policies and practices are enforced to resolve cyber security threats (Maximilian, Markl & Mohamed, 2018).

Secure Communication

The data in the Industry 4.0 environments is very sensitive. Hence it is very essential to secure the transfer of data between layers. Secure communication prevents data confidentiality and integrity adversaries and prevents data leakage and manipulation attacks like man in the middle, eavesdropping and other vulnerabilities against data communication channels (Bakhshi, etal., 2018).

Access Management

In an IIoT environment each device, end-user, process, application and so forth is known as users and should be authenticated and rules of authorization of which should be regulated for each component in order to prevent intruders to access data or do malicious modifications (Misra, & Vaish, 2011).

Data Encryption

Security limitations are very important in Industry 4.0 environment. Various types of IIoT data transfer protocols are required to solve issues which are related to identity and authentication. It is essential to protect huge data against attacks in an IIoT environment. Hence cryptographic algorithms are developed. Efficient encryption techniques are developed to protect data communication and transfer.

Privacy

In IIoT environment, preserving data privacy is a big challenge. Malicious users steal the data or the data gets corrupted. General Data Protection Regulation (GPDR) is introduced to ensure data privacy in industries and organisations. This regulation protects the unauthorised access to industrial data by the malicious users. Industries are encouraged to follow GPDR which ensures data security and privacy in IIoT.

Best Practices to Ensure Security in IIoT

The infrastructure in IIoT incorporates interconnected mechanical devices and computer controlled devices accessing the Internet. The complexity of IIoT environment leads to the subsequent increase of security threats. Efficient security measures are needed to protect information from intruders in IIoT environment. These issues will not be solved by the existing antivirus and firewall software solutions. To resolve these issues enhanced firewall and antivirus software solutions are required. Certain practices can prevent security threats.

- Types of security threats can be listed and identified. The list of security threats can be updated in the organizations' infrastructure.
- Vulnerable equipment should be spotted out. Components of IIoT environment should be recorded in a device status register. If any one of the components starts malfunctioning, the status register should be updated. Hence it is easy to trace out the vulnerable equipment and resolving the security issues.
- Apart from password security techniques, an access mechanism should be created. Implementation of face recognition techniques, voice recognition techniques and biometrics ensure authorized access in IIoT environment.
- Suspicious activities have to be continuously monitored. In general all the equipment in IIoT environment follows specific operation standards. Any deviation in the operation standards will locate the suspicious activity. Then the suspicious device is isolated from the IIoT environment.
- To monitor and analyse the data transfer between devices advanced software utilities are used in IIoT environment. Using these software utilities data from firewall must be captured, processed and analysed for unknown security hazards. It is possible to detect any threat and alert automatically since advanced software utilities set new standards for IIoT environment.

TRUST MANAGEMENT IN IIOT

Trust management in IIOT is an efficient technology to provide security service. Its outcome has been used in many applications such as P2P, Grid, and ad hoc networks and so on. Trust management (TM) is a vital role in IIOT for reliable data communication and mining, enhanced user privacy and information security. IIOT contains smart objects which have heterogeneous characteristic features and need to work jointly and cooperatively. These smart objects are used to public areas and communicate through wireless networks and vulnerable to malicious attacks. IIOT provides a proper trust service according to the request of the clients. This will be translated to trust-related request.

Trust Relationship and Decision (TRD)

Trust management provides an efficient way to evaluate trust relationships between IOT entities and make sure to take wise decision to communicate and collaborate with each other (Neeraj & Singh, 2016).

Trust relationship evaluation focuses all IIoT system entities in all layers. It contributes a vital role for intelligent and autonomic trust management.

Data Perception Trust (DPT)

Data perception trust objective property of trust management in IIoT should be achieved at physical perception layer. Data sensing and collection should be reliable in IIoT. In this aspect, serious attention may be given to the trust properties like sensor sensibility, confidentiality, security, reliability, persistence, preciseness, data collection and integrity.

Privacy Preservation (PP)

Privacy preservation deals with user privacy. This includes user data, and preservation of user location according to the policy and expectation of IIoT users.

Data Fusion and Mining Trust (DFMT)

The data collected in IIoT is in huge amount. This should be processed and analysed with regard to data privacy and data accuracy as a whole. It also relates to trusted social computing in order to mine user demands based on their social behaviours and social relationship exploration and analysis. DFMT focuses the objective properties of the data processor in the IIoT network layer.

Data Transmission and Communication Trust (DTCT)

Private data of others cannot be accessed by illegal system entities in data communications and transmission. This objective is related to the security and privacy properties of IIoT system wherein light security/trust/privacy solution is needed.

POWER SAVING AND GREEN IIOT

Power Saving

There are a number of important challenges that need to be considered, since Industrial IoT (IIoT) systems are becoming more complicated. One such challenge is increasing energy consumption. IIoT was adopted to reduce resource consumption and carbon emissions of industrial systems. IIoT systems include a range of computer systems, electronic devices and other supporting accessories. Large amounts of energy have led to an increasing carbon footprint. IIoT systems normally consist of low-power devices supported by batteries, which is a constraint to the continuous operations.

In the IIoT domain, the massive sensor nodes and smart devices facilitate data collection. Optimization in processing, sensing, and communications may effectively minimize energy consumption for IIoT devices. The prime platform of energy consumption is wireless sensor networks (WSNs) also called the backbone of IIoT. The four types of topological structures are mesh, hybrid, hierarchical and plane.

This facilitates the implementation of large scale WSNs. Both ad hoc and mesh WSNs suffer from a limited overall lifetime.

The method for hierarchical WSNs, places nodes in a tiered framework and limits communications among sensor nodes. Thereby it can improve routing efficiency dramatically and make the network more scalable and extensible. Network lifetimes may be prolonged and the traffic loads can be balanced by categorising nodes as sense nodes, control nodes, and gateway nodes.

Remarkable technology development in the field of Industrial Internet of Things (IIoT) has changed the society we live and work. Apart from the benefits of IIoT to the society, it should be kept in mind that the IIoT also consumes energy, E-waste and generates toxic pollution. These lead to stress on the environment and society.

Green IIoT

Green IIoT is becoming an increasing desire to maximize the benefits and minimize the harm of IIoT. Green IIoT will be environmentally friendly in future. To achieve this, it is necessary to follow a lot of measures to conserve fewer resources, minimize carbon footprint and promote efficient techniques for energy usage. Since it minimizes carbon emission substantially there is an urge to move towards Green IIoT.

Green IIoT has three concepts, namely, design technologies, leverage technologies and enabling technologies (Wang, et al, 2016). Design technologies refer to the communication protocols, energy efficiency of devices, interconnections and network architectures. Leverage technologies reduce carbon emissions and enhance the energy efficiency. By minimizing hazardous emissions, energy, pollution and resource consumption Green IIoT becomes more efficient. Subsequently, Green IIoT leads to minimizing the technology impact on the environment, preserving natural resources and human health and reducing the cost significantly. Hence Green IIoT focuses on green manufacturing, green utilization, green design, and green disposal.

1. **Green Use:** Minimizing power consumption of computer systems and other information.
2. **Green Disposal:** Reusing old computers and recycling discarded computers and other electronic equipment.
3. **Green Design:** Devising energy efficient for green IIoT sound components, computers, and servers and cooling equipment.
4. **Green Manufacturing:** Manufacturing electronic components, computer system and other associated accessories with minimal or no impact on the environment.

COST

While planning on their IIoT initiatives, it is essential for industries to make an accurate calculation of the daily basis operational costs, the future revenue from new service offerings, the operational cost savings, and the period it will take to attain return on investment (ROI). There are many costs incurred in the development and operation of a connected service business.

There are recurring and non-recurring expenses borne by industries. Initial non-recurring costs are application design, security and implementation cost, hardware sourcing and manufacturing. Recurring operational expenses (OpEx) include administrative labour, network communication, and technical support. The categories of costs incurred in running a connected service business are retrieved from https://www.cisco.com/c/dam/m/en_ca/never-better/manufacture/pdfs/hidden-costs-of-delivering-iiot-services-white-paper.pdf (CISCO-JASPER, 2017).

Network Communication Costs

Connecting devices to the Internet is the key factor of IIoT. However, connecting all of the end devices, wherever they may be, poses an important challenge and is dependent on multiple requirements including technical, security, performance, reliability, and location. There are a lot of options available when choosing connectivity. Recently, many industries are moving to mobile networks (or cellular) to give them better control over the IIoT experience. Given the growing adoption of mobile network connectivity, it should be considered as the primary channel for network communication. Hence payments to mobile network operator (MNO) typically account for 33-50 percentages of the overall recurring operational expenses.

Administrative Labour Costs

IIoT services enable industries to drive continuous customer engagement, gain real-time product insights, and tap into new revenue streams. Planning appropriate staff and labour is the prime factor to achieving this. Administrative labour cost includes the expenditure associated with the tasks of maintaining connected devices. The administrative cost is proportional to the scale of IIoT deployment, the complexity of device lifecycle, the number of devices deployed, and the supporting infrastructure. Industries which are recently establishing IIoT need to follow the effective cost management and efficient operation management, and keep costs in check to ensure a successful business model.

Technical Support Costs

Manually supporting IIoT services is a very complicated task. When a problem is reported, a support engineer must first carry out the route cause analysis to determine the reason for failure. Trouble shooting technical issues and their associated costs vary across industries depending on the number of devices and complexity. With a full pledged understanding of the total cost of delivering an IIoT service, industries can make smart, profitable and sustainable business decisions.

CONCLUSION

The main focus of this chapter is to identify and analyse various key challenges in the implementation of IIoT. IIOT requires a proper action plan and changes in existing Industry environment. IIoT platform and architecture need to be built for security from the scratch. Currently, Industries are not implementing enough to prevent security challenges at the design stage. Deploying automated intelligence and security procedures are mandatory for the implementation of IIoT in industries. Automation minimizes vulnerability by improving response time. This chapter also focuses on various issues related to device management, connectivity, data integration, interoperability, cost factors, power saving and green IIoT. This chapter makes an initial attempt to identify the challenges to IIoT adoption. It provides valuable guidelines to academicians, researchers, managers/practitioners and policymakers in understanding the various challenges to IIoT adoption.

REFERENCES

Bakhshi, Z., Balador, A., &Mustafa, J. (2018). Industrial IoT Security Threats and Concerns by Considering Cisco and Microsoft IoT reference Models. *2018 IEEE Wireless Communications and Networking Conference Workshops,* 173-178.

Boyes, H., Hallaq, B., Cunningham, J. & Watson, T. (2018). The industrial internet of things (IIoT): An analysis framework. *Computers in Industry, 101.*

Chen, A. &Stegner, Z. (2018). Rethinking Connectivity: Considerations for Designing Industrial IoT Networks. *IoT Now Magazine.*

Chen, B., Wan, J., Shu, L., Li, P., Mukherjee, M., & Yin, B. (2017). Smart factory of industry 4.0: Key technologies: application case, and challenges. *IEEE Access: Practical Innovations, Open Solutions, 6,* 6505–6519. doi:10.1109/ACCESS.2017.2783682

CISCO-JASPER. (2017). *The hidden costs of delivering Internet of Things (IoT) Services.* White paper. Retrieved from https://www.cisco.com/c/dam/m/en_ca/never-better/manufacture/pdfs/hidden-costs-of-delivering-iiot-services-white-paper.pdf

Gudlur, V. V. R., Shanmugan, V. A., Perumal, S., & Mohammed, R. M. S. R. (2020). Industrial Internet of Things (IIoT) of Forensic and Vulnerabilities. *International Journal of Recent Technology and Engineering, 8*(5).

Hatzivasilis, G., Askoxylakis, I., Alexandris, G., & Anicic, D. (2018). The Interoperability of Things: Interoperable solutions as an enabler for IoT and Web 3.0. *Conference Paper.* 10.1109/CAMAD.2018.8514952

Kagermann, H., Helbig, J., Hellinger, A., &Wahlster, W. (2013). *Recommendations for implementing the strategic initiative INDUSTRIES 4.0.* Securing the future of German manufacturing industry; final report of the Industries 4.0 Working Group, Forschungsunion.

Kamel, S. O., & Hegazi, N. H. (2018). A Proposed Model of IoT Security Management System Based on A study of Internet of Things (IoT) Security. *International Journal of Scientific & Engineering Research, 9*(9).

Kusters, D., Praß, N., &Gloy, Y. S. (2017). Textile Learning Factory 4.0–Preparing Germany's Textile Industry for the Digital Future. *Procedia Manufacturing, 9,* 214-221. doi:10.1016/j.promfg.2017.04.035

Lu, S., Xu, C., Zhong, R. Y., & Wang, L. (2017). A RFID-enabled positioning system in automated guided vehicle for smart factories. *Journal of Manufacturing Systems, 44,* 179–190. doi:10.1016/j.jmsy.2017.03.009

Maximilian, L., Markl, E., & Mohamed, A. (2018). Cyber security Management for (Industrial) Internet of Things: Challenges and Opportunities. *Inform Tech Software Eng, 2018*(8), 5. doi:10.4172/2165-7866.1000250

Misra, S., & Vaish, A. (2011). Reputation-based role assignment for rolebasedaccess control in wireless sensor networks. *Computer Communications, 34*(3), 281–294. doi:10.1016/j.comcom.2010.02.013

Moura, R. L., Ceotto, L. L. F., Gonzalez, A., & Toledo, R. A. (2018). Industrial Internet of Things (IIoT) Platforms: An Evaluation Model. *International Conference on Computational Science and Computational Intelligence (CSCI)*. DOI 10.1109/CSCI46756.2018.00194

Mourtzis, D., Vlachou, E., & Milas, N. (2016). Industrial Big Data as a result of IoT adoption in Manufacturing. *Procedia CIRP, 55*, 290–295. doi:10.1016/j.procir.2016.07.038

Nagy, J., Olah, J., Erdei, E., Mate, D., & Popp, J. (2018). The Role and Impact of Industry 4.0 and the Internet of Things on the Business Strategy of the Value Chain—The Case of Hungary. *Sustainability, 2018*(10), 3491. doi:10.3390u10103491

Neeraj & Singh, A. (2016). Internet of Things and Trust Management in IoT – Review. *International Research Journal of Engineering and Technology, 3*(6).

Phan, L., & Kim, T. (2020, May 14). Breaking down the Compatibility Problem in Smart Homes: A Dynamically Updatable Gateway Platform. *Sensors (Basel), 20*(10), 2783. doi:10.339020102783 PMID:32422946

Schroder, C. (2016). *The challenges of industry 4.0 for small and medium-sized enterprises*. Friedrich-Ebert-Stiftung.

Sengupta, J., Ruj, R., & Bit, S. D. (2020). A Comprehensive Survey on Attacks, Security Issues and Block chain Solutions for IoT and IIoT. *Journal of Network and Computer Applications, 149*, 102481. doi:10.1016/j.jnca.2019.102481

Sethi, P., & Sarangi, S. R. (2017). Internet of Things: Architectures, Protocols, and Applications. *Journal of Electrical and Computer Engineering, 2017*, 9324035. Advance online publication. doi:10.1155/2017/9324035

Umar, B., Hejazi, H., Lengyel, L., & Farkas, K. (2018). Evaluation of IoT Device Management Tools. IARIA, 2018.

Varney, M. (2018). *Why Machine Intelligence Is the Key to Solving the Data Integration Problem for the IIOT?* Bit Stew Systems Inc.

Wang, K., Wang, Y., Sun, Y., Guo, S., & Wu, J. (2016, December 16th). Green Industrial Internet of Things Architecture: An Energy-Efficient Perspective. *Article in IEEE Communications Magazine, 54*(12), 48–54. doi:10.1109/MCOM.2016.1600399CM

Wang, S., Wan, J., Li, D., & Zhang, C. (2016). Implementing smart factory of industries 4.0: An outlook. *International Journal of Distributed Sensor Networks, 12*(1), 3159805. doi:10.1155/2016/3159805

Wurm, J., Hoang, K., Arias, O., Sadeghi, A., & Jin, Y. (2016).Security analysis on consumer and industrial IoT devices. *21st Asia and South Pacific Design Automation Conference (ASP-DAC)*, 519–524. 10.1109/ASPDAC.2016.7428064

Yang, H., & Kim, Y. (2019). Design and Implementation of High-Availability Architecture for IoT-Cloud Services. *Sensors (Basel), 2019*(19), 327. doi:10.339019153276 PMID:31349629

Zeiid, A., Sundaam, S., Moghaddam, M., Kamarthi, S., & Marion, T. (2019). Interoperability in Smart Manufacturing: Research Challenges. *Machines, 2019*(7), 21. doi:10.3390/machines7020021

Zhou, K., Liu, T., & Zhou, L. (2015): Industry 4.0: Towards future industrial opportunities and challenges. In *12th International Conference on Fuzzy Systems and Knowledge Discovery (FSKD)* (pp. 2147-2152). IEEE.

Chapter 10
Industrial Internet of Things:
Benefit, Applications, and Challenges

Sam Goundar
https://orcid.org/0000-0001-6465-1097
British University, Vietnam

Akashdeep Bhardwaj
https://orcid.org/0000-0001-7361-0465
University of Petroleum and Energy Studies, India

Safiya Shameeza Nur
The University of the South Pacific, Fiji

Shonal S. Kumar
The University of the South Pacific, Fiji

Rajneet Harish
The University of the South Pacific, Fiji

ABSTRACT

This chapter focused on the importance and influence of industrial internet of things (IIoT) and the way industries operate around the world and the value added for society by the internet-connected technologies. Industry 4.0 and internet of things (IoT)-enabled systems where communication between products, systems, and machinery are used to improve manufacturing efficiency. Human operators' intervention and interaction is significantly reduced by connecting machines and creating intelligent networks along the entire value chain that can communicate and control each other autonomously. The difference between IoT and IIoT is that where consumer IoT often focuses on convenience for individual consumers, industrial IoT is strongly focused on improving the efficiency, safety, and productivity of operations with a focus on return on investment. The possibilities with IIoT is unlimited, for example, smarter and more efficient factories, greener energy generation, self-regulating buildings that optimize energy consumption, smart cities that can adjust traffic patterns to respond to congestion.

DOI: 10.4018/978-1-7998-3375-8.ch010

INTRODUCTION

The idea of a world where systems with local processing, sensors and controllers are interconnected with each other and to the larger network and cloud to share data and information is captivating within every single industry. These systems will be connected at a global level with each other and its end users to help entities and users make better-informed decisions based on the data retrieved from these systems. This idea has been given many labels so far, but ubiquitous is the Internet of Things (IoT). The IoT includes everything from smart cities to smart homes, everyday smart appliances, and connected toys to the Industrial Internet of Things (IIoT) with smart agriculture, smart factories, and the smart grid.

The Industrial Internet of Things (IIoT) is often presented as a revolution that is changing the face of the industry in an innovative and rapid manner. However, as it may take a bit of time for global standards to be generalised, the full benefits of IIoT is still a few years away. End users though will still be able to take advantage of the available new IIoT technologies and leverage their existing investment in technologies and people. Introducing IIoT solutions using "Wrap & Re-use'' approach, rather than a "Rip & Replace" approach will enable greater business control. In addition, this measured approach will drive the evolution towards a smart manufacturing enterprise that is more efficient, safer, and sustainable.

The IIoT vision for the world is one where smart connected machinery and equipment operate as part of a much bigger system that make up the smart manufacturing enterprise. The machinery and equipment, or the "things" will posses' different levels of intelligent functionality, ranging from sensory functions, control mechanisms, optimisation, and full autonomous operations.

The smart manufacturing plant comprises of smart equipment, machinery, and operations, all of which have high levels of intelligence embedded at the core. The automated and linked systems use various internet and cloud technologies that ensure secure access to devices and information. New and advanced analytics tools allow for Big Data to be processed efficiently to deliver greater business value.

Think of industrial machineries or systems that can sense their own environments and health and make appropriate adjustments. Instead of working until breakdown, the machines schedule their own regular maintenance or adjust control algorithms dynamically to compensate for the troubled part and the communicate this shortcoming to other machines in the system as well as users of these machines. IIoT can solve problems that were previously thought impossible. However, as the saying goes, "if it was any easier, everyone would be doing it". As IIoT, innovation grows so does the complexity, which makes the IIoT a very large challenge that no company alone can meet. In a recent report on Forbes (), it is estimated that the Industrial Internet of Things could create a total value of up to $11.1 trillion on an annual basis by 2025 and about 70% of this would be captured by business-to-business solutions-leaving the value of the consumer Internet at about $3.5 trillion. In other words, the Industrial Internet will be worth more than twice the consumer Internet will as illustrated in Figure 1 below.

LITERATURE REVIEW

According to Khan, et al. (2020), "the adoption of emerging technological trends and applications of the Internet of Things (IoT) in the industrial systems is leading towards the development of Industrial IoT (IIoT). IIoT serves as a new vision of IoT in the industrial sector by automating smart objects for sensing, collecting, processing, and communicating the real-time events in industrial systems. The major objective of IIoT is to achieve high operational efficiency, increased productivity, and better management

Figure 1. Global Field of Opportunity for IIoT.

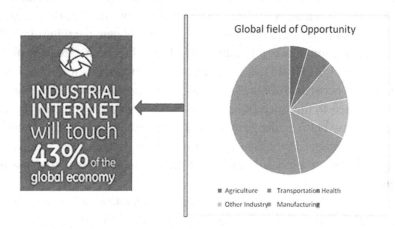

of industrial assets and processes through product customization, intelligent monitoring applications for production floor shops and machine health, and predictive and preventive maintenance of industrial equipment. In this paper, we present a new and clear definition of IIoT, which can help the readers to understand the concept of IIoT. We have described the state-of-the-art research effort s in IIoT. Finally, we have highlighted the enabling technologies for IIoT, and recent challenges faced by IIoT." (Khan, et al., 2020).

Industry 4.0 and its main enabling information and communication technologies are completely changing both services and production worlds (Aceto, Persico & Pescape, 2020). "This is especially true for the health domain, where the Internet of Things, Cloud and Fog Computing, and Big Data technologies are revolutionizing eHealth and its whole ecosystem, moving it towards Healthcare 4.0. The authors by selectively analysing the literature systematically surveyed how the adoption of Industry 4.0 technologies (and their integration) applied to the health domain in changing the way to provide traditional services and products. In their paper, Aceto, Persico & Pescape, 2020 provide (i) a description of the main technologies and paradigms in relation to Healthcare 4.0 and discuss (ii) their main application scenarios; and then provide an analysis of (iii) carried benefits and (iv) novel cross-disciplinary challenges; finally, extract (v) the lessons learned". (Aceto, Persico & Pescape, 2020).

As claimed by Jaidka, Sharma & Singh (2020) "The emerging Internet of Things (IoT) provides a wide range of platform to different technologies by connecting different devices and are automated by using sensors. It builds a platform and is responsible for functioning of various smart devices over a range. After installing IoT in devices they can communicate with each other without involving human and computer interaction, it is also being widely used in all the fields, as no human intervention is required in any IoT based applications. In a paper we present a survey of how IoT is transformed into IIoT (Industrial Internet of Things), what is Industry 4.0, what are key differences between IoT and IIoT, various application of IoT in different sectors, powerful features and advance characteristics of IoT. Since the demand of IoT systems is increasing in various fields day by day and ease of life after using smart devices but then also it possesses certain limitations too. The later part of paper focuses on security threats and issues addressed by IoT based systems along with ways to overcome the limitations possessed in the system". (Jaidka, Sharma & Singh, 2020)

In their paper titled "Challenges and Recommended Technologies for the Industrial Internet of Things: A Comprehensive Review", Younan, et al. 2020 states that the "physical world integration with cyber world opens the opportunity of creating smart environments; this new paradigm is called the Internet of Things (IoT). Communication between humans and objects has been extended into those between objects and objects. Industrial IoT (IIoT) takes benefits of IoT communications in business applications focusing in interoperability between machines (i.e., IIoT is a subset from the IoT). Number of daily life things and objects connected to the Internet has been in increasing fashion, which makes the IoT be the dynamic network of networks. Challenges such as heterogeneity, dynamicity, velocity, and volume of data, make IoT services produce inconsistent, inaccurate, incomplete, and incorrect results, which are critical for many applications especially in IIoT (e.g., health-care, smart transportation, wearable, finance, industry, etc.).

Discovering, searching, and sharing data and resources reveal 40% of IoT benefits to cover almost industrial applications. Enabling real-time data analysis, knowledge extraction, and search techniques based on Information Communication Technologies (ICT), such as data fusion, machine learning, big data, cloud computing, blockchain, etc., can reduce and control IoT and leverage its value. Their research presents a comprehensive review to study state-of-the-art challenges and recommended technologies for enabling data analysis and search in the future IoT presenting a framework for ICT integration in IoT layers. Their paper surveys current IoT search engines (IoTSEs) and presents two case studies to reflect promising enhancements on intelligence and smartness of IoT applications due to ICT integration". (Younan, et al. 2020)

A study by Leminen, et al., (2020) analyses "Industrial Internet of Things (IIoT) business models in the machine-to-machine (M2M) context. Thereby, it develops a conceptual framework to categorize different types of business model innovation for companies operating in the M2M business space. Business model innovations tend to cross multiple industries and drive ecosystems in which smart objects facilitate business models and service applications that are incrementally or radically novel in terms of their modularity or architecture. Our framework identifies four distinct types of IIoT business models: (I) Company-specific business models, (II) Systemic business models, (III) Value designs, and (IV) Systemic value designs. Moreover, it sheds light on different abstraction levels of business model building blocks and exposes the characteristics and differences in the value potential between the four business models. Finally, we advance the idea of 'value design' referring to business models of multiple actors coupled together, ultimately resulting in complex networks and ecosystems of diverse things, processes, and companies". (Leminen, et al., 2020)

According to Kim & Tran-Dang, 2019, "recently, the technological advancement in embedded systems and wireless communication has enabled interconnection of massive electronic devices to support innovative services and promises better flexibility and efficiency. Such paradigm is referred to as the Internet of Things (IoT) that promises ubiquitous connection to the Internet, turning common objects into connected devices. An emerging class of IoT-enabled industrial production systems is called the Industrial Internet of Things (IIoT) that, when adopted successfully, provides huge efficacy and economic benefits to industrial system installation, maintainability, reliability, scalability, and interoperability. Their paper introduced the concept of IIoT from technological and practical application aspects". (Kim & Tran-Dang, 2019).

Digital technologies have changed the way supply chain operations are structured (Radanliev, et al. 2019). In their article, (Radanliev et al. 2019), "develop design principles to show determining factors for an Internet-of-Things approach within Supply Chain Management. From the design principles, the article derives a new model and a transformational roadmap for the Industrial Internet of Things in Industry 4.0 supply chains of Small and Medium Enterprises (SMEs). Their focus is on SMEs. Their literature review included 173 academic and industry papers and compared the academic literature with the established supply chain models. Taxonomic review was used to synthesise existing academic and practical research. Subsequently, case study research was applied to design a transformational roadmap. That was followed by the grounded theory methodology, to compound and generalise the findings into a theoretical model. Their research design resulted in a new process of compounding knowledge from existing supply chain models and adapting the cumulative findings to the concept of supply chains in Industry 4.0. The findings from their study present a new model for Small and Medium Sized companies to transform their operations in the Industrial Internet of Things and Industry 4.0". (Radanliev, et al. 2019)

Balaji, Nathani, & Santakumar, 2019, posits that "Internet of things (IoT) is a very unique platform which is getting very popular day by day. The very reason for this to happen is the advancement in technology and its ability to get linked to everything. This feature of getting linked has in itself provided multiple opportunities and a vast scope of development. The fact that technology in various fields has evolved through the years, is the reason why we observe a rapid change in the shape, size and capacity of various instruments, components and the products used in daily life. And this benefit of simplified technology when accompanied by a platform like IoT eases the work as well as benefits both the manufacturer and the end user. The Internet of Things gives us an opportunity to construct effective administrations, applications for manufacturing, lifesaving solutions, proper cultivation and more. This paper proposes an extensive overview of the IoT technology and its varied applications in life saving, smart cities, agricultural, industrial etc. by reviewing the recent research works and its related technologies. It also accounts the comparison of IoT with M2M, points out some disadvantages of IoT. Furthermore, a detailed exploration of the existing protocols and security issues that would enable such applications is elaborated. Potential future research directions, open areas and challenges faced in the IoT framework are also summarized". (Balaji, Nathani, & Santakumar, 2019).

Rehman, et al., 2019 looks at "big data production in industrial Internet of Things (IIoT) and claim that it is evident due to the massive deployment of sensors and Internet of Things (IoT) devices. According to them, big data processing is challenging due to limited computational, networking and storage resources at IoT device-end. Big data analytics (BDA) is expected to provide operational- and customer-level intelligence in IIoT systems. Although numerous studies on IIoT and BDA exist, only a few studies have explored the convergence of the two paradigms. In their study, they investigate the recent BDA technologies, algorithms and techniques that can lead to the development of intelligent IIoT systems. They devise a taxonomy by classifying and categorising the literature on the basis of important parameters (e.g. data sources, analytics tools, analytics techniques, requirements, industrial analytics applications and analytics types). They present the frameworks and case studies of the various enterprises that have benefited from BDA. They also enumerate the considerable opportunities introduced by BDA in IIoT". (Rehman, et al., 2019).

Industrial production plays an important role for achieving a green economy and the sustainable development goals (Beier, Niehoff, & Xue, (2018). Therefore, the nascent transformation of industrial production due to digitalization into a so-called Industrial Internet of Things (IIoT) is of great interest from a sustainable development point of view. This paper discusses how the environmental dimension of a sustainable development can potentially benefit from the IIoT—focusing especially on three topics: resource efficiency, sustainable energy and transparency. It presents a state-of-the-art literature analysis of IIoT-enabled approaches addressing the three environmental topics. This analysis is compared with the findings of a survey among Chinese industrial companies, investigating the sustainability-related expectations of participants coming along with the implementation of IIoT solutions. China has been chosen as a case study because it brings together a strong industrial sector, ambitious plans regarding industrial digitalization and a high relevance and need for more sustainability. The survey was conducted with the means of a questionnaire which was distributed via email and used for direct on-site interviews. It focused on large and medium sized companies mainly from Liaoning Province and had a sample size of 109 participants. (Beier, Niehoff, & Xue, (2018).

Major challenges identified by Xu & Li, 2014 in their paper were security and privacy whereas traceability, visibility and controllability were identified as the benefits of Industrial IoT. Also mentioned was the idea that success of IoT is influenced by calibration of standards which would improve interoperability, reliability, compatibility, and global connectivity.

The future research direction mentioned by Breivod & Sandstorn, 2015 in their paper is interoperability and compatibility. Basically, by what means will the long-lived systems and new services which are based on fast-paced technologies integrate with each other. The paper discusses general IoT challenges, automation domain specific constraints, and industrial IoT challengers then lastly managing those Industrial Internet of Things challenges.

Gurtov & Korzun (2016), outlines in their paper some of IIoT challenges as integration with 5G wireless networks, Software Defined Machines, ownership and smart processing of digital sensor data. It explains how these challenges affect IIoT and it outlines some proposals for tackling these issues. The outlined proposals are secure communication architecture for the Industrial Internet based on Smart Spaces and Virtual Private LAN Services.

Hartmann & Halecker (2015), describe IIoT as the industrial application of IoT. They state that IIoT is technologically focused, however the cyber-physical systems have not been defined clearly. Based on the critical review of evolutionary cycles regarding IIoT, (Hartmann & Halecker, 2015) suggest a perspective on the changing rules in IIoT and define strategic challenges and imperatives for IIoT innovation management.

The next big industrial revolution that we are looking at is Industry 4.0. IoT is set to play a very big part in this. Some of the benefits of IoT adoption in energy management process discussed by the authors in their paper includes: Finding and reducing energy waste sources, Improving energy-aware production scheduling, Reduction in energy bill, Efficient maintenance, and its management. They also discussed a few challenges in implementing IIoT solutions within their systems. The paper did well in presenting on how IIoT can be utilized to solve many environmental issues currently faced in the industry. We looked at how these concepts and solutions can be applied in industries locally and signal out constraints or benefits we may encounter here. (Pacis, Subdio & Bugtai, 2017)

Niranjan, et al. 2017 discusses the use of Industrial Internet of Things to connect industrial machines with sensors and controllers over the internet and collecting data that can be used by authorized users in the useful manner. It further investigates how IoT based industry automations will enable the control and monitoring of in-house production systems from anywhere in the world. The paper makes a brief comparison of automation in the industry through time. The journal however falls short in including some evident data from factories in the real world to further support the ideologies discussed in the paper.

Kettuen & Salmela, 2017 study focused on determining the state and trend of digital transformation in manufacturing industry in Finland. It also focuses its studies on the impact on a business's competitive edge the use of Internet of Things can give and if digital transformation is becoming a necessity. A research was conducted with numerous companies where data was generally collected through semi-structured interviews. Some challenges of introducing IIoT specifically within manufacturing industry were also discussed in this journal.

Canizares & Valero, 2018 in their paper focused on a company in the metal-mechanical sector. It aimed to show the improvements that could be obtained from the applications of Internet of Thing in this company. It reported findings on how some improvements on the company's processes made these processes more efficient and at the same time reducing costs. It also defined KPIs and the application of existing technologies such as RFID and Wi-Fi are utilized to aid in the rollout of IoT technology within the company. The research quantified some of its results in term of process efficiency which is a key to smart factory automation.

Industry 4.0. It investigates the security challenges and inherent risks that result from the Internet of Things and Industry 4.0 within the context of digital transformation. With the widespread use of devices and data, it discusses the lack of standards, and ubiquitous adoption without due understanding of the need to consider the shift in implications. The use of IoT will enable companies to take advantage of the rich data streams that can be collected and analysed from myriad inexpensive sensors and devices. (Bligh-Wall, 2017)

Abdelhafidh & Fourati, 2017 in their paper, briefly discuss the development of various IoT applications in different industries with the rise of Industry 4.0 such as IoT in healthcare Industry, Smart Agriculture, IoT in manufacturing and Industry Security. Most of the paper, however, focused on the use of IoT for a Fluid Distribution Monitoring System, which includes Water Distribution networks and Oil & Gas Distribution networks. IoT becomes a source of big quantity of data obtained from heterogeneous and dynamic connected objects. In our research we explore and build upon on the idea of using IoT based systems to monitor real time information in the industry and how we can incorporate the use of cloud storage to collect and store real-time and historical data.

RESEARCH METHODOLOGY

To gather information on the studied topic, the research was conducted in two parts. The first part focused on research based on published materials such as journal articles, website data and books. This research helped in understanding the concept of Internet of Things and how it is being integrated in the industrial world (Industrial Internet of Things).

The second part of the research focused on exploring the Technology Acceptance Model (TAM) and its use to identify the drawbacks of integrating Industrial Internet of Things concepts to the Fijian manufacturing industries. This research was conducted through surveys which were designed after consulting the variables studied in the Technology Acceptance Model. Two surveys were designed, and they were focused on collaboration from two sets of contributors. The first sample had a size of fifty participants, and these were comprised of managers from the information technology and operations departments of manufacturing industries in Fiji. The second survey was intended towards a more general user base consisting of information technology officers and other potential users of the integrated technology. This sample focused on a group of hundred participants.

We have looked at five applications of Industrial Internet of Things (IIoT). Starting from industrial automation to predictive maintenance, smart logistics management, enhanced product quality and lastly smart inventory management. To begin with; **industrial automation**, the usage of coordinated systems like computers or mechanical devices and information expertise systems for management of diverse processes and equipment's in a manufacturing industry to substitute personnel's. Succeeding further from mechanization to the opportunities of industrial development. These systems can be categorized into three basic categories been fixed automation, programmable automation and flexible automation. "Industrial automation spans over many different types of control systems, e.g., motion control, protection systems, and digital control. Furthermore, industrial automation can be found in many different domains that execute different processes" (Breivold & Sandström, 2015).

- **Predictive Maintenance** is an approach which ensures cost savings for routine or time-based preventive maintenance due to smart metered connections. Procedures are aimed to aid in concluding the state of in-service equipment in order to forecast in what period maintenance ought to be executed. With the innovations in big data analytics and cloud computing predictive maintenance is continuing to grow. "Predictive maintenance relies heavily on data, and IoT provides the best methodology for analysing data and connecting all users" (Bayoumi & McCaslin, 2016).

- **Smart Logistics Management** is made up of smart products from smart services, basically having the precise product at the accurate time at a true place of need and want in the prefect condition. They are invisible, calm and therefore transparent in management and it releases personals from controlling of logistics. "Smart means that planning and scheduling, ICT infrastructure, people and governmental policymaking need to be efficiently aligned." (Kawa, 2012).

- **Smart Logistics** is the coordinated relationship of these four fundamental areas. ICT infrastructure provides right resources at the exact time for more fast and detailed information which permits enhanced planning and scheduling. As we are moving into augmented intelligence, people and machines work together for the benefit of all. People are provided training before they commerce working with these smart yet complex machinery at first. Lastly, policies which are important as they govern everything in the business.

- **Enhanced product quality** is critical for both the customer satisfaction and the business growth. Factors to consider for enhanced product quality is design, production, inspection and traceability. Adjustments are done in product design to close gaps identified and quality inspection while in production with latest technologies. Unknown causes of defects that delay inspection are identified and enhancement in traceability to increase quality and speed. "Zhang et al. designed an intelligent monitoring system to monitor temperature/humidity inside refrigerator trucks by using RFID tags, sensors, and wireless communication technology" (Xu, He, & Li, 2014).

- **Smart inventory management** links all aspects of inventory management flawlessly, from raw materials to finished goods. Such management provided visible effect using frameworks and interconnected intelligent technology systems in real time. There is real time inventory updates, alerts of issues and updates of latest process the inventory is in and also materials utilized to get output. Radio frequency identification tags, barcodes and intelligent sensors are used to identify, trace and track all required objections and devices. Reduction of expiry products and improvement in workflow. "As more and more physical objects are equipped with bar codes, RFID tags or sensors, transportation and logistics companies can conduct real- time monitoring of the move of physical objects from an origin to a destination across the entire supply chain including manufacturing, shipping, distribution, and so on" (Xu, He, & Li, 2014).

Benefits and Challenges of IIoT

- **Facility management** is the interconnectivity of nearly all the systems in communication to each other and with personnel via interface whilst keeping hardware's connected. These physical systems are progressively able to competent to control and connect themselves automatically within an information network. "This is far more than just installing a robot or processing a fully automated conveyor belt. These physical systems will be able to behave as autonomous systems in a changing environment" (Hartmann & Halecker, 2015) Sensors can also be used to monitor alarm vibrations, temperature changes and other dynamics that can be future reasons for less operational conditions.

- **Real-Time data** is mobile information when compared to cloud-based or decentralized. A large data center placed someplace in the world is a cloud and accessed on need basis. Whereas real data is communicated same time as the operations are happening for these manufacturing industries where data is crucial for the success. Integration with IIoT enables companies to collect data from assets and make informed decisions in real time.

The biggest industry concern is safety and is very much minimized when IIoT is adopted by industries. Potential issues are identified prior therefore reducing risks taken by individuals and actual hazards can be avoided. Enhanced industrial safety is enabled through efficient procedures and remote real time troubleshooting leading to less dangerous travel for personnel. Lastly, environmental footprint, enabled by smart meters that monitor consumption of industry resources such as fuels, time, water, electricity, etc. Resource utilization will be identified through the meters and changes will be made according. From increased efficiency to lessened safety risk and reduced travel, IIoT adoption can significantly reduce the environmental impact. Using less energy, avoiding oil spills and other accidents, and emitting less carbon are significant enough to pay attention to IIoT. IIoT also allows for clearer monitoring of energy and resource usage.

The biggest challenge of IIoT devices are security vulnerabilities as private information can be communicated with the automatically shared information. "Protecting privacy in the IoT environment becomes more serious than the traditional ICT environment because the number of attack vectors on IoT entities is apparently much larger" (Xu, He, & Li, 2014) Adoption of IIoT can lead to more security vulnerabilities and challenges in the absence of a properly secure and encrypted network. IIOT permits tracking, monitoring and connectivity of industrial data. "Reasonable efforts in technology, law, and regulation are needed to prevent unauthorized access to or disclosure of the privacy data" (Xu, He, & Li, 2014) "Therefore, the software architecture of an IoT solution needs to protect the interconnected devices from intrusions and interference from new attack vectors (i.e., coming from the communication channels) from entering the system so as to ensure secure operations" (Breivold & Sandström, 2015).

Second challenge is the absence of IIoT standards. For the success of manufacturing industries, it requires standardization in terms of capability, interoperability and reliability. "In order to provide high-quality services to end users, IoT's technical standards need to be designed to define the specification for information exchange, processing, and communications between things", (Xu, He, & Li, 2014). All in all, IIoT technology and innovations will spread widely from standardization.

An essential challenge for the industry is the cost of implementing IIoT which cannot be underestimated. Return of investment is not very promising in terms of cost if a detailed proposal is not prepared. Advantages of integration with the current or move to the new infrastructure should be listed with all the information possible. A lot of possible future issues and challenges can be avoided if such a step is taken in the present.

Last challenge identified is skills set, just having the approval with funding from higher management is not enough. We have to keep in mind who's going to make it happen and take the project till end, to design, develop, implement, fine tune and maintain the new structure. This is where skill set challenge comes in. To have the correct skill ready for the correct merge from current structure to the new one with IIoT. "Analyzing or mining massive amounts of data generated from both IoT applications and existing IT systems to derive valuable information requires strong big data analytics skills, which could be challenging for many end users (Xu, He, & Li, 2014).

PROPOSED TECHNOLOGY ACCEPTANCE MODEL (TAM)

The technology acceptance model was used to study the integration and use of industrial internet of things in the Fijian context. A total of eight factors contributing to the technology acceptance model were employed to understand and outline the results of the tests. The variables used for this study were perceived risk, cost, self-efficacy, compatibility, perceived usefulness, perceived ease-of-use, behavioural intentions, and actual use.

Perceived risk is the risk factor of integrating IIoT concepts into the business operations. The stakeholders may be wary whether these integrations would work since there are not many successful examples locally. The employees may also misinterpret the integration of such technology to affect their jobs. This can be thought of as having a negative impact on the implementation and usage of IIoT in local industries.

Cost is one of the major factors of hindrance when it comes to technology integration. The cost of equipment, setup, training and integration into the business is always posed as a challenge. Though these costs are generally one-off, they are worth a significant monetary value and stakeholders may feel that if the IIoT project does not fulfil its promised functionalities, then the capital investment will be lost. This is seen as another negative impact to the IIoT integration.

Self-efficacy relates to the efficiency of the proposed technology system. This variable determines how efficient the technology is in terms of its use and cost effectiveness. Self- efficacy is seen as a positive impact towards the integration of IIoT.

Compatibility determines if the proposed technology will be easily adapted in the organisation given its current infrastructure and workforce. If the proposed technology can fit into the operations seamlessly without changing much of the equipment already used and if it is easy to use with little to no training required by the current workforce, then it poses itself as an advantage.

Perceived usefulness is considered as a cognitive construct (Park, 2009). As defined by Davis in his paper, perceived usefulness is "the degree to which a person believes that using a particular system would enhance his or her job performance" (Davis, 1989). In IIoT, the perceived usefulness would be the how the employees would see the innovations as support to their daily functions. Would it make their work easier and efficient? From the stakeholder's perspective, they would see the perceived usefulness as how integrating the IIoT technologies would benefit the organisation by increasing productivity, work efficiency and provide data for futuristic growth.

Perceived ease-of-use "is an assessment of the mental effort involved in the use of the system" (Park, 2009). It concerns the perception of a user's interaction with a proposed system. The introduction of IIoT technologies to local industries will be largely impacted by employees' perception on the ease-of-use of the system. If they feel it is too difficult to operate, they will have rejection towards the implementation. The stakeholder's perspective also matters. If they feel the technology is not fit for operation by their current workforce, then they may hesitate to integrate such technologies into the organization's operations.

Behavioural intentions as defined by the consumer health informatics research resource is "a person's perceived likelihood or subjective probability that he or she will engage in a given behaviour" (Consumer Health Informatics Research Resource - Behavioural Intention, n.d.). The behavioural intentions of the stakeholders and the employees towards IIoT integration defines how they perceive IIoT to be beneficial and do they need to adapt such to a change for the organisation to prosper?

Actual use simply means the actual integration of the IIoT technologies into the business. It attempts to question whether the actual system will be implemented or if this is just a perception that the systems will be beneficial towards the organizations overall performance. Figure 2 describes the proposed model

Figure 2. Technology Acceptance Model for IIoT in the Fijian Context

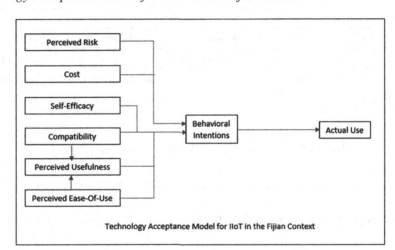

diagram for technology acceptance of Fijian industries.

HYPOTHESIS BASED ON THE TECHNOLOGY ACCEPTANCE MODEL

There were four hypothesis statements which were drawn from the technology acceptance model. These are outlined as follows:

H1. The actual use of IIoT is affected by the behavioural intention to use
H2. Perceived Risk and Cost of IIoT projects negatively affect Behavioural Intentions of decision makers

H3. The Perceived Use of IIoT is affected by Compatibility and Perceived Ease-Of-Use

H4. The Perceived Ease-Of-Use of IIoT, its Self-Efficacy, Compatibility and Perceived Usefulness

Table 1. Research Questions developed from the Technology Acceptance Model

Construct	Definition
Actual use (AU)	Are there any current or future IIoT projects outlined for your organisation?
Behavioural intention to use (BI)	Are the decision makers willing to accept IIoT integration into the organizational operations?
Behavioural intention to use (BI)	How likely is it that the decision makers will integrate IIoT into the organizational operations?
Cost (C)	Is the cost associated with IIoT project implementation a hinderance to the organisation?
Cost (C)	In your opinion, do you think the benefits of IIoT will outweigh its implementation costs?
Perceived risk (PR)	How likely is it that the IIoT project implementation will be successful?
Perceived usefulness (PU)	Do you see the implementation of these technologies as a benefit to the organisation?
Perceived ease-of- use (PEOU)	Are the current employees ready to accept the implementation of such technologies?
Self-Efficacy (SE)	How efficient do you think the IIoT technologies are when compared with legacy systems?
Compatibility (CO)	With the current infrastructure available, how likely is it able to support the implementation of IIoT projects?

directly affect the Behavioural Intentions of Stakeholders and employees

Table 1 depicts the research questions developed from the Technology Acceptance Model.

Enabling IIoT: Opportunities and Challenges

IIoT revolution has been made possible by the advancements in hardware, network connectivity, data analytics and storage and cloud infrastructure. Powerful and affordable hardware is empowering smart sensors, wireless networks, and gateways. Ability to carry large volumes of data over networks for processing had made possible greater analytics. These analytics for a range of data sets in real-time as well as offline has given rise to new business models and more efficient workflows. Getting all these components to work together in harmony is what makes the whole greater than the sum of parts.

The overall IIoT adoption started gradually and seems to be accelerating now. Just like all other revolutions, the 4th industrial revolution is also not without its own resistance and challenges. There are several factors that slowed its expected adoption rate, ranging from technical to human aspects. Some of the most significant are:

Factors Affecting Adoption of IoT

As depicted in Figure 3 below:

- **Security and Safety Concerns**: For the entire end-to-end deployment and at each stage and interfaces in between, including physical security of the deployment, data security and safeguards

- **Integration Difficulties**: Legacy hardware, lack of inter-operability, mix of multitude of technolo-

Figure 3. Factors Affecting Adoption of Internet of Things

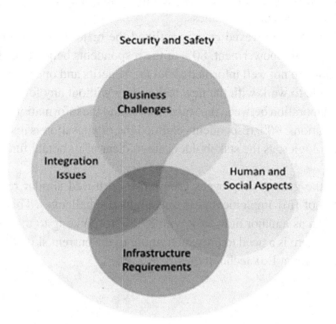

gies and protocols, lack of IT infrastructure on the shop floor
- **Infrastructure Requirements:** Reliable and consistent network connectivity, storage space and analytical processing power required to handle very large data volumes
- **Business Related Challenges**: Complexity of deployments, Initial investment, not very attractive

- **Return on Investment**: unless full potential is realized using analytics, system down-time and testing for deployments
- **Human/Social issues**: Lack of trained resources with technology and domain skills, worker concerns about job security, reluctance for change, retraining
- **Ecosystem issues**: Too many platforms and vendors to choose from, lack of interoperability and accepted standards

Apart from these, all industrial deployments must adhere to the established standards related to confidentiality, availability, integrity, and safety as well as typical enterprise IT security practices.

RESEARCH FINDINGS

From the responses collected in the surveys out of the fifty organizations that responded, 80% mentioned there is some form of IIoT engagement being carried out by their respective organizations. Out of these, 90% responded saying their organizations is interested in further integrating IIoT technologies into the production line and other aspects of the business.

The responses from both the surveyed groups outlined the major issues in implementing IIoT into Fijian industries as employee empowerment. 80% of the respondents believe that the current set of employees in the organisation are not well informed about the benefits and operations of IIoT technologies and/or they will not be able to work with the new technology without any form of training. 70% stated that there is a lack of collaboration between the stakeholders and the information technology department to discuss the implementations. 90% respondents claimed the organisation is not ready to bear the costs of integrating IIoT technologies, as the stakeholders are unclear of its actual functional benefits and its returns on investment.

The responses from the general employees group also portrayed similar results where 87.5% responded that the benefits of IIoT implementation outweighs its challenges. This group also thought of the cost of implementation as a major drawback with 87.5% responding to this as a challenge. 58% of these employees thought there is a need to provide training to the current skill set of the organization in order to successfully implement IIoT technologies.

CONCLUSION

From this research, it was found that IIoT is an important aspect, which needs to be adapted by the local industries in order for them to maximize efficiency and profits. This would in turn help the consumers as the prices of items may decrease as well as providing a maintained supply. From the research, it was quite evident that the major drawback for IIoT in local industries is due to the lack of pioneering organizations in these technological advancements. Through further research based on international adaptation and benefits of localization of IIoT concepts, this gap can be bridged.

The technology acceptance model portrayed positive results towards the hypothesis statements. It provides a platform for conclusion that states there are mixed reactions from employees and stakeholders towards the actual use of IIoT. There are positive impacts on the behavioural intentions towards IIoT through the perceived ease of use of the system, its usefulness to the current organizational operations, its compatibility with the current systems and people and the proposed efficiency. There are, however, negative impacts as well such as the cost of the project implementation and the perceived risks of the IIoT integration, which lead to a negative mindset towards the actual integration of these technological systems. These challenges may be overcome by outlining the benefits of IIoT and overall acceptance on the global scale. A further proposal on the benefits per cost analysis will help improve on the behavioural intentions.

REFERENCES

Abdelhafidh, M., Fourati, M., Fourati, L. C., & Chouaya, A. (2017). *Internet Of Things In Industry*. Academic Press.

Aceto, G., Persico, V., & Pescapé, A. (2020). Industry 4.0 and health: Internet of things, big data, and cloud computing for healthcare 4.0. *Journal of Industrial Information Integration*, *18*, 100129. doi:10.1016/j.jii.2020.100129

Balaji, S., Nathani, K., & Santhakumar, R. (2019). IoT technology, applications and challenges: A contemporary survey. *Wireless Personal Communications*, *108*(1), 363–388. doi:10.100711277-019-06407-w

Bayoumi, A., & McCaslin, R. (2016). *Internet of Things – A Predictive Maintenance Tool for General Machinery, Petrochemicals and Water Treatment*. Sustainable Vital Technologies in Engineering & Informatics.

Beier, G., Niehoff, S., & Xue, B. (2018). More sustainability in industry through industrial internet of things? *Applied Sciences (Basel, Switzerland)*, *8*(2), 219. doi:10.3390/app8020219

Bligh-Wall, S. (2017). Industry 4.0: Security imperatives for IoT — converging networks, increasing risks. *Cyber Security: A Peer-Reviewed Journal*, 61-68.

Breivold, H. P., & Sandström, K. (2015). Internet of Things for Industrial Automation – Challenges and Technical Solutions. In *2015 IEEE International Conference on Data Science and Data Intensive Systems*. Sydney: IEEE. 10.1109/DSDIS.2015.11

Cañizares, E., & Valero, F. A. (2018). Analyzing the Effects of Applying IoT to a Metal-Mechanical Company. *Journal of Industrial Engineering and Management*, *11*(2), 308–317. doi:10.3926/jiem.2526

Consumer Health Informatics Research Resource - Behavioral Intention. (n.d.). Retrieved October 25, 2018, from Consumer Health Informatics Research Resource: https://chirr.nlm.nih.gov/behavioral-intention.php

Davis, F. D. (1989). Perceived Usefulness, Perceived Ease of Use and User Acceptance of Information Technology. *Management Information Systems Quarterly*, *13*(3), 319–340. doi:10.2307/249008

Gurtov, A., Liyanage, M., & Korzun, D. (2016). Secure Communication and Data Processing Challenges in the Industrial Internet. *Baltic Journal of Modern Computing*, 1058-1073.

Hartmann, M. H., & Halecker, B. (2015). Management of Innovation in the Industrial Internet of Things. *XXVI ISPIM Conference – Shaping the Frontiers of Innovation Management*.

Jaidka, H., Sharma, N., & Singh, R. (2020). *Evolution of IoT to IIoT: Applications & Challenges*. Available at SSRN 3603739

Kawa, A. (2012). *SMART logistics chain. Intelligent Information and Database Systems*. ACIIDS.

Kettunen, K., & Salmela, E. (2017). Internet of Things as a Digital Transformation Driver in the Finnish Manufacturing Technology Industry. *Journal of Innovation & Business Best Practice*.

Khan, W. Z., Rehman, M. H., Zangoti, H. M., Afzal, M. K., Armi, N., & Salah, K. (2020). Industrial internet of things: Recent advances, enabling technologies and open challenges. *Computers & Electrical Engineering*, *81*, 106522. doi:10.1016/j.compeleceng.2019.106522

Kim, D. S., & Tran-Dang, H. (2019). An Overview on Industrial Internet of Things. In *Industrial Sensors and Controls in Communication Networks* (pp. 207–216). Springer. doi:10.1007/978-3-030-04927-0_16

Leminen, S., Rajahonka, M., Wendelin, R., & Westerlund, M. (2020). Industrial internet of things business models in the machine-to-machine context. *Industrial Marketing Management*, *84*, 298–311. doi:10.1016/j.indmarman.2019.08.008

Niranjan, M., Madhukar, N., Ashwini, A., Muddsar, J., & Saish, M. (2017). IOT Based Industrial Automation. *IOSR Journal of Computer Engineering (IOSR-JCE)*, 36-40.

Pacis, D. M., Subido, E. D. Jr, & Bugtai, N. T. (2017). *Research on the Application of Internet of Things (IoT) Technology towards a Green Manufacturing Industry: A Literature Review*. DLSU Research Congress, Manila.

Park, S. Y. (2009). An Analysis of the Technology Acceptance Model in Understanding University Students' Behavioral Intention to Use e-Learning. *Journal of Educational Technology & Society*, 150–162.

Radanliev, P., De Roure, D. C., Nurse, J. R., Montalvo, R. M., & Burnap, P. (2019). The Industrial Internet-of-Things in the Industry 4.0 supply chains of small and medium sized enterprises. University of Oxford.

Rehman, M. H., Yaqoob, I., Salah, K., Imran, M., Jayaraman, P. P., & Perera, C. (2019). The role of big data analytics in industrial Internet of Things. *Future Generation Computer Systems*, *99*, 247–259. doi:10.1016/j.future.2019.04.020

Xu, L. D., He, W., & Li, S. (2014). Internet of Things in Industries: A Survey. *IEEE Transactions on Industrial Informatics Vol*, *10*(4), 2233–2243. doi:10.1109/TII.2014.2300753

Younan, M., Houssein, E. H., Elhoseny, M., & Ali, A. A. (2020). Challenges and recommended technologies for the industrial internet of things: A comprehensive review. *Measurement*, *151*, 107198. doi:10.1016/j.measurement.2019.107198

Chapter 11
Knowledge–Driven Autonomous Robotic Action Planning for Industry 4.0

Ajay Kattepur
Ericsson Research, India

ABSTRACT

Autonomous robots are being increasingly integrated into manufacturing, supply chain, and retail industries due to the twin advantages of improved throughput and adaptivity. In order to handle complex Industry 4.0 tasks, the autonomous robots require robust action plans that can self-adapt to runtime changes. A further requirement is efficient implementation of knowledge bases that may be queried during planning and execution. In this chapter, the authors propose RoboPlanner, a framework to generate action plans in autonomous robots. In RoboPlanner, they model the knowledge of world models, robotic capabilities, and task templates using knowledge property graphs and graph databases. Design time queries and robotic perception are used to enable intelligent action planning. At runtime, integrity constraints on world model observations are used to update knowledge bases. They demonstrate these solutions on autonomous picker robots deployed in Industry 4.0 warehouses.

INTRODUCTION

Advances in robotics, cyber-physical systems and industrial automation has come to the forefront with Industry 4.0 Lasi et al. (2014), with the following key requirements:

1. **Interoperability:** Machines, Internet of Things (IoT) Greengard (2015) enabled devices and humans connected and coordinating with each other.
2. **Information transparency:** Physical systems enhanced with sensor data to create added value information systems.
3. **Technical Assistance:** Use of intelligent devices to aid in informed decision making. Robotic automation may be identified to perform repetitive, unsafe or precise tasks.

DOI: 10.4018/978-1-7998-3375-8.ch011

4. **Decentralized Decisions:** The ability of such systems to make autonomous decisions; only critical cases will involve human intervention.

A fundamental characteristic required in Industry 4.0 deployments is the ability of autonomous robotic devices to self-configure in dynamic goal and deployment conditions. Autonomic computing Huebscher and McCann (2008) models have been proposed to create self-aware robotic systems that respond to both high level goals as well as external stimuli Faniyi et al. (2014). This has led to the development of *Cognitive Robotic Architectures* Levesque and Lakemeyer (2010)Beetz et al. (2010), that are at the intersection of robotics, IoT and Artificial Intelligence Russell and Norvig (2015).

Cognitive robots are able to intelligently execute tasks based on high level *goals*, dependent on *world model* knowledge and sensory *perceptions* to generate efficient *actions* Levesque and Lakemeyer (2010). In order to be deployed in dynamic Industry 4.0 environments, the robots must be autonomous and adaptive to runtime changes. Given a high level task such as *"pick ball from warehouse rack"*, the autonomous robot must identify appropriate *action plans* to perform this task. As the robots are intended to be learning world models, *knowledge bases* are needed to populate information about the world, object, perception and action sequences needed. Any runtime anomalies are dealt with through further queries and eventual exception handling.

Distilling these high level requirements, an autonomous planning module for robots should include:

- *Knowledge Bases* that efficiently capture relationships between world models, objects, robot actions and tasks
- *Action Plans* that are efficiently decomposed from a high level goal task; this involves querying the knowledge base as well as triggering perceptions in case of knowledge mismatch
- Techniques to *Reconfigure* actions at runtime, when plans cannot be executed due to constraints
- Rules for consistent *Updates* to the world model, which allows multiple robots to coordinate or analyze exceptions during execution.

While individual modules may have been developed in the robotic and embedded software communities, integrating these features into a common framework for industrial deployments remains a challenge.

In this paper, we propose *RoboPlanner*, a structured technique to generate design time action plans for autonomous robots. In order to enable autonomy in deployments, we integrate *knowledge bases, design time action planning* and *runtime adaptation* modules. Knowledge representation and queries are enabled using efficient graph database technologies Angles and Gutierrez (2008). Design time action plans as provided using the formal concurrent programming knowledge Orc Kitchin et al. (2009), that allows structured composition of action plans. To take care of runtime adaptation, we provide general rules for triggering perception and exception handling. An integrity check is also provided to update the graph database with runtime knowledge. This framework is implemented over a realistic industrial use case involving autonomous picking robots employed in Industry 4.0 warehouses Wurman et al. (2008).

Principal contributions of this chapter:

1. *RoboPlanner* Knowledge Base module that formally models robotic world models, capabilities, object descriptions and task templates.
2. *RoboPlanner* Action Planner that uses design-time queries/updates to knowledge graph databases, including exception handling.

3. *RoboPlanner* Runtime simulation, adaptation and performance analysis of action plans using graph queries. This may be used to generate executable task templates for physical robots.
4. *RoboPlanner* integrity checks for runtime updates to the knowledge base.
5. Demonstration of the framework over an Industry 4.0 warehouse automation task.

The rest of this chapter is organized as follows: Firstly, we provide a detailed overview of the state of the art. Next, we provide an overview of Industry 4.0 warehouse automation and the autonomous robots deployed in them. The *RoboPlanner* modules are described in with details of knowledge base representation using graph databases are covered. This follows with description of the techniques used for action plan generation. Simulation, performance analysis and knowledge updates in autonomous robot deployments are finally presented. The chapter ends with conclusions.

RELATED WORK

Industry 4.0 Automation

Industry 4.0 deployments Lasi et al. (2014) propose the use of autonomous robotic entities to complete complex tasks. Commercial deployments have been used in warehouses Bartholdi and Hackman (2016) Zhang and et al. (2016) to improve throughput of automated tasks. Amazon[1] has deployed hundreds of autonomous robots to aid in reducing costs of warehouse logistics Wurman et al. (2008). Inspiration is drawn from the use of autonomic computing technologies Huebscher and McCann (2008), that allow robotic runtime reconfiguration and adaptation. Architectures with self-aware, self-configuring and self-optimizing capabilities have also been proposed Faniyi et al. (2014), that may be applied to such automation frameworks.

Autonomous / Cognitive Robotics

The need for autonomy in robots has led to recent research on cognitive robotic systems Levesque and Lakemeyer (2010), with architectures such as RoboEarth Tenorth and Beetz (2013) and CRAM Beetz et al. (2010) being proposed. While a few of these make use of semantic ontologies to represent knowledge, others make use of biological memory models to cache information. A review of cognitive architectures applied in multiple domains such as vision, learning, memory models and robotics have been provided in Kotseruba and Tsotsos (2018).

Table 1 provides a detailed comparison between *RoboPlanner* and other cognitive/autonomous robotic architectures. We notice that OWL based ontologies Grimm et al. (2007) and queries using SPARQL/ Prolog are heavily used, which suffer from performance deterioration when the knowledge base is large. Automated planners such as ROSPlan Cashmore and et al. (2015) make use of logical PDDL transitions at task design time, rather than runtime executions. In particular, runtime exception handling and consistent model updates have not been fully considered in these frameworks.

Table 1. State of the Art Autonomous / Cognitive Robotic Architectures.

Feature Modules	CRAM Beetz et al. (2010) RoboEarth Tenorth and Beetz (2013)	ACT-R/E Trafton et al. (2013)	SOAR Laird et al. (2012)	OpenRobots Ontology (ORO) Lemaignan et al. (2010)	RoboPlanner
Application Domains	Cognitive Service Robots, Knowledge Sharing among Robots	Human–Robot Coordination	Autonomous Mobile Robots	Cognitive Service Robots	Autonomous Robots
Knowledge Base	KnowRob Tenorth and Beetz (2013) OWL Ontologies	Declarative knowledge (factbased memories); Procedural knowledge (rulebased memories)	Semantic Memory Models – Symbolic and Episodic	ORO OWL and RDF Triplestore	Graph Databases with World Model, Robot Capabilities, Algorithms and Task Templates
Knowledge Queries	Knowrob (Prolog) queries, that can be extended to other ontology queries	High level model interactions	STRIPS Russell and Norvig (2015) like decision procedures	SPARQL Queries	Gremlin Knowledge Graph Queries
Action Planning	CRAM Plan Language (CPL) allowing concurrent, parallel processes	Intentional (Goal) module	Procedural memory module	CRAM integration with logical rules	Formal Concurrent Orc Specifications with Knowledge Base queries
Runtime Exception Handling	COGNITO reasoning system that processes failure traces	Utility based rewards; Visual and Aural modules	Reinforcement Learning	Human expert intervention	Adaptation and exception handling modules
Runtime Knowledge Base Updates	No Explicit Mention	Knowledge chunks updated	Chunks of memory data updated	RDF Triple updates with consistency checks	Graph database update with integrity checks
Performance Evaluation	No Explicit Mention	Accuracy of Actions with respect to World Model changes	Cognitive Reactivity measured	ORO Server performance evaluation (updation, queries)	Graph database performance, exception handling delay

Knowledge Driven Action Planning

In *RoboPlanner*, we propose the use of graph databases Angles and Gutierrez (2008) for knowledge representation, which maintain graph relationships within the database. Efficient graph queries are useful in dialogue and chatbot engines as presented in M. Maro and Origlia (2017). We also propose using the Orc concurrent programming language, that may be use in conjunction with industrial workflow specifications (redacted for double blind review). Aspects of the Orc framework are similar to Hierarchical Task Networks Erol et al. (1994), with complex expressions being sub-divided into atomic tasks. Orc further provides granularity in controlling concurrency, temporal actions and runtime behaviour, that is more suited for action planning in robotics. A related programming approach is the GOAL agent programming language Hindriks and Dix (2014), that makes use of belief-desire-intention approaches to programming intelligent agents. Aspects of knowledge modeling, action templates and goal functions may be mapped to similar axioms provided in our framework. Such an approach may be extended to multiple autonomous robotic deployments.

This chapter is an extended version of the short conference paper Kattepur and Balamuralidhar (2019). The main extensions with respect to Kattepur and Balamuralidhar (2019) are:

1. More detailed description of state of the art with a comparison Table 1.
2. An example use case description in Industry 4.0 warehouses, that further motivates the use of knowledge driven robotic actions.
3. More detailed analysis of knowledge graphs, graph databases and queries.
4. Algorithm 2 detailing the action planning and adaptation mechanism.
5. Newer more detailed simulation results covering runtime adaptation, ROS code generation, latency analysis and graph database queries.

WAREHOUSE AUTOMATION

In this section, we introduce Industry 4.0 warehouse automation tasks that may be fulfilled by autonomous robots. A high level description of autonomous robots is also introduced, which is used to build the *RoboPlanner* framework in proceeding sections.

Figure 1. Automation for Warehouse Pick & Place Tasks.

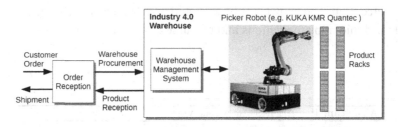

Industry 4.0 Warehouses

Industrial warehouses are employed as buffers in supply-chains to maintain excess product, when there are variations in procurement/customer demand Bartholdi and Hackman (2016). Considerable effort has gone in reducing the stowing and procurement times in such warehouses, with automated picking robots Zhang and et al. (2016) being throughput of *pick & place* tasks.

Fig. 1 presents a high level view of operations taking place in automated warehouses. Once a delivery order is received, the products are procured from the warehouse. As shown in Fig. 1, autonomous Picker robots (such as KUKA KMR Quantec[2]) are being proposed for Industry 4.0 automating pick & place tasks. The robots are intended to be autonomous, with adaptation seen for varying pick-up locations, product dimensions and rates of procurement. When the required products are procured, they are collated and checked for final packing and product shipment.

In order to successfully integrate robotic entities into complex industrial deployments, it is crucial to develop a unified modeling framework for autonomous robots.

Autonomous Robots

To model the robotic components in warehouses, we make use of the *Autonomous Robot* abstraction, inspired by intelligent agents Russell and Norvig (2015). Typical activities, for instance with a pick & place robot in a smart warehouse, include:

1. *Goals*: Understanding goals of each task and subtask, such as, placing correct parts into correct bins within the given time constraints.
2. *Perception*: Object identification and obstacle detection using camera and odometry sensors that sense the environment. This aids the robot in object detection and identification. Robot location, view and environment may also be perceived.
3. *Actions*: Identifying granular actionable subtasks, such as, moving to particular location, picking up parts of orders or sorting objects. Constraints may be placed on the robot capabilities, motion plans and accuracy in performing such actions.
4. *Knowledge Base*: Using domain models of the world for goal completion, such as warehouse environment maps, rack type and product features. The robot capabilities and necessary algorithms should enable completion of goals.

Algorithm 1 presents an overview of an intelligent robot's perception and action via a *Knowledge Base* Russell and Norvig (2015). The knowledge base coordinates the appropriate action in relation to an individual robot's perception. The knowledge base should also include descriptions of domain ontology, task templates, algorithmic implementations and resource descriptions.

Algorithm 1.

Algorithm 1: Stateful Intelligent Robotic Agent.

1 **Input:** Robot Perception; Knowledge Base; Robot State;
2 **Output:** Robot Action;
3 Robot State ← *Interpret*(Perception);
4 Knowledge Base ← *Update*(Knowledge Base, Perception);
5 Action ← *Choose-Best-Action*(Knowledge Base);
6 Robot State ← *Update*(State, Action);
7 Knowledge Base ← *Update*(Knowledge Base, Action);

To integrate the above elements into robotic interactions for Industry 4.0, we propose the *RoboPlanner* autonomous architecture framework.

ROBOPLANNER MODULES

In this section, we provide details about the various modules to be integrated within *RoboPlanner*. These modules cover the principal requirements of cognitive robotic architectures Levesque and Lakemeyer

(2010)Beetz et al. (2010), including knowledge representation, action planning, reconfiguration and knowledge updates. Fig. 2 provides an overview of the modules that are integrated within *RoboPlanner*:

- **Design Time Action Planning Module**: This module is responsible for generating efficient action plans, when input with a high level goal. The module decomposes the goal into atomic tasks, and applies workflow specification languages (such as Orc Kitchin et al. (2009)) to complete the goal task. Action planning involves querying the *Knowledge Graph Database Module* to ascertain requirements for goal completion. *Robot perception* may also be triggered to acquire further information for action planning.

- **Knowledge Graph Database Module**: An integral part of all autonomous/cognitive robotic architectures is the knowledge base. We model this using graph databases Angles and Gutierrez (2008), that maintain relationships between data in a graphical form. Entities such as the world model, robotic algorithms and task templates are stored in the database. The knowledge database is queried both at design time for action generation and at runtime for knowledge updates.

- **Runtime Execution Module**: The action plans are executed by one or multiple autonomous robots to complete the task. Translation of the action plan to a robot specific middleware language such as ROS[3] may be done. The execution module may be aided by robotic perception. Knowledge that is gained during the execution is to be updated to the graph knowledge database, after satisfying some *integrity constraints*.

- **Adaptation Monitoring Module**: This modules monitors runtime deployments of intelligent robots to estimate plan completion. While robotic perception may be used to aid in unforeseen circumstances, more severe exceptions may require re-planning. Performance degradation (leading to non-completion of plans), may also trigger re-planning. Knowledge of instances that trigger re-planning are learnt and updated.

The following sections dive further into the modeling and implementation of these modules.

Figure 2. RoboPlanner Design/Runtime Execution Modules.

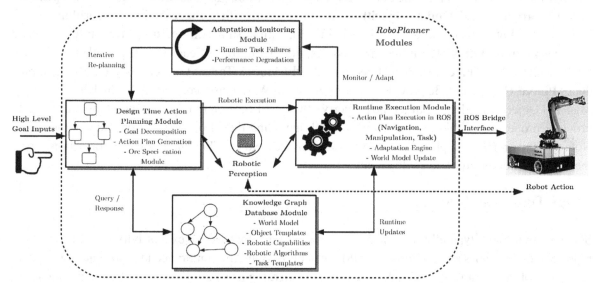

ROBOTIC KNOWLEDGE BASE

The robotic knowledge base is modeled using property graphs, with data stored in graph databases. Queries using the Gremlin graph query language are also studied.

Knowledge Graphs

In order to model knowledge bases inherent in intelligent automation, we make use of property graphs Angles and Gutierrez (2008). Property graphs are attributed, labelled, directed graphs. This is an alternative to semantic ontologies Grimm et al. (2007) and tuple data-stores that are use in implementations such as Knowrob Tenorth and Beetz (2013) and CRAM Beetz et al. (2010). Our knowledge base has the following knowledge graphs included:

- **World Models:** Describes the environment map and layout, including object locations.
- **Object Templates:** Describes the target objects of interest, including shape, size, colour and location.
- **Robot Capabilities:** Provides robot models, capabilities, sensors and actuators that are integrated to perform tasks.
- **Robotic Algorithms:** Navigation, manipulation and task allocation algorithms that are used within robotic actions.
- **Task Templates:** High level task requirements and corresponding outputs are provided.

Fig. 3 provides the property graph models for world models, task templates, object templates, robot capabilities and robot algorithms. To describe properties between edges, we limit ourselves to four relations: **isOfType**, **hasProperty**, **requires** and **produces**. **isOfType** provides hierarchical sub-class relationships; **hasProperty** extends property descriptions using key–value pairs; **requires** provides preconditions to extract knowledge from the graph; **requires** provides post condition effects of executing the node. These relationships may be queried to extract information from the knowledge base.

Fig. 3a provides the capabilities of a **Pick Robot** that **Robot Model**, **Capabilities**, **Perception**; it **requiresTarget**, **World Model**, **Algorithms** and produces the **Pick**, **Place** Actions. Algorithms necessary for the robotic executions are provided in Fig. 3d, with **path planning**, **image template matching** and **grasp manipulation** algorithms included. Explicit definitions of each task is provided in Fig. 3b, for instance with the **Place** task, which requires **World Model**, **Target Object**, **Picker Robot** and produces **Placed Object**. Fig. 3e provides an example of the Warehouse world model, which **hasProperty** Map and Object. In order to extract the property of Object Location requires a Map of the area. Fig. 3c provides properties of objects in the world model, including their Location, Shape and Contour Map. Note that the property graph modeling approach provides extensibility and reuse of information across multiple autonomous robotic deployments.

Graph Database Queries

Semantic ontologies typically store data in tuple data-stores that reduce expressivity provided in graph representations Angles and Gutierrez (2008). Scalability is another hindrance in representation, update and query of large ontologies. Graph databases are emerging as an appropriate tool to model intercon-

Figure 3. Knowledge Property Graphs for Autonomous Robots.

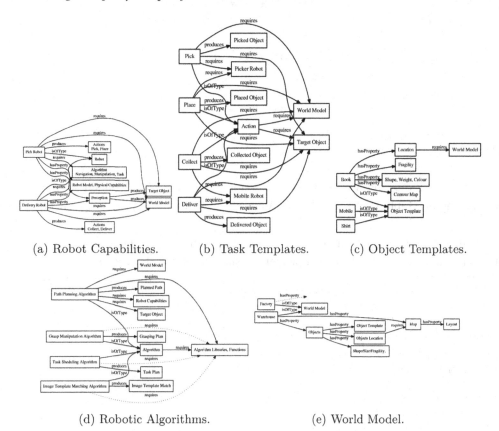

(a) Robot Capabilities.   (b) Task Templates.   (c) Object Templates.

(d) Robotic Algorithms.   (e) World Model.

nectivity and topology of relationships among large knowledge data sets Angles and Gutierrez (2008). Principal advantages include: (i) Being able to keep all the information about an entity in a single node and show related information by arcs connected to it; (ii) Queries can refer directly to this graph structure, such as finding shortest paths or determining certain subgraphs; (iii) Graph databases provide efficient storage structures for graphs, thus reducing computational complexity in operations. Graph databases are also emerging as high-performance back end stores when making use of complex dialogue and chatbot engines M. Maro and Origlia (2017).

To implement the property graphs, we make use of the multimodal *OrientDB* database[4]. OrientDB uses a generic vertex persistent class *V* and a class for edges *E*. Unlike ontologies that store data using triple stores, graph databases maintain the graphical structure with vertices and edges. In the graph data model, nodes are physically connected to each other via pointers, thus enabling complex queries to be executed faster and more effectively than in a relational data model Angles and Gutierrez (2008). Properties are represented as *Key–Value* pairs that may be queried.

An example graph database (of the World Model in Fig. 3e) with vertices, edges and properties in OrientDB is presented below:

```
gremlin> g.V.map
==>{Name=WorldModel}
```

```
==>{Name=Objects, Properties=ObjectProperties,
Location=ObjectLocation}
==>{Name=Warehouse}
==>{Name=Map, Rack=RackConfig, Layout=WarehouseLayout,
Aisles=AislesConfig} gremlin> g.E
==>e[#26:0][#10:0-isOfType->#9:0]
==>e[#29:0][#10:0-hasProperty->#11:0]
==>e[#30:0][#10:0-hasProperty->#9:1]
==>e[#33:0][#9:1-requires->#11:0]
```

In order to query this graph, we use *Gremlin*[5], a domain-specific (DSL) open source programming language focusing on graph traversal and manipulation. The following types of queries may be made:

1. **Filtering:** Filter out vertices or edges according to given property labels. For instance, the query may **g.v().Name** be used to filter out properties such as **Name** of a vertex.

```
gremlin> g.v('10:0').Name
==>Warehouse gremlin> g.v('10:0').bothE
==>e[#26:0][#10:0-isOfType->#9:0]
==>e[#29:0][#10:0-hasProperty->#11:0]
==>e[#30:0][#10:0-hasProperty->#9:1]
```

2. **Complex Queries:** Queries can combine multiple vertices, edges and properties. Queries can also provide range or equality constraints to numeric property values. For instance, the complex query **g.V.has('Name', 'Warehouse').out('hasProperty'). map** matches the vertex with property key–value pair (**Name, Warehouse**), output edge with property **hasProperty** and produces an output of the vertices.

```
gremlin> g.V.has('Name', 'Warehouse').
out('hasProperty').map
==>{Rack=RackConfig, Layout=WarehouseLayout,
Aisles=AislesConfig, Name=Map}
==>{Properties=ObjectProperties, Name=Objects, Location=ObjectLocation}
```

3. **Graph Traversal:** Another advantage of storing data using graph databases is the ability to traverse graphs. For instance, the query **g.v().outE.inV.name.path** traverses the output edges (**outE**) of a vertex, and provides the path traversed.

```
gremlin> g.v('10:0').outE.inV.name.path
==>[v[#10:0], e[#26:0][#10:0-isOfType->#9:0], v[#9:0], null]
==>[v[#10:0], e[#29:0][#10:0-hasProperty->#11:0], v[#11:0], null]
```

While we have made use of Gremlin as the language for explicit graph database querying, this can also be a backend for an efficient dialogue/chatbot implementation M. Maro and Origlia (2017). Questions such as "Where is the target?" or "What are the target's properties?" or "Can the robot lift this?" can be translated into efficient knowledge base queries as defined above. It is of interest to translate this knowledge to efficient action plans for the robot to act upon, which is explored next.

ACTION PLAN GENERATION

In order to study the design time action planning module, we formalize the interaction between the planner and knowledge base. An overview of the concurrent programming language Orc is also provided, that is later used to simulate action planning.

Orc Language

In order to implement robotic action plans, we make use of the formal specification language *Orc*. The Orc concurrent programming language is grounded on formal process-calculi to specify complex distributed computing patterns Kitchin et al. (2009). The execution of programs in Orc makes use of *Expressions*, with the atomic abstraction being a site. To create complex expressions based on site invocations, Orc employs the following *Combinators*:

- **Parallel Combinator (|):** Given two sites s_1 and s_2, the expression $s_1 \mid s_2$ invokes both sites in parallel.
- **Sequential Combinator (>x>, >>):** In the expression $s_1 > x > s_2$ (shorthand $s_1 >> s_2$), site s_1 is evaluated initially, with every value published by s_1 initiating a separate execution of site s_2.
- **Pruning Combinator (<x<, <<):** In the expression $s_1 < x < s_2$ (shorthand $s_1 << s_2$), both sites s_1 and s_2 execute in parallel. If s_2 publishes a value, that value is bound to x and the execution of s_2 is terminated.
- **Otherwise Combinator (;):** The expression $s_1 ; s_2$ first executes site s_1. If s_1 publishes no value and halts, then s_2 is executed instead.

The val declaration in Orc binds variables to values. The def declaration defines a function. Orc further contains built-in sites incorporating distributed computing paradigms such as channels, semaphores and synchronization primitives (further details available in the Orc website[6]).

Action Planning Module

As specified in Fig. 2, action planning involves interacting with the knowledge base to efficiently plan manipulation, navigation and task planning actions. However, perception and exception handling must also be built in to take care of insufficient knowledge.

To formalize the process of generating *action plans* required to satisfy *goals*, we present Algorithm 2. Given an input *goal* (e.g. pick ball from rack using picker robot), the first step (lines 3, 4 in Algorithm 2) is to verify and subdivide goals from the *task descriptions* available (pick target, being an atomic subgoal). For each of these *subgoals*, there are pre-conditions to be satisfied (lines 6–8 in Algorithm 2): *subgoals* require *(actions, targets)*, *actions* require *(targets, object attributes, capabilities)*, *targets* require *(object attributes)*. The *object attributes* of interest (environment rack, locations) can either be derived from the world model knowledge base or by querying *robot perception* (robot sensor observation and interpretation, environment point cloud). The *target* of interest (ball) can either be identified from the object templates knowledge base or by querying *robot perception* (robot sensor observation and interpretation, perception algorithms). The *capability* to complete goal (robot model, arm length, battery state) is also extracted from the robot capability knowledge base. Finally, the *action* (pick ball) needed to satisfy the *subgoal* is derived, dependent on the specified *target*, *object attributes* and *capability* (line 12 in Algorithm 2). The *actions* consist of both navigation (path planning) and manipulation (grasping, lifting) procedures. This process is used iteratively for each *subgoal* to derive the *action plan* needed to enact the goal (line 13, 14 in Algorithm 2). In case there are *Exceptions* observed within the subgoal planning, re-planning is triggered.

An example of such an action plan in presented in Fig. 4, wherein the high level input task of: picker | pick | ball | rack is decomposed iteratively to complete the task. Queries to the knowledge base enable generating information to identify targets (target?) or atomic actions (action?). Perception triggering and re-planning in case of exceptions are also provided. Such a process of decomposing high level expressions to actionable tasks has also been employed in the automated planning community with Hierarchical Task Networks Erol et al. (1994).

To further generalize this action planning framework, we provide instances of multiple artifacts in Table 2. We emphasize that the procedure outlined in Algorithm 2 and Fig. 4 is structured to be generic, allowing action plans to be generated with various world models, robotic capabilities and task templates.

Such a structured way of planning actions will prove valuable across multiple deployments.

Human Centric Knowledge Elicitation

Note that though we have represented this via graph queries, another view would be request-responses with a dialogue agent A. Bordes and Weston (2017), representing the knowledge base. The dialogue agent could be used to further clarify queries that may be ambiguous. A typical conversation instance could be:

```
user: pick ball
RoboPlanner Dialogue Agent: recognize two target:= ball; colour? red colour?
blue. Which color?
User: pick red ball
RoboPlanner Dialogue Agent: target:= ball; colour:= red; action:= pick; pro-
ceed?
User: yes
```

Such a system unifies the modeling of both knowledge acquisition from a central repository, robotic updates and queries that may be made to human participants. The action plans that are formulated result in valid goal fulfillment, due to varied knowledge sources incorporated.

Algorithm 2.

Algorithm 2: Generating Action Plans for Goals via Knowledge Bases.

---

1 **Input:** Input Goal; Knowledge Base[World Model, Object Templates, Task Descriptions, Robot Capability, Algorithms];
2 **Output:** Action Plan;
3 Goal ← *Verify*(Input Goal, Knowledge Base[Task Descriptions]);
4 Subgoals ← *Decompose*(Goal, Knowledge Base[Task Descriptions]);
5 **for** *each Subgoal* **do**
6   (Action?, Target?) ← *Requirements*(Subgoal);
7   (Target?, Object Attributes?, Capability?) ← *Requirements*(Action);
8   Object Attributes? ← *Requirements*(Target);
9   **if** *Object Attributes? is a member of Knowledge Base[World Model]* **then**
      | Object Attributes ← *Query*(Object Attributes?, Knowledge Base[World Model]);
   else
      if *Object Attributes? can be obtained by Perception* **then**
         | Object Attributes ← *Perception*(Object Attributes?, Knowledge Base[World Model, Robot Capability, Perception Algorithms]);
      else
         | Exception ← Object Attributes?

10   **if** *Target? is a member of Knowledge Base[Object Templates]* **then**
      | Target ← *Query*(Target?, Knowledge Base[Object Templates]);
   else
      if *Target? can be obtained by Perception* **then**
         | Target ← *Perception*(Target?, Knowledge Base[World Model, Robot Capability, Perception Algorithms]);
      else
         | Exception ← Target?

11   Capability ← *Query*(Capability?, Knowledge Base[Robot Capability]);
12   **if** *Capability satisfies Action* **then**
      | Action ← *Query*(Action?, Target, Object Attributes, Capability, Knowledge Base[Navigation/Manipulation Algorithms]);
   else
      | Exception ← Action?
13   **if** *Exception is null* **then**
      | Action Plan ← *Update*(Action);
   else
      | Trigger Re-planning of Subgoal

14 **return** Action Plan satisfying Input Goal;

While we have demonstrated the use of dialogues to aid in knowledge elicitation for robotics, there are other associated processes available as described in K. Jokinen et al. (2019). Through the use of demonstration, gestures and feedback, the accuracy of knowledge and actions that have been deployed by robots can be improved. In other applications such as smart homes or helper robots, it is crucial to incorporate these aspects to ensure safety and human-centricity of planned actions.

Figure 4. Action Planning for a pick task.

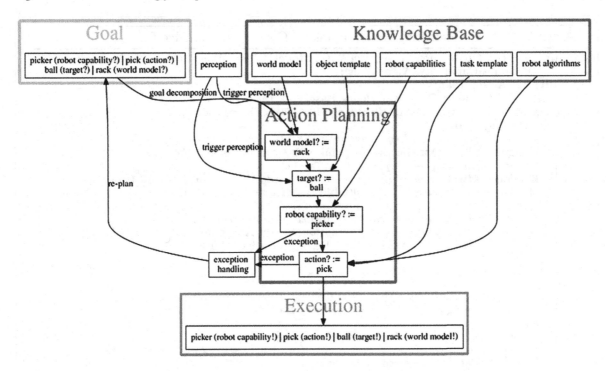

SIMULATION AND ANALYSIS

In this section, we provide an end-to-end simulation of the design time planning and runtime adaptation process, with further analysis on performance aspects. Further constructs to ensure graph database integrity with knowledge base updates are provided.

Table 2. A (Non-exhaustive) List of Artifacts in the Action Planning Framework.

Artifacts	Instances
world model	warehouse | factory | home environment | shipyard
object template	ball | box | obstacle | component
robot capabilities	picking | movement | obstacle avoidance | detection
perception	depth camera | odometry sensor | gyroscope
task templates	delivery | scheduling | monitoring
robot algorithms	localization & mapping | edge detection | path planning
target	ball | bin | book | package | conveyor belt
action	pick | grasp | move | follow | drop | hold
goal	action? | target? | world model? | robot capability?

Action Planning Simulation

Given a high level goal task such as "Pick Ball from Rack using Picker Robot", the first step is to decompose goals into appropriate sub-tasks. The tasks are mapped to appropriate Knowledge Bases (using the member function in Orc), depending on whether they represent actions, targets, robotic components or properties. The following code provides a map of the atomic terms to individual knowledge base elements (Line 12 in Knowledge Resolution). For instance, the term rack is located as a member of the World model knowledge base (Line 15 in Knowledge Resolution).

Box 1.

```
1   +++ Knowledge Resolution +++
2   --Knowledge Base Pointers
3   include "KB.inc"
4   val b=[World_model, Object_template, Robot_cap, Task_template, Robot_algo]
5
6   --Mapping Process
7   def search(a,b) = Ift member(a,head(b)) >> Println(a+" in "+head(head(b))) | Iff member(a,head(b)) >> search(a,tail(b))
8   def plan(a) = search(a,b)
9
10  --Input Goals
11  map(plan,["pick","ball","rack","picker"])
12
13  --Orc Output-------------------------------------
14  rack in World_model
15  ball in Object_template
16  picker in Robot_cap
17  pick in Task_template
18
```

Once the appropriate knowledge base elements are recognized, Gremlin queries are used to obtain dependencies from the Graph database. We assume that the knowledge base is pre-populated with property.

Terms used in Fig. 3, such as hasProperty and requires are used in conjunction with Gremlin graph database filtering and complex queries, to populate local robotic knowledge bases (Lines 8–20 in Knowledge Query). While we represent this as explicit queries, alternate implementations may use dialogue engines to extract necessary information from the knowledge base via question-answers A. Bordes and Weston (2017) M. Maro and Origlia (2017). We make use of the def class declaration that allows us to implement sites within Orc Kitchin et al. (2009), which provides encapsulation similar to classes in object-oriented programming. we make use of the Ref site in Orc, that creates a rewritable storage location. The following Orc code presents these aspects:

Action planning can now be performed, with the high level goals being enacted through decomposition. Queries are made to the knowledge base to determine if the query terms are located in the world or target models, the absence of which triggers perception (Lines 8– 11 in Action Planner). Similarly, queries for the robot and action models are triggered, which can trigger runtime exceptions such as lack of robot capabilities (Lines 14–19 in Action Planner). We also introduce a function to trigger replanning replan action, that looks for exceptions and may add capabilities such as a new robot model or action template (Lines 21 in Action Planner). The following Orc code presents these aspects:

Box 2.

```
1    +++ Knowledge Query +++
2    --Reference store for retrieved data
3    val World_model = Ref([])
4    def append_model(v) = World_model? >m>
5    append([v],m) >q> World_model:= q
6
7    --Gremlin Query site
8    def class gremlin()=
9    def find(v,D) = g.V.has(v,D).map >v>
10   append_model(v)
11   def hasProperty(v,D) = g.V.has(v,D).outE
12   ('hasProperty').inV.map >v> append_model(v)
13   def requires(v,D) = g.V.has(v,D).outE
14   ('requires').inV.map >v> append_model(v)
15   def isOfType(v,D) = g.V.has(v,D).outE
16   ('isofType').inV.map >v> append_model(v)
17   def produces(v,D) = g.V.has(v,D).outE
18   ('produces').inV.map >v> append_model(v)
19   stop
20
21   --Searching Dependencies for "rack"
22   val gremlin = gremlin()
23   gremlin.find("Name","rack") | gremlin.hasProperty("Name","Objects") | gremlin.requires("Name","Map") | gremlin.
24   isOfType("Name","WorldModel") >> World_model?
25
26   --Output----------------------------------------
27   ["WorldModel", ("Map", "WarehouseLayout"),("Objects", "ObjectLocation", "ObjectProperties"), "rack"]
28
```

The output of a typical action plan is now presented, which is input goals that are similar to those planned at design time. Once the query results from various knowledge models are obtained, the action

Box 3.

```
1    +++ Action Planner +++
2
3    --Knowledge Base, Perception and Exception Pointers
4    include "KB.inc"
5    val perception = Dictionary()
6    val exception = Dictionary()
7
8    --Queries for world and object templates, with perception
9    def query1(v,db) = Ift member(v,db) >> (v,db) | Iff member(v,db) >> perception.p := v >> perception.p?
10   def world(w) = query1(w, world_model)
11   def target(o) = query1(o, object_template)
12
13   ---Queries for robot capabilities and actions, with exceptions
14   def query2(v,db) = Ift member(v,db) >> (v,db) |
15   Iff member(v,db) >> exception.ex := v >> add_capabilities(v,db)
16   def robot(r) = query2(r, robot_cap)
17   def action(a) = query2(a, task_template) >> (query2(("navigation","task","manipulation"),robot_algo)) | replan_action(a))
18
19   --Replanning procedures for runtime exceptions
20   def add_capabilities(v,db) = merge(db,[v])
21   def replan_action(a) = Ift (exception.ex? = null) >> stop | Iff (exception.ex? = null) >> exception.ex := null >> action(a)
```

can be performed that includes navigation, manipulation and task completion (Lines 12–14 in Design Time Simulation). Such an execution is straightforward as neither external perception or exceptions are triggered. The following Orc code presents these aspects:

Box 4.

1	**+++ Design Time Simulation +++**
2	--Input goals
3	**robot**("picker") ǀ action("pick") ǀ object("ball") ǀ world("rack")
4	
5	--Output--
6	**Target Query Triggered** for ball
7	**World Model Query Triggered** for rack
8	**Robot Capability Query Triggered** for picker
9	**Action Query Triggered** for pick
10	("ball", ["ball", "cube"])
11	("picker", ["picker"])
12	("rack", ["warehouse", "rack"])
13	(("navigation", ["navigation", "manipulation", "task"]), ("task", ["navigation", "manipulation", "task"]), ("manipulation",
14	["navigation", "manipulation", "task"]))
15	**Action Completed** pick

Runtime Adaptation Simulation

An important aspect of autonomous robotic deployment is runtime adaptation to changes. The goals are modified with the robot type replaced by mover, action collect and the target object replaced by cylinder. As these requirements are not pre-populated into the graph knowledge base, adaptation and exception handling procedures are triggered in Algorithm 2. We notice that perception is triggered to identify the target cylinder (Lines 10–11 in Runtime Adaptation Simulation). Exceptions are also triggered for the lack of collect actions and mover robot capabilities, that are further added into the knowledge base (Lines 14–19 in Runtime Adaptation Simulation). Post this adaptation, the action execution is completed. The following Orc code presents these aspects:

ROS Smach Code Generation

To deploy the action plans to physical/virtual robots, we make use of the open source ROS Smach [7] framework. This is a finite state machine where states and transition of the robot may be described with respect to complex tasks. We autogenerate this from the Orc task list, by referencing robot capabilities, world model and task templates seen in Fig. 3. An example of the ROS Smach code generated is presented below, that produces the output of each state transition as succeeded, aborted or preempted. The PERCEPTION task, if successful is followed by ROBOT ARM MOVEMENT; else an abort or pre-emption is triggered:

```
ActivityDiagram PickPlace produces outcomes (succeeded,aborted,preempted) has
activities{
Activity PERCEPTION { inputData: {WORLD} requireCapability: {robot.camera}
```

Box 5.

1	**+++ Runtime Adaptation Simulation +++**
2	--Input goals
3	**robot**("mover") | action("collect") | target("cylinder") | world("rack")
4	
5	--Output--
6	**Action Query Triggered** for collect
7	**Target Query Triggered** for cylinder
8	**Robot Capability Query Triggered** for mover
9	**World Model Query Triggered** for rack
10	("rack", ["warehouse", "rack"])
11	**Perception Trigged** for cylinder}
12	"cylinder"
13	**Action Replan Triggered**
14	**Exception Trigged** for mover
15	**Adding** knowledge **of** mover
16	**Updated KB** ["mover", "picker"]
17	**Action Query Triggered** for collect
18	**Exception Trigged** for collect
19	**Adding** knowledge **of** collect
20	**Updated KB** ["collect", "pick", "drop", "assign"]
21	(("navigation", ["navigation", "manipulation", "task"]), ("task", ["navigation", "manipulation", "task"]), ("manipulation",
22	["navigation", "manipulation", "task"]))
23	**Action Completed** collect

```
conditions { if (outcome is succeeded) nextActivity:
ROBOT_ARM_MOVEMENT, if(outcome is preempted) final outcome:
PickPlace.preempted, if(outcome is aborted) final outcome: PickPlace.aborted}}
Activity ROBOT_ARM_MOVEMENT { inputData: {WORLD ROBOT TASK TARGET} requireCa-
pability: {robot.movement} conditions { if (outcome is succeeded) nextActiv-
ity:
ROBOT_GRIPPER_GRASP, if(outcome is preempted) final outcome:
PickPlace.preempted, if(outcome is aborted) final outcome: PickPlace.aborted}
}...}
```

Integrating *RoboPlanner* with physical robotic simulator for action planning, such as that shown in Fig. 5, is then done using ROS API calls mapped to each ROS Smach task. ROS also provides ROS bridge interfaces to call physical robot sensor-actuator APIs via the task planning framework. This presents an end-to-end system for autonomous robot action planning (refer to Fig. 2), with knowledge integration, design time action planning, runtime execution and adaptation.

Performance Analysis

Given that we propose the use of knowledge bases and Gremlin graph queries to retrieve the language, performance impact of the queries must be analyzed. This is specially important in the case of Industry 4.0 deployments, where automation is intended to improve throughput. To estimate query and update times in OrientDB graph databases, we run the following stress test on a Linux workstation with 4 core i5-6200U CPU @ 2.30GHz, 4 GB RAM, which simulates the hyper-connected graph traversal over 50 nodes:

Figure 5. KUKA Picker Robot API Call Integration via ROS Smach.

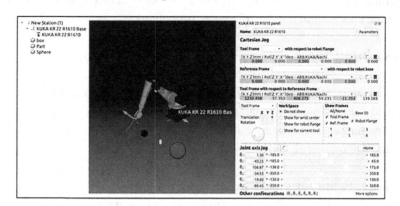

```
Starting workload GSP (concurrencyLevel=4)...
-          Workload in progress 100% [Shortest paths blocks
(block size=50) executed: 50/50] - Total execution time: 2.768 secs
-          Executed 50 shortest paths in 2.762 secs
-          Path depth: maximum 8, average 5.286, not connected 0
-          Throughput: 18.103/sec (Avg 55.240ms/op)
-          Latency Avg: 211.996ms/op (58th percentile) - Min: 55.838ms 99th
Perc: 576.653ms - 99.9th Perc: 576.653ms Max: 576.653ms - Conflicts: 0
```

The average graph traversal latency is seen to be around 211 milliseconds, that outperforms conventional perception and object recognition algorithms (2300 milliseconds in Zhang and et al. (2016)). Using these mean values for exponentially distributed latency outputs, Monte-Carlo runs are performed over 20,000 runs. Fig. 6 demonstrates outputs for various cases with the Knowledge Base having 100%, 90% and 70% of the action planning information (triggering perception and exception handling in case of missing knowledge). For instance, over the base case of 70% plan information in the Knowledge Base, the 95% percentile latency improves by 56.5% (90% queries answered by knowledge base) and by 73.9% (90% queries answered by knowledge base). This demonstrates that continuous learning and runtime

Figure 6. Latency Measurements dependent on Knowledge Base Queries.

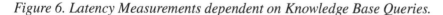

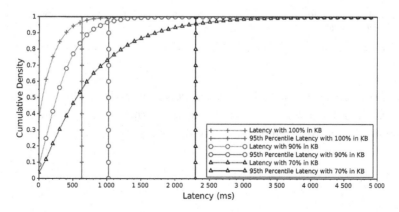

updates have a significant impact on autonomous robotic performance. Thus, it is crucial to maintain an updated knowledge base within the *RoboPlanner* framework.

Graph Database Integrity

While the multi-modal OrientDB satisfies ACID (Atomicity, Consistency, Isolation, Durability) properties for databases, integrity checks are to be maintained when updating the databases. Integrity constraints are rules which define the set of consistent database states or changes of state. Typically, three types of checks are performed Rabuzin et al. (2016):

1. **Schema Instance:** Entity types and type checking integrity.
2. **Referential Integrity:** This checks that the nodes and edges are uniquely named and that the edges are provided with labels and start/end vertices.
3. **Functional Dependencies:** Value restrictions on particular attributes. Defining minimum and maximum property value.

These checks are incorporated into the below Orc code for knowledge base updates. We notice that type checking (Line 4 in Database Update Integrity), redundancy of input data (Line 10 in Database Update Integrity) and valid range of properties (Line 12 in Database Update Integrity) are included. When a robot produces a runtime update, the site update node(key,value) checks for integrity before pushing it to the knowledge base (Line 16 in Database Update Integrity).

Box 6.

```
1    +++ Database Update Integrity +++
2    --Type information
3    type world_model = { . Name:: String, Colour:: String, Location:: (Number,Number,Number) .}
4    val new_world_model = Dictionary()
5
6    --Integrity check site
7    def class integrity()=
8    val range = range
9    def redundancy_check(key,value) = Ift(key = value) >> false | Iff(key = value) >> true
10   def value_check(key,range) = Ift(member(key,range)) >> true | Iff(member(key,range)) >> false
11   stop
12
13   def class update()=
14   def update_node(key,value) = Read(key) >aa> (integrity.redundancy_check(key,value),integrity.value_check(key,value)) >>
15   new_world_model.aa := value
16   stop
```

Such integrity checks and superior performance aspects can prove useful in other applications such as intelligent chatbots and dialogue engines, where updated knowledge bases and real time responses are crucial. In **summary**, our work demonstrates the following:

1. *RoboPlanner* Knowledge Base module that formally models robotic world models, capabilities, object descriptions and task templates – Fig. 3 and inputs to Knowledge Resolution/Knowledge Query.
2. *RoboPlanner* Action Planner that uses design-time queries/updates to knowledge graph databases, including exception handling – Algorithm 2, Fig. 4 and Action Planner/Design Time Simulation.
3. *RoboPlanner* Runtime simulation, adaptation and performance analysis of action plans using graph queries – Runtime Adaptation Simulation example and Fig. 6. Executable task templates as ROS Smach codes.
4. *RoboPlanner* integrity checks for runtime updates to the knowledge base – Database Update Integrity example.

Such modules will prove useful across a host of Industry 4.0 deployments invoking autonomous robots.

CONCLUSION

Autonomous robots are being increasingly used in Industry 4.0 deployments to solve problems via intelligent adaptive mechanisms. A central tenet in such deployments is eliciting efficient action plans that may be executed at runtime. To this end, it is necessary to have proper semantic representations of domains and actions of interest, queries to the knowledge base that may be efficiently parsed and adaptive action executions.

In this paper, we generate action plans through graph knowledge base queries via the *RoboPlanner* framework. Knowledge about robotic world models and capabilities are encoded in efficient graph database models, that may be efficiently queried to extract information for task completion. Using the concurrent programming language Orc, action plans are generated that can handle robotic runtime exceptions and perception information. End-to-end design/runtime simulations and performance analysis demonstrate the advantages of maintaining the robotic knowledge base.

In future, we see development in Industry 4.0 environments along these lines. Robots are evolving from those designed for a specific task to those that can handle multiple tasks. In order to complete these, cognitive abilities such as situation analysis, knowledge evaluation, planning, execution, monitoring and learning will be needed. Areas such as knowledge graphs, AI planning and reinforcement learning and being incorporated further into these areas. The material provided in this chapter provides a roadmap to incorporate some of these aspects into Industry 4.0 deployments. For other robotics use cases such as smart homes or remote robotics, other areas may be of critical importance, that would lead to multiple research foci.

ACKNOWLEDGMENT

This research received no specific grant from any funding agency in the public, commercial, or not-for-profit sectors.

REFERENCES

Angles, R., & Gutierrez, C. (2008). Survey of graph database models. *ACM Computing Surveys*, *40*(1), 1–39. doi:10.1145/1322432.1322433

Bartholdi, J., & Hackman, S. (2016). Warehouse and Distribution Science. The Supply Chain and Logistics Institute, School of Industrial and Systems Engineering, Georgia Institute of Technology.

Beetz, J., Mosenlechner, L., & Tenorth, M. (2010). Cram: A cognitive robot abstract machine for everyday manipulation in human environments. *Intl. Conf. on Intelligent Robots and Sys (IROS)*, 1012-1017. 10.1109/IROS.2010.5650146

Bordes, A. Y. B., & Weston, J. (2017). Learning end-to-end goal-oriented dialog. *Intl. Conf. on Learning Representations (ICLR)*, 1-15.

Cashmore, M. (2015). Rosplan: Planning in the robot operating system. *Proc. of the Intl. Conf. on Automated Planning and Scheduling (ICAPS)*, 1-9.

Erol, K., Hendler, J., & Nau, D. (1994). Htn planning: Complexity and expressivity. AAAI, 1123-1128.

Faniyi, F., Lewis, P. R., Bahsoon, R., & Yao, X. (2014). Architecting self-aware software systems. *IEEE/IFIP Conf. on Software Architecture*, 91-94.

Greengard, S. (2015). *The Internet of Things*. MIT Press. doi:10.7551/mitpress/10277.001.0001

Grimm, S., Hitzler, P., & Abecker, A. (2007). *Knowledge representation and ontologies*. Springer Semantic Web Services.

Hindriks, K., & Dix, J. (2014). Goal: A multi-agent programming language applied to an exploration game. *Springer Agent-Oriented Software Engineering*, 112-136.

Huebscher, M., & McCann, J. (2008). A survey of autonomic computing – degrees, models, and applications. *ACM Computing Surveys*, *40*(3), 1–28. doi:10.1145/1380584.1380585

Jokinen, K., Nishimura, S., Watanabe, K., & Nishimura, T. (2019). *Human-Robot Dialogues for Explaining Activities*. Academic Press.

Kattepur, A., & Balamuralidhar, P. (2019). Robo-planner: Autonomous robotic action planning via knowledge graph queries. *Proc. of the 34th ACM/SIGAPP Symposium on Applied Computing*, 953-956. 10.1145/3297280.3297568

Kitchin, D., Quark, A., Cook, W., & Misra, J. (2009). The orc programming language. *Proc. of FMOODS/FORTE*, 1-25.

Kotseruba, I., & Tsotsos, J. (2018). 40 years of cognitive architectures: Core cognitive abilities and practical applications. *Springer Artificial Intelligence Review*, *53*(1), 17–94. doi:10.100710462-018-9646-y

Laird, J., Kinkade, K., Mohan, S., & Xu, J. (2012). Cognitive robotics using the soar cognitive architecture. *AAAI Tech. Report*, 46-54.

Lasi, H., Fettke, P., Kemper, H., Feld, T., & Hoffmann, M. (2014). Industry 4.0. *Business & Information Systems Engineering*, *6*(4), 239–242. doi:10.100712599-014-0334-4

Lemaignan, S., Ros, R., Mosenlechner, L., Alami, R., & Beetz, M. (2010). Oro, a knowledge management platform for cognitive architectures in robotics. *Intl. Conf. on Intelligent Robots and Systems*, 3548-3553. 10.1109/IROS.2010.5649547

Levesque, H., & Lakemeyer, G. (2010). Cognitive robotics. *Dagstuhl Seminar Proc., 10081*, 1-19.

Maro, M., Valentino, M. A. R., & Origlia, A. (2017). Graph databases for designing high-performance speech recognition grammars. *Proc. of the 12th Intl. Conf. on Computational Semantics*, 1-9.

Rabuzin, K., Sestak, M., & Konecki, M. (2016). Implementing unique integrity constraint in graph databases. *Intl Multi-Conf. on Computing in the Global Information Technology*.

Russell, S., & Norvig, P. (2015). *Artificial Intelligence: A Modern Approach* (3rd ed.). Pearson.

Tenorth, M., & Beetz, M. (2013). Knowrob – a knowledge processing infrastructure for cognition-enabled robots. *The International Journal of Robotics Research*, *32*(5), 566–590. doi:10.1177/0278364913481635

Trafton, G., Hiatt, L., Harrison, A., Tamborello, F., Khemlani, S., & Schultz, A. (2013). Act-r/e: An embodied cognitive architecture for human-robot interaction. *J. of Human Robot Interaction*, *2*(1), 30–55. doi:10.5898/JHRI.2.1.Trafton

Wurman, P., D'Andrea, R., & Mountz, M. (2008). Coordinating hundreds of cooperative, autonomous vehicles in warehouses. *AAAI Artificial Intelligence Mag., 29*, 9–19.

Zhang, H., & ... . (2016). Dorapicker: An autonomous picking system for general objects. *IEEE Intl. Conf. on Automation Science and Engineering (CASE)*, 721-726. 10.1109/COASE.2016.7743473

ENDNOTES

[1]   https://www.amazonrobotics.com/
[2]   https://www.kuka.com
[3]   https://www.ros.org/
[4]   https://orientdb.com/
[5]   http://tinkerpop.apache.org/
[6]   https://orc.csres.utexas.edu/
[7]   http://wiki.ros.org/smach

Chapter 12
Industrial Internet of Things 4.0:
Foundations, Challenges, and Applications – A Review

Vishwas D. B.
NIE Institute of Technology, India

Gowtham M.
NIE Institute of Technology, India

Gururaj H. L.
Vidyavardhaka College of Engineering, India

Sam Goundar
https://orcid.org/0000-0001-6465-1097
British University, Vietnam

ABSTRACT

In the era of mechanical digitalization, organizations are progressively putting resources into apparatuses and arrangements that permit their procedures, machines, workers, and even the products themselves to be incorporated into a solitary coordinated system for information assortment, information examination, the assessment of organization advancement, and execution improvement. This chapter presents a reference guide and review for propelling an Industry 4.0 venture from plan to execution, according to base on the economic and scientific policy of European parliament, applying increasingly effective creation forms, and accomplishing better profitability and economies of scale may likewise bring about expanded financial manageability. This chapter present the contextual analysis of a few Industry 4.0 applications. Authors give suggestions coordinating the progression of Industry 4.0. This section briefly portrays the advancement of IIoT 4.0. The change of ubiquitous computing through the internet of things has numerous difficulties related with it.

DOI: 10.4018/978-1-7998-3375-8.ch012

INTRODUCTION

Industry 4.0 - a term that has fallen more regularly than generally others. At driving exchange fairs, the point has been assuming a focal job for quite a while. The main modern transformation, which kept going from around 1760 to 1840, was activated by the development of railroads and the innovation of the steam motor. It introduced the period of mechanical creation. (Anita, & Bodla, 2017).

The second mechanical unrest started in the nineteenth century and proceeded into the mid twentieth century. Drivers were the presentation of power and the sequential construction system in the car business by Henry Ford in 1913. Thus, creation turned out to be a lot quicker, as every worker focused on just one work unit appear in in Figure 1.

Figure 1. Industrial Revolutions
(Anita, & Bodla, 2017).

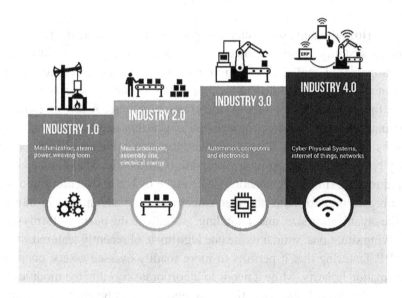

The third modern insurgency started during the 1960s and was fundamentally impacted by the advancement of semiconductors, centralized server PCs, PCs and the Internet.

The fourth modern insurgency was fundamentally molded by physical and advanced patterns. Klaus Schwab names four material signs. (Kagermann, Wahlster, & Helbig, 2013).

1. Autonomous engine automobiles
2. 3D printing
3. Advanced robotics
4. Novel (new) materials

Industry 4.0 makes the idea of savvy/insightful processing plant wherein the digital physical frameworks settle on decentralized choices. Machines speak with one another, educate each other about deformities in the creation procedure, distinguish and reorder exhausting material inventories. Computerization and mechanical autonomy ascend to a next degree of digital physical frameworks by joining IoT (Internet of things), and distributed computing. Digital physical frameworks speak with one another and with human administrators utilizing the Internet of Things. (Kagermann, Wahlster, & Helbig, 2013).

There are four structure standards of industry 4.0, which must be followed:

1. **Interoperability:** Human administrators, machines, sensors, gadgets must speak with one another by means of IOT.
2. **Information Transparency:** The data frameworks means to have the option to make a simulated copy of all physical data with sensor information.
3. **Technical Assistance:** The capacity of the frameworks used to help people for taking care of pressing many issues. The capacity of digital physical frameworks to genuinely bolster people for assignments which are danger or risks for people.
4. **Decentralized Decision Making:** The capacity of the digital physical structure to play out their assignments as self-sufficient as could be considered.

Internet of Things (IoT) the biggest advanced megatrend that connects the physical and virtual universes. The expanding systems administration of individuals, articles and machines with the Internet is prompting the development of new plans of action. Among the various advances describing the development procedure of an organization, the ERP appropriation is one of the most significant, because of the critical impacts that it could have on the whole creation process. A few investigations feature that organizations, contingent upon their size, approach ERP executions in various ways, and need to confront heterogeneous issues. Simultaneously, it has been likewise shown that advantages originating from the ERP selection vary based on the organization size itself. (Chun-Chin, Mao-Jiun, & Wang, 2004).

When all is said in done, littler organizations advantage of better execution in assembling and co-ordinations, while bigger organizations show enhancements and advantages in money related fields. Notwithstanding the organization size, understanding, and enhancing business forms speaks to a triumph factor in quick evolving situations, similar to the one legitimate of recently featured situations. Improved the meaning of ERP, featuring that it permits to more readily oversee assets, coordinate procedures, and authorize information honesty, while it needs to incorporate coordinated modules for bookkeeping, money, deals and circulation, HR, materials the board and different business capacities dependent on a typical design that interfaces the undertaking to the two clients and providers. (Ahel, et al., 2018)

Also, ERPs permit a consistent joining of procedures across useful regions with improved work process, normalization of different strategic approaches, and improved request of the executives, precise bookkeeping of stock, and better gracefully chain the executives. (Chun-Chin, Mao-Jiun, & Wang, 2004).

Specifically, there are contrasts among enormous and little endeavors over a scope of issues, including:

1. Motivation that prompts the reception of an ERP framework in the processing plant.
2. Kind of embraced framework.
3. Implementation techniques viably received.
4. Customization and re-science's level of the framework working with the ERP.

Specifically in regards to the connection between auxiliary factors legitimate of an organization (e.g., industry type, size, and authoritative structure) and their effect on different tasks, having that the hierarchical size is the most of the time analyzed issues related with advancement, R&D uses, and market power. Proposes that little organizations will in general fall behind huge production lines in executing new advances, additionally utilizing various practices, though bigger makers are bound to be early adopters of data innovation developments, being likewise increasingly fit to have an inward IT office dealing with all the systems administration activities. (Sarah, Abdellah, & Mustapha, 2018).

The various kinds of organizations overviewed and those among them that have an incorporated framework appear in Figure 2.

Figure 2. Integrated System.
(Sarah, Abdellah & Mustapha, 2018).

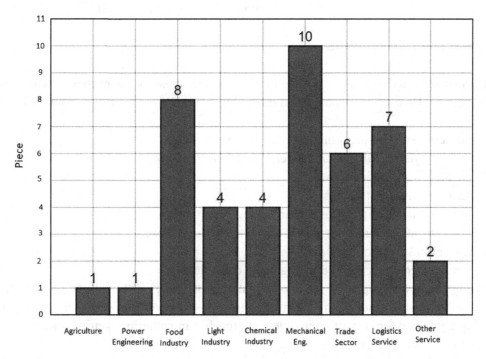

Different angles that must be remembered in picking an ERP framework are:

Programming establishment multifaceted nature—this, in any case, ordinarily speaks to only a piece of the procedure, because of the way that a fruitful ERP execution includes more than having modern programming and propelled processing innovations. (Chun-Chin, et al., 2004).

1. Motivation directing an organization to actualize an ERP
2. Type of ERP that will be actualized and the subsequent framework setup
3. Implementation techniques embraced by the assembling organization
4. Degree of customization required to the ERP designers
5. Implementation cost required by the ERP appropriation, with the subsequent speculation incomes and returns.

This prompts the way that any change to the framework can have significant ramifications. By and large, adjustments lead to greater expenses, longer usage times and increasingly convoluted executions.

CHALLENGES, BENEFITS, AND ADVANTAGES

We are amidst the fourth modern unrest, or Industry 4.0 as it's all the more normally known. Obviously, with an incredible chance, comes extraordinary difficulties. In any case, before we dig any more profound, how about we take a gander at what Industry 4.0 is basically. Industry 4.0 is the robotization data and information trade in assembling advances. The idea incorporates digital physical frameworks, the Internet of things (IoT) and distributed computing. (Baotong, et al., 2018).

Industry 4.0: The Challenges

As Industry 4.0 keeps on changing the manner in which we associate with our general surroundings, new difficulties emerge. Here are the fundamental difficulties you may look not long from now:

1. New business models—the significance of another strategy
2. Reconsidering your affiliation and systems to help new outcomes
3. Understanding your business case
4. Leading productive pilots
5. Helping your relationship to grasp where action is required
6. Change organization, something that is over and over overlooked
7. Assessment of association culture
8. The genuine interconnection everything being equivalent
9. Enrolling and developing new capacity The Internet of Things (IoT) will interface machines and structures and award unsurprising information transmission over all pieces of a work environment, opening up open doors for all around new approaches in gathering, getting ready, and different associations. (Baotong, et al., 2018).

Industry 4.0: The Benefits

Here are the key points of interest of the new present day change.

1. Upgraded gainfulness through progress and robotization.
2. Ongoing data for nonstop smoothly chains in a consistent economy.
3. More prominent business rationality through bleeding edge upkeep and watching possibilities.
4. More excellent things in light of constant watching, IoT-enabled quality improvement.
5. Better working conditions and transcendent reasonability.
6. Personalization open entryways that will pick up the trust and endurance the front line purchaser.

Table 1. SWOT Table (Strengths, Weaknesses, Opportunities and Threats) for IIoT 4.0.

Strengths	Weaknesses
• Improved profitability, productivity, seriousness, income. • Growth in high-skill employees. • Improved consumer reliability – new markets. • Production adaptability and control.	•Highly dependence on strength of advance and systems: little disturbances can have significant effects. • Dependence on a scope of achievement factors. Devices, cognizant structure, work gracefully with fitting aptitudes, venture and R&D. • Costs of improvement and usage • Semi-skilled employees become joblessness. • Need to import talented work and incorporate settler networks.
Opportunities	Threats
• Develop new top markets for items and administrations. • Improve EU socioeconomics • Lower section obstructions in new markets, connections to new flexibly chains.	• Cyber-security, protected innovation, information security • Workers, SMEs, ventures, and national economies without the mindfulness and additionally intends to adjust to Industry 4.0 and who will truly fall behind • Vulnerability to and unpredictability of worldwide worth chains • Adoption of Industry 4.0 by remote contenders killing EU activities.

(Parsa, Najafabadi, & Salmasi. 2017).

Industry 4.0 Advantages

1. **Customization:** Creating a customer organized market that is versatile and will address the masses-sissues and creating demands speedy and profitably. It will decimate the gap between the creator and the customer and correspondence will happen authentically between them.
2. **Streamlining:** Production upgrade is an essential piece of breathing space of Industry 4.0. A 'Clever Factory' containing hundreds and thousands of sharp contraptions that can self-improve will provoke directly around zero creation individual time. This is incredibly fundamental for endeavors which use exorbitant and high collecting equipment as the apparatus in the semiconductor organizations.
3. **Pushing Research:** The task of industry 4.0 degrees of progress will influence research in various fields like IT security and will have an effect expressly o the direction business. Another industry will require new extents of limits. Thusly, 'Direction and Training' will consider another shape which takes such undertakings which require skilled work. (Baotong, et al., 2018).

Stages

A wide scope of business segments, including yet not restricted to human services, vitality, transport, aviation, self-sufficient vehicles, mechanical hardware and brilliant gadgets have used diverse IoT stages The goal of this area is to audit and break down the consequences of IoT stages and locate a basic yet reasonable IoT stage for fast improvement, specifically, changing traditional apparatuses to brilliant (Jehoon, & Sung, 2018).

IoT Stands (platform) for Smart Systems and Products.

1. AWS IoT.
2. Azure IoT Suite.
3. Google Cloud IoT.
4. IBM Watson IoT.

5.   Kaa IoT Platform.
6.   PTC Thing.

CASE STUDIES

Manufacturing Enterprise

Contextual analysis in Industrial Enterprise the accompanying results acquired from a contextual analysis with an Austrian assembling venture with around 400 representatives which plans and produces aviation segments and test hardware are introduced. To guarantee exactness of results, we have picked an association that as of now is occupied with Industry 4.0 and along these lines has required fundamental information and comprehension about its essential ideas. (Shohin, et al., 2020). The organization got a poll for every email to take into account reflected appraisal of their inside circumstances individually. Although self-evaluation of the development things is a legitimate strategy and simple to lead, we know that most organizations at the time don't have the necessary information about Industry 4.0 to self-survey the development of their own organization. Following the second period of the appraisal system (see

Figure 3. Assessment procedure.
(Maroua, et al., 2019).

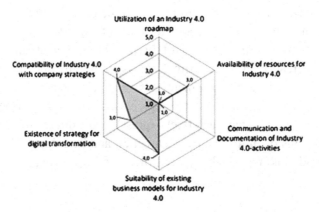

Figure 3), the reaction at that point was embedded into the product device to compute the development levels and to make the development report. In Figure 4 the development level in nine measurements is pictured. A radar graph is utilized to delineate the general outcome initially. Figure 4: Radar graph picturing Industry 4.0 developments in nine measures. (Alessandro, et al., 2019).

The apparently high development level in the measurement "Items" (see Figure 4) is reasonable, as the made aviation parts normally show profoundly developed qualities as to Industry 4.0. For instance, estimated development things in this measurement were (among others) "the likelihood to coordinate items into different frameworks", "the self-rule of items'', the "adaptability of item qualities" or "the likelihood to digitize items". It is conceivable that segments for the application in aviation have item qualities that lead to a high appraising of these development things. The high development levels in dif-

Figure 4. Radar chart visualizing Industry 4.0.
(Maroua, et al., 2019).

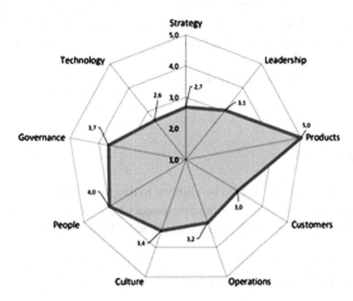

ferent measurements appear to be exact too, as the picked organization is viewed as an early connector of the ideas of Industry 4.0 by a few specialized diaries. (Sarah, Abdellah, & Mustapha, 2018).

To empower evaluation in five levels complete information about the capability of Industry 4.0 with respect to all things is required. Along these lines, either the people inside the organization have or increase enough information, or an outer examiner is welcome to help basic evaluation. A powerful way to deal with help reflected evaluation of the organization's own circumstance is to introduce propelled industry cases in the separate measurement (benchmarking). Investigation of the applied evaluation system in the exploration group prompted the distinguishing proof of essential enhancements and potential further improvements which will be examined in the following segment. (Aleksandr, et al., 2018).

Home Appliance

Relevant examination to show how Industry 4.0 developments explicitly, IoT, Cloud, Edge figuring, and gigantic data could change a standard cooler into an insightful thing for encouraging extra worth. This assessment gives a structure thing progression from perceiving requirements, structure, model, and test to commercialization and undertakings. The purpose of using IoT in an adroit home is to upgrade power use, offer personalization, enabling impelled upkeep, and sensible system refreshes with higher purchaser steadfastness. (Parsa, Najafabadi, & Salmasi, 2017).

This effort is a bit of a portfolio, including a couple IoT-enabled home mechanical assemblies' adventures in a developed modern office with over 70 years of exercises. Changing over normal home machines to insightful ones is outstandingly mentioned concerning the base impact on current creation lines. Notwithstanding the way that this paper fixates on coolers as a common home device, distinctive devices, for instance, garments washers can get sharp through a comparative approach.

This cutting edge relevant examination can be loosened up to all the sagacious home machines and the proposed framework can be gotten by other home devices as a result of comparable qualities, for instance, organize affiliations, obliged undertakings, data volume, data structure, response time, and the most basic chance to embed a revamp IoT-enabled load up inside each home machine. Finally, yet basically, utilizing a single IoT-middleware and versatile application for solitary home machines. This fragment presents a couple of parts of from beginning to end IoT-enabled sharp machine system. (Hugh, et al., 2018).

The system in 5 includes three rule divisions.

1. Wise thing.
2. Correspondence show and
3. Steady checking and control.

Directly off the bat, the ice chest ought to be climbed to transform into an electronic asset instead of just a physical thing. Likewise, suitable remote correspondence is required for building a CPS. Taking everything into account, an IoT structure with portrayal and Cloud capacities is key. The entire terri-

Figure 5. Home Automation System (End to End IoT).
(Parsa., Najafabadi & Salmasi, 2017).

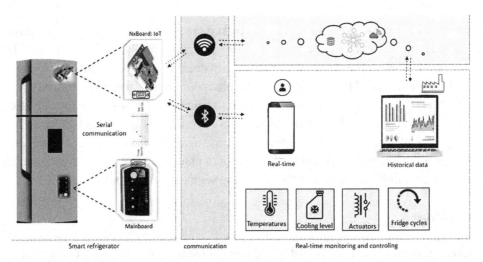

tory inspects current challenges, PCB setup, orchestrate affiliations edge enrolling, gear plan, firmware building, IoT stage, progressing checking/dashboard, programming including firmware and versatile application headway. (Parsa, Najafabadi, & Salmasi, 2017).

IOT for Fluid Distribution Monitoring System (FDMS)

A couple of works are completed to screen the Fluid structure yet concerning IoT based responses for such issue barely any works are explored. In this paper, we will offer a composing review about IoT-based responses for Fluid Distribution Monitoring structure which joins Water Distribution frameworks and Oil and Gas Distribution sort out. The general IOT-based Architecture for FPMS is outlined in figure

Figure 6. The general IOT-based Architecture for FDMS.
(Maroua, et al., 2017).

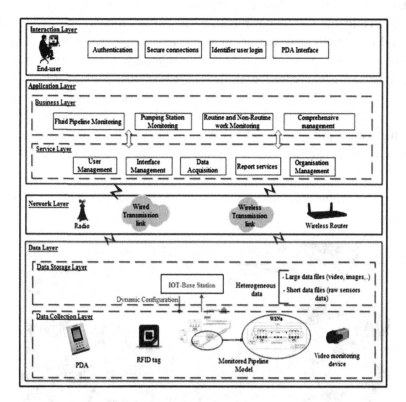

6 that presents the structure configuration got together with the item modules made out of four layers (the Data layer joins data variety and data accumulating layers, the Network layer, the Application layer fuses organization layer and business layer and the Interaction layer). (Maroua, et al., 2017).

A disseminated Database rather than a focal one guarantees the coherence of the framework and gives high accessibility and adaptation to non-critical failure.

The significance of an on-line information the executive's framework to permit a straightforward approval to information access in-time. (Ramalingam, et al., 2019).

Smart Farming Blooms in South Korea

One of the world's biggest mushroom-cultivating activities — a South Korean organization creating in excess of 23 million kilograms of mushrooms every year — perceived the need to drastically change its rambling association in the midst of rising work costs and furious rivalry in the worldwide market. (Sehan, Meonghun, & Changsun, 2018).

To drive new degrees of significant worth and intensity in the advanced economy, a strong change toward 'brilliant cultivating' was considered inescapable. The organization left on an i4.0 venture in association with KPMG's demonstrated group of computerized specialists. With worth and execution boundaries set up and an exhaustive i4.0 guide set up, the customer worked together intimately with the KPMG group to drive its driven change venture forward. Continuous creation information was gathered and investigated utilizing IoT innovation, advanced sensors and a cutting edge information examination stage. Manual assignments over the cultivating, arranging and pressing biological system were to a great extent motorized utilizing present day robotization gear and robots. The organization moved from a substantial dependence on physical work to an automated computerized undertaking. The outcome? Information examination has uncovered another creation approach that essentially improves crop yields by advancing the compost to-soil proportions utilized in mushroom development — producing a surprising 30-percent expansion in yield. (Sehan, Meonghun, & Changsun, 2018).

Likewise, the organization of robotization hardware and robots — joined with investigation of laborers' developments in the creation procedure — uncovered new business bits of knowledge. The organization has thusly redesigned its plant format, further improving efficiency and bringing down work costs. The

Figure 7. Smart farming blooms in South Korea, Korea Bizwire in Agriculture, Business, Industries, Policies and Regulation

organization's comprehensive methodology has drastically improved its worth, execution and intensity as far as profitability, productivity and quality — and its future possibilities for accomplishment in the advanced economy. (Chen, & Zhou, 2018).

Transportation and Logistics

These days, vehicles just as streets and moved merchandise, are furnished with progressively modern innovative gadgets, for example, on close to handle correspondence (NFC) labels, radio-recurrence ID (RFID) labels, actuators, sensors and so forth. (Nallapaneni, Manoj, & Archana, 2017). IoT advancements can be utilized to upgrade the capability of these frameworks and streamline their utilization in the areas of transportation, coordination's and providers, which are viewed as basic segments to the profitability of numerous enterprises. Savvy transportation frameworks (ITS) can impart, offer and trade strategic data and information instantly, ideal and precisely. Subsequently, they are utilized to guarantee that the transportation arrange is proficiently observed and controlled (Azab, et al., 2016).

IoT offers a few contemporary applications and administrations and in blend with the omnipresent 5G versatile systems can furnish enterprises with insightful transportation and coordination's frameworks. These frameworks give arrangements which are structured explicitly for specific needs and objectives, subsequently quickening efficiency, benefit and activities. Also, they offer constant checking and fol-

Figure 8. Transportation and logistics.
(Ghodrati, Hoseinie & Garmabaki, 2015).

lowing all through the whole gracefully chain, in this manner helping endeavors increment start to finish deceivability just as keep up proficient transportation control and financially savvy the executives. Furthermore, they lead progressively compelling course arranging and enhancement, take into account better vitality effectiveness and diminish the general framework personal time. (Ghodrati, Hoseinie, & Garmabaki, 2015).

Figure 9. Agricultural robots.
(Jehoon, & Sung, 2018)

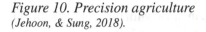

Figure 10. Precision agriculture
(Jehoon, & Sung, 2018).

The Fourth Industrial Revolution Changes in Agriculture

The idea of Agriculture 4.0 draws on the expression "Industry 4.0" and alludes to its expanded mix and correspondences innovation with horticultural creation organized frameworks, consolidating different various kinds of information from numerous sources, guarantee to expand profitability and proficiency. This unrest changes the devices utilized in agribusiness divisions replaces prompts brilliant cultivating. High normal time of rural apparatuses emerges the interest to coordinate it into the computerized world. At present these devices with strong, widespread and interoperable and not required with no unique additional preparation. Because of deficient media communications framework in provincial regions they create devices in horticulture 4.0 that work even where there is no cell phone signal across parts of the cultivation areas. (Jehoon, & Sung, 2018)

1. **Agricultural Robots:** Agriculture robots in Figure 9 will work in various field like creation, preparing, dissemination, and utilization. These robots recognize the administration environment and independently offer keen work. The fourth unrest brings about usage of robots in many cultivating gear's for choice of fitting item and appropriate conveyance of vermin for bug control. This method additionally fixed with elevated vehicle used to control the wellbeing with ordinary checking of natural products, vegetables and creatures in farming field. Robots uncommonly structured under horticulture 4.0 are first is Open-field robots use in quite a while like water system and development of harvests, the subsequent robot is known as office robots used to screen the yield of harvests and controlling cultivating exercises, the third robot named domesticated animals robots used to deal with creatures utilized in farming divisions. This unrest in horticulture division with the target to improve profitability through robotization, unmanned cultivating and the ecofriendly cultivating advancement. (Sishi, & Telukdarie, 2017).

2. **Precision Agriculture:** The keen cultivating innovation under prescient farming used to gauge crop yield and soundness additionally, these insurgencies in horticulture segments utilized for watching various kinds of yield, its development during supporting and post collecting periods. The farming industry 4.0 ideas build up the accuracy horticulture which is utilized for overseeing various

exercises in farming. (Hermann, Pentek, & Otto 2016). The greetings tech association actualized accuracy farming for enormous scope through a choice emotionally supportive network (DSS) in different fields of agribusiness segments. This assists with upgrading comes back from agribusiness and builds request of these innovation. The fourth upheaval in agribusiness is separated in three division as appeared in figure 10. The main division including sensor-based innovation for assortment of a few boundary identified with harvests, land and whether conditions are attainable for compelling development. Additionally, these strategies include handling the information with improvement of dynamic capacity continuously application in field of horticulture. (Fabio Gregori, et al., 2017).

The subsequent division is invigorated based on first division investigation with respect to prerequisite by the agribusiness crops in regards to water substance and manures on proper planning. Digitized cultivating hardware is utilized to play out the activity expressed by arrangement of dynamic in the main division. The third division comprises of control frameworks of different ranch hardware is inputted by handling database gathered from mechanized geological data and ranchers' information. Absence of third dynamic procedure, it is hard to do accuracy farming however the initial two divisions are very much evolved. (Kamilaris, et al., 2016).

3. **RFID in Agriculture:** Technology utilized for recognizing animals at programmed taking care of machine in RFID (Radio Frequency Identification). This framework is outfitted with transponder and it is associated with IT gadgets. This framework comprises of programmed, contactless, ID and limiting of items and creatures. (Ramya Krishnan, et al., 2016). As of late these frameworks are utilized for following of Cereal yields in horticulture areas. RFID plays out the assignment of Identification, situating of article in agribusiness areas and has applications like Identifying and restricting of domesticated animals, oat bunches, and hardware in spread. (Colombo, et al., 2017).

THE FUTURE OF EMPLOYMENT

Present day change 4.0 ensures a lot with respect to wander, pay, and mechanical movements. In any case, work remains one of the most astounding pieces of the business 4.0 change. Thusly, it is fundamentally harder to check or measure the normal pace of work. What sort of new position will be introduced? What does a 'Splendid Factory' master require to battle in the ever-changing condition thusly? Will such an advancement lay-off various masters? These are all in all significant requests of the ordinary expert. (Hampel, et al., 2019). Industry 4.0 could be the zenith of mechanical improvement in gathering. Notwithstanding the way that it in spite of everything appears like the machines are over-taking the business. In this way, it is fundamental to research the system further to close the work and economics later on. This will set up the workers of today for a not so far future change. Given the possibility of the business, it will introduce new openings in robot pros, huge data examination, and a colossal part of mechanical architects. While attempting to choose such an occupations which the business 4.0 will introduce or such a skilled works they would require. BCG has disseminated a report to show off how ten of the key use-cases for the foundation of Industry 4.0 will be impacted. (Sharma, et al., 2017). Ranking of challenges identified based on a survey carried out in 2013 appear in Figure 11.

Following are a portion of the basic changes that will affect the socioeconomics of future work:

Figure 11. Ranking of challenges identified based on a survey carried out in 2013.
(Sharma, et al., 2017).

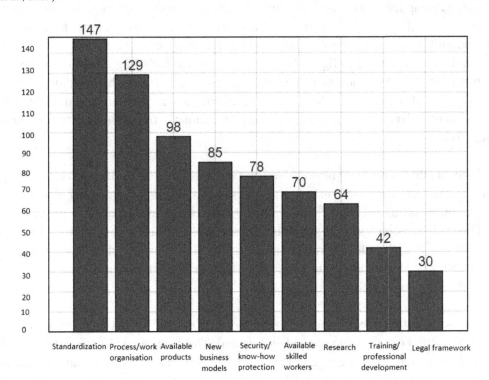

1. **Mechanical Helped Production:** The new business totally relies upon wise contraptions having the choice to talk with the general condition. It suggests that the workers who help the creation line packaging will be laid off and displaced with a shrewd device outfitted with sensors, actuators, and cameras that can separate the thing and pass on fundamental changes for it. In this way, the enthusiasm for such a specialists will be superseded with "Robot-coordinators". (Hugh, et al., 2018).

2. **Big Data-Driven Quality Control (QC):** In the standard term, quality control focuses in lessening the unavoidable assortments between things. QC depends by and large upon genuine methodologies to show whether a particular segment of a thing, for instance, weight, size, or shape is changing in a way that can be seen for instance. While, such a strategy by and large, depends after social affair obvious or ceaseless data as for the thing. Thusly, since Industry 4.0 will in a general sense rely upon gigantic data, the prerequisite for QC workers will reduce. On the other hand, the enthusiasm for data scientists is extending dependably. (Jay Lee, et al., 2014).

3. **Product offering Simulation:** The prerequisite for headway for transportation is declining; the necessity for mechanical draftsmen who go after smoothing out and reenactment to copy creation lines will increase. Having the development for creation line amusement before the affiliations could open up occupations for mechanical fashioners increasing down to earth involvement with the cutting edge fields. (Gowtham, Banga, & Mallanagouda, 2020).

4. **Predictive Maintenance:** The presentation of shrewd gadgets will empower producers to foresee machine disappointments well ahead of time. Insightful machines can freely have the option to

look after themselves. Thusly, the prerequisite of the customary support specialists or professionals will drop, and they will be supplanted with progressively specialized and profoundly talented ones. (Hugh, et al., 2018).

5. **Self-Driving Logistics Vehicles:** The hugest point of convergence of progress is transportation. Fashioners use straight programming methods, for instance, the transportation model to utilize the use of transportation. (Puiu, Bischof, & Serbanescu, et al., 2017). In any case, with the free or self-driving vehicles, and with the data help, various drivers will lose their positions. In addition, one moved vehicles will consider restriction free working hours with higher utility rate. (Mora, et al., 2017).

6. **The 'Machine as a Service' (MaaS):** The business 4.0 will enable creators to sell a "Machine as a Service." This infers, as opposed to offering the absolute machine to the client, the machine will be set-up and kept up by the maker while the client abuses the organizations it gives. This will open up openings for work in the upkeep and would require an expansion in the business work for selling the MaaS. (Ghodrati, Hoseinie, & Garmabaki, 2015).

In future creation forms, man-machine and item will speak with one another, taking into account a self-composed creation process. The item itself will turn into a vital data transporter. By utilizing industry 4.0 advances, they will independently choose where and how they are delivered.

All through (Ghodrati, Hoseinie, & Garmabaki, 2015). Germany organizations and exploration establishments are cooperating to make programming correspondence between individuals, machines, coordination, and items, must be accomplished continuously. Along these lines, all creation, the procedures going from gracefully to conveyance can be advanced and modified. Toward one side of the chain, the business have the savvy manufacturing plant improving creation quality at the other they have the shrewd item and information driven administrations. While they are very much situated with respect to keen items, they need to additionally create on know-how in the field of information driven administrations.

Comparison of IIOT 4 and Industrial Internet consortium appear in Table 2.

MECHANICAL CHANGES AND EFFECTS MAY RESULT FROM INDUSTRY 4.0

1. Enormous ability gains by accomplishing without one moment to spare preservation and close to zero individual time.
2. 3d printing will make changed, near to creation conceivable Social Innovation Policy for Industry 4.0.
3. Machine thriving may enlarge by virtue of self-streamlining.
4. The worth chain for creation can be smoothened over the creation cycle as thing parts pass on when they have been done and following stages, for example, development can be readied.
5. Virtual industrialization: before new plant or creation lines are set up, it will be conceivable to structure and test these in detail in the automated world.
6. Pushed mix of working over the whole worth chain.
7. Vertical coordination and composed gathering structures. (Judit, et al., 2018).

Table 2. Comparison of IIOT 4.0 and industrial internet consortium.

IIOT 4.- and Industrial Internet Consortium – Comparison	
Industry 4.0	**Industrial Internet Consortium**
Key Stakeholders	
Government services: R&D, Economics and Technology Academia: National Academy for Science and Engineering, German Research Center for Artificial.	12 staff individuals arranged in the USA and Europe. Enlistment is open, charges are assessed by firm size, around 170 people including Bosch, SIEMENS, SAP.
Support Platforms	
"A definitive objective of Industry 4.0 is to defend a feasible upper hand of Germany's assembling base … we should prepare German industry to manufacture and introduce CPS, and … make these stay serious worldwide" "Germany sits at the head of world exchange with its examination concentrated products".	To empower and quicken the selection of the web on any kind of business procedure, fabricating or something else. Organizations join to profit by sharing accepted procedures. The emphasis is on the interoperability of items and innovations and "proving grounds"
Sectorial focus	
Assembling - worry about falling down in the marriage of equipment German quality) and programming/advanced advances considered outside Germany. (22% GDP – including agribusiness and foundation).	Manufacturing concentrate however more extensive degree (65-70% GDP – including agribusiness and foundation).
Technological Focus	
Inserted frameworks, mechanization, mechanical technology – CPS that can associate with a gracefully chain.	Anything that can be associated with the web, huge information, with information input and raising productivity.
Geographical Focus	
Concentrated on Germany, and utilizing German citizens' cash keeping that in mind. Given the significance of Germany's assembling area it has more extensive EU impacts.	World-wide center, which could prompt a progressively quick shutting of the efficiency hole among cutting edge and rising countries.
Corporate Focus	
Concentrating on conveying the message to and instructing Germany's SMEs.	Dominated by enormous organizations (for example ABB, Siemens, China Telecom, Mitsubishi) - open to all, has SME individuals.
Optimization Focus	
CPS centers around assembling effectiveness	Return on any advantages
Standardization Focus	
Creating norms is focal yet so far it isn't clear how this ought to continue.	To give direction to measures associations
Economic Approach	
Will in general be a hypothetical portrayal of future assembling - nonexclusive change process throughout the following 10-20 years.	Solid direction to the present and working with what is as of now accessible

(Ghodrati, Hoseinie & Garmabaki, 2015).

CONCLUSION

This part succinctly portrays the advancement of IIoT 4.0. The change of universal registering through the internet of things has numerous difficulties related with it. This chapter gives a brief explanation about the challenges and benefits presently there in IIoT 4.0. And it gives the Strength, Weakness, Opportunities and Threats in IIoT 4.0. Taking everything into account, having spread out the foundation including a review of related terms, we gave a study of existing meanings of IIoT 4.0 and built up our own definition which we trust enhances those. Also the chapter gives a brief discussion about various case studies of IIoT 4.0 like Manufacturing Enterprise, Home Appliance, IoT for Fluid Distribution Monitoring System (FDMS), Smart Farming Blooms in South Korea, Transportation and Logistics and Fourth Industrial Revolution changes in agriculture. Likewise we gave diverse contextual investigations of IIoT alongside the future workforce.

REFERENCES

Ahel, Eroff, Shohin, Zhong, & Runyang. (2018). IoT-Enabled Personalisation for Smart Products and Services in the Context of Industry 4.0. *Proceedings of International Conference on Computers and Industrial Engineering, CIE.*

Aleksandr, Babkin, Vladimir, Plotnikov, Yulia, & Vertakova. (2018, January). The Analysis of industrial cooperation models in the context of forming digital economy. *SHS Web of Conferences, 44.*

Alessandra, Papetti, Pandolfi, Peruzzini, & Germani. (2017). Digital Manufacturing Systems: A Framework to Improve Social Sustainability of a Production Site. In *50th CIRP Conference on Manufacturing Systems*. Elsevier.

Alessandro, Marzocca, Alfredo, Liverani, Cees., & Bil. (2019, October). Maintenance in aeronautics in an Industry 4.0 context: The role of Augmented Reality and Additive Manufacturing. Journal of Computational Design and Engineering, 6(4), 516-526.

Anita, Bodla, & Abhinav. (2017). Internet of Things (IoT)—It's Impact on Manufacturing Process. International Journal of Engineering Technology Science and Research.

Azab, Ahmed, Noha, Mostafa, & Park. (2016). OnTimeCargo: A Smart Transportation System Development in Logistics Management by a Design Thinking Approach. *20th Pacific Asia Conference on Information Systems (PACIS 2016), 44.*

Baotong, Wan, Lei, Shu, Peng, Li, Mithun, Mukherjee, & Yin. (2018, September). Smart Factory of Industry 4.0: Key Technologies, Application Case, and Challenges. In Special Section on Key Technologies for Smart Factory of Industry 4.0. IEEE.

Behrad, Bagheri, & Kao. (2014, December). A Cyber-Physical Systems architecture for Industry 4.0-based manufacturing systems. In Society of Manufacturing Engineers (SME). Elsevier Ltd.

Chen, J., & Zhou, J. (2018, July). Revisiting Industry 4.0 with a Case Study. In *2018 IEEE International Conference on Internet of Things (iThings) and IEEE Green Computing and Communications (GreenCom) and IEEE Cyber, Physical and Social Computing (CPSCom) and IEEE Smart Data (SmartData)* (pp. 1928-1932). IEEE. 10.1109/Cybermatics_2018.2018.00319

Chong, Li, & Li. (2019, May). IEEE 5G Ultra-Reliable Low-Latency Communications in Factory Automation Leveraging Licensed and Unlicensed Bands. IEEE Communications Magazine, 57(5).

Colombo, A. W., Karnouskos, S., Kaynak, O., Shi, Y., & Yin, S. (2017). Industrial Cyberphysical Systems: A Backbone of the Fourth Industrial Revolution. In *Special Section on Key Technologies for Smart Factory of Industry 4.0* (pp. 6–16). IEEE. doi:10.1109/MIE.2017.2648857

Ghodrati, Hoseinie, & Garmabaki. (2015). Reliability considerations in automated mining Systems. *International Journal of Mining Reclamation and Environment, 29*(15), 404-418.

Gowtham, Banga, & Patil. (2019, July). Secure Internet of Things: Assessing Challenges and Scopes for NextGen Communication. In *2nd IEEE International Conference on Intelligent computing, Instrumentation and Control Technologies (ICICICT-2019)*. IEEE. 10.1109/ICICICT46008.2019.8993327

Gowtham, Banga, & Patil. (2020, April). Cyber-Physical Systems and Industry 4.0 Practical Applications and Security Management an Intelligent Traffic Management System. In *An Intelligent Traffic Management System*. CRC Press.

Hallaq, Cunningham, & Watson. (2018, October). The industrial internet of things (IIoT): An analysis framework. Computers in Industry, 101, 1-12.

Hamdi, Abouabdellah, & Oudani. (2018, December). Disposition of Moroccan SME Manufacturers to Industry 4.0 with the Implementation of ERP as A First Step. In *Sixth International Conference on Enterprise Systems (ES)*. IEEE.

Hermann, M., Pentek, T., & Otto, B. (2016). Design principles for industry 4.0 scenarios. In *Proceedings of the 49th Hawaii International Conference on System Sciences (HICSS)*, (pp. 3928–3937). 10.1109/HICSS.2016.488

Jehoon & Sung. (2018, March). The Fourth Industrial Revolution and Precision Agriculture. In *Automation in Agriculture - Securing Food Supplies for Future Generations*. INTECH.

Judit, Olah, Erdei, & Mate. (2018, September). The Role and Impact of Industry 4.0 and the Internet of Things on the Business Strategy of the Value Chain—the Case of Hungary. Sustainability.

Kagermann, H., Wahlster, W., & Helbig, J. (2013, April). Recommendations for Implementing the Strategic Initiative INDUSTRIE 4.0 -- Securing the Future of German Manufacturing Industry. National Academy of Science and Engineering, Munchen.

Kamilaris, A., & Gao, F. (2016). A semantic framework for internet of things-enabled smart farming applications. In *Proceedings of the 2016 IEEE 3rd World Forum on Internet of Things (WF-IoT)*, (pp. 442–447). 10.1109/WF-IoT.2016.7845467

Maroua, Fourati, Fourati., & Chouaya. (2017, November). Internet of Things in Industry 4.0 Case Study: Fluid Distribution Monitoring System. *9th International Conference on Networks & Communications*.

Meonghun, Lee, & Shin. (2018). IoT-Based Strawberry Disease Prediction System for Smart Farming. Sensors 2018.

Mora, H., Gil, D., Terol, R. M., Azorín, J., & Szymanski, J. (2017). An IoT-based computational framework for healthcare monitoring in mobile environments. Sensors (Basel). doi:10.339017102302

Nallapaneni, Kumar, Archana, & Dash. (2017 November). The Internet of Things: An Opportunity for Transportation and Logistics. In *IEEE International Conference on Inventive Computing and Informatics (ICICI)*. IEEE.

Noha, Hamdy, Hisham, & Alawady. (2019 March). Impacts of Internet of Things on Supply Chains: A Framework for Warehousing. Social Sciences.

Parsa, Najafabadi, & Salmasi. (2017). Implementation of smart optimal and automatic control of electrical home appliances (IoT). In *IEEE Proceedings 2017 Smart Grid Conference. SGC* (pp. 1–6). IEEE.

Puiu, Bischof, Serbanescu, Nechifor, Parreira, & Schreiner. (2017). A public transportation journey planner enabled by IoT data analytics. In *20th Conference on Innovations in Clouds, Internet and Networks (ICIN), Paris, 2017* (pp. 355-359). Academic Press.

Ramalingam, M., Puviarasi, R., Chinnavan, E., & Foong, H. K. (2019). Self-monitoring framework for patients in IoT-based healthcare system. International Journal Innovation Technology and Engineering, 3641-3645.

Ramya, Krishnan, Renuka, Swetha, & Ramakrishnan. (2016). Effective Automatic Attendance Marking System Using Face Recognition with RFID. *International Journal of Scientific Research in Science and Technology*, 2(2).

Sharma, N. K., Singh, R. J., Mandal, D., Kumar, A., Alam, N. M., & Keesstra, S. (2017). Increasing farmer's income and reducing soil erosion using intercropping in rainfed maize-wheat rotation of Himalaya, India. *Agriculture, Ecosystems & Environment*, 247, 43–53. doi:10.1016/j.agee.2017.06.026

Sishi, M. N., & Telukdarie, A. (2017, December). Implementation of Industry 4.0 Technologies in the Mining Industry: A Case Study. In *IEEE International Conference on Industrial Engineering and Engineering Management (IEEM)*. IEEE. 10.1109/IEEM.2017.8289880

Wei, Mao-Jiun, & Wang. (2004). A comprehensive framework for selecting an ERP system. International Journal of Project Management.

Xun, Xu, Lu, Aristizabal, Velásquez, Yesid, & Valencia. (2020 January). IoT-enabled smart appliances under industry 4.0: A case study. Advanced Engineering Informatics.

Chapter 13
Blockchain–Enabled Decentralized Reliable Smart Industrial Internet of Things (BCIIoT)

Chandramohan Dhasarathan

(iD) https://orcid.org/0000-0002-5279-950X

Department of Computer Science and Engineering, Madanapalle Institute of Technology and Science, India

Shanmugam M.

Department of Computer Science and Engineering, Vignan's Foundation for Science, Technology, and Research, India

Shailesh Pancham Khapre

ASET-CSE, Amity University, Noida, India

Alok Kumar Shukla

School of Computer Science and Engineering, VIT-AP University, Amaravati, Andhra Pradesh, India

Achyut Shankar

ASET-CSE, Amity University, Noida, India

ABSTRACT

The development of wireless communication in the information technological era, collecting data, and transfering it from unmanned systems or devices could be monitored by any application while it is online. Direct and aliveness of countless wireless devices in a cluster of the medium could legitimate unwanted users to interrupt easily in an information flow. It would lead to data loss and security breach. Many traditional algorithms are effectively contributed to the support of cryptography-based encryption to ensure the user's data security. IoT devices with limited transmission power constraints have to communicate with the base station, and the data collected from the zones would need optimal transmission power. There is a need for a machine learning-based algorithm or optimization algorithm to maximize data transfer in a secure and safe transmission.

DOI: 10.4018/978-1-7998-3375-8.ch013

INTRODUCTION

The development of wireless communication in the information technological era, collecting data, and transfer it from unmanned systems or devices could be monitored by any application while it is online. Due to direct and aliveness of countless wireless device in a cluster of the medium could legitimate unwanted users to interrupt easily in an information flow. It would lead to data loss and security breach. Many traditional algorithms are effectively contributed to the support of cryptography-based encryption to ensure the user's data security. IoT devices with limited transmission power constraints have to communicate with the base station the data collected from the zones would need optimal transmission power. There is a need for a machine learning-based algorithm or optimization algorithm to maximize data transfer in a secure and safe transmission.

Past few decades IIoT based service is collectively targeting open-loop environments in the competitive customer service satisfaction. Statistical techniques are adopted to check plenty of historical data for evaluating service mechanisms with extensive research for the empowerment of industrial service management.

A significant decomposition with clustering of multi-purpose granularity realization for splitting of data from various IoT information tasks deals for unified data storing and transfer. In industries high band-width wireless devices are capable enough to handle the behavior of smart manufacturing and comprehensive service monitoring. Supply chain management would be more flexible and secure if it would gain the support of protective mechanisms and algorithms.

Attributes such as contract extension into the smart contract, data ownership are climbed to shared ownership with mutual benefits, service management could be concentrated more on customer's feedback and support, technical support depends on deployment and maintenance feasibility, and it would increase the yield of an industry. The performance of IoT implication in manufacturing industries would directly depend on service reliability, economy, and assurance with collaborative partners. The reliability of any communication medium is always questionable until there would be no compromise in data delivery. The economy in implementing current technology into the business era is quit complicated and directly implicated with company budget and considerable checking is always required for maintenance cost and resource requirements.

IIoT attacks are improving in all perspectives to discover the characteristic behavior of optimized cases and pre-analyzing the features for a back entry. Many more protocols and techniques are imposed on the diverse manufacturers to bring them into a centralized control. Intra-organization communication is more important for the effective supply chain management. Digital service increased the risk factor and reduces the time factor in terms of goods delivery and getting orders. It also relays on the business model of an organization to analyze the ecosystem to inherit the interconnectivity of the public and industry. Cyber security providers developed a business eco-system with the interdependent hierarchy to maintain the loosely coupled business cluster analysis. The similarity of business financial services, aggregator infrastructure, would lead the instantiation mechanism with consumers more easily. Conceptual modeling of digital transformation with a peripheral of an emerging model to recombine the organizations with a single contract would help each other in sharing the best and worst.

This paper is organized with an introductory section followed by related articles as a literature review. A separate session is designed to express the approach of Blockchain-enabled Internet of Things (BCI-IoT) for secure and efficient data transmission is proposed. The experimental result evaluation section is described in the proposed BCIIoT with comparative configuration under certain simulation range is tested for its effectiveness. Finally, the paper is concluded with future research directions as a section.

LITERATURE REVIEW

IoT utilization by(Mukherjee et al.,2018) designing a method based on power demand by clustering Industrial pieces of equipment and IoT-based applications for assuring security to the information flow in production units. The network is associated with clustered systems for predefined demands. Wang proposed manufacturing services with a fuzzy analytic network process to (Shijie & Yingfeng, 2020) employ the manufacturing unit under the control of the Cross-Entropy Method. The service scoring mechanism is introduced for credit evaluation in real-time for arriving at a quick decision. Sung to apply in a heterogeneous environment and influence different program with fuzzing test technique (Kim, Cho, Lee, & Shon, 2020) is proposed to analyze the vulnerability and develops trust from various manufacturers. It improves the productivity of the physical system and the need for a secure system emerges. Tobias described the proposition between the data and maintenance of industrials experts (Shijie & Yingfeng, 2020) value to prevent the ecosystem from examining current trends with respective cluster digital systems.

To state the digitalization of industrials benefits of intelligent communication to help manufacturers into (Saadaoui et al., 2020) binary error and industrial waste. It is proposed to develop a multi-layered impulse ratio wavelet packet with convolution coding. Device shows smart sensors transparency (Rathee, Balasaraswathi, Chandran, Gupta, & Boopathi, 2020) for extorted storage by protecting it with blockchain-based security transmission is adapted. The probability of attack gets reduced by authenticating delay of IoT devices and falsification attacks are identified. In cloud fog hybrid network is designed with distributed (Liu, Huang, Zheng, & Liu, 2020) computing to minimize the processing delay by simulated annealing and PSO. Its latency gets improved by efficient utilization of industrial ecosystem with the help of cloud computing heterogeneous access is monitored by the future centric system.

Knowledgebase concentrated (Vimal, Khari, Dey, Crespo, & Robinson, 2020) on the reinforcement of swarm intelligent learning of behavioral approach for IIoT. The predictive process affects the delay of neighboring devices based on their resource allocation with an enhanced ant colony optimization. Authors created a multi-objective whale optimization algorithm that is used to identify the device communication (Sharma & Saini, 2020)and an elliptical curve cryptography technique is applied for secure transmission. Researchers proposed a technique(Li, Yue, & Wang, 2020)to minimize the problem in mechanism to overhead the communication by delay optimized decryption is wrapped with deep learning for privacy protection. An approach(N. et al., 2019) designed to design model-based engineering based on multiple views to perform cloud-based desirable features.

Wenbo secure data communication n(Zhang, Wu, Han, Feng, & Shu, 2020) is achieved by a lightweight consensus algorithm to ensure the IIoT data transmission securely with high reliable data delivery and comparatively less probability of data loss. A data model for (Oberländer, Übelhör, & Häckel, 2019) effective business in the IIoT context with transparency in manufacturing and deliver products globally. The possibility of a business model is pay-per-user is tested under industrial practice. An IIoT security issue is discussed (Dakhnovich, Moskvin, & Zegzhda, 2019) and designed architecture for it to restrict network issues while globally connected. A must need situation to design a Petri net privacy-preserving framework to ensure cloud users' data in the storage area and by multi-agent algorithms (Chandramohan, Vengattaraman, Rajaguru, & Dhavachelvan, 2013a, 2013b, 2016).

Multimedia based routing is (Al-Turjman & Alturjman, 2018) designed for IIoT problems with unmanned aerial vehicles by selecting K-disjoined paths. An intelligent automation system is designed (Gawron-Deutsch, Diwold, Cejka, Matschnig, &Einfalt, 2018) to control the cyber-physical system to solve a high degree of optimization to be more flexible and intelligent. An approach to reduce the data theft happens in cloud storage (Han, Que, Jia, & Zhang, 2018) area's energy consumption of active devices by balancing the availability and utilization of the resource. A seamless communication with an actuator motivating (Hatzivasilis et al., 2018) by designing a framework for managing devices communicating based on QoS aware sustainability and it is tested with less memory for its effectiveness. A design for an e-healthcare privacy framework (Hossain & Muhammad, 2016)to collect and process the ECG information and secure it by watermarking to reduce private data that leads to professional clinical errors.

A discrete state machine is designed for real-time scheduling to validate it in a software-defined (Jha, Babiceanu, & Seker, 2019) network to test for low-performance invariants. Network embedding algorithm to measure the node-to-node distance for diagnostic information, that is computationally (Kan, Yang, & Kumara, 2018) expensive. A priority aware secure system is designed (Kharb & Singhrova, 2019) for IIoT to provide low cost and low power communication to improve the wireless device battery power. Service architecture is designed for industrial management (Lee, Lee, & Lee, 2019)to improve the reliability of the production process and customer decision system. An encryption algorithm is used to protect the customer's and owner's information. An end to end communication is characterized (Liao, Shen, Sun, Dai, & Wan, 2019) to promote the robustness and reliability to reduce the complexity of channel. An intelligent machine monitoring system is designed (Mcninch, Parks, Jacksha, & Miller, 2019) for lockout targets present in an industry to ensure the safety of workers and employers with concrete sensors. There is a comfortable design for a prototype to implement in the IT and telecom sector (Rizvi, Zubair, Ahmad, Hashmani, & Khan, 2019) with the use of IoT and IoUT with even 5G technology.

An environmental uncertainty design is discussed to test the questioner for influencing the top management to (Arnold & Voigt, 2019) balance logistic regression of management authority to handle IIoT. By proposing a business model necessity creates an opportunity (Laudien & Daxböck, 2019), to develop a multi-case contribution that leads constructional and manufacturing industry adequately to support technological development. IIoT need valuable insights with a semi-structured approach(Arnold, Kiel, & Voigt, 2019)for the industry-specific business model to update the workforce and automotive suppliers.

A structured extraction of triple bottom line picture benefited (Daniel, 2019) sustainable incorporation within industrial dimension to improve the social benefits in unexpected typical cases. Individual contribution is evaluated to the industrial effects of IIoT (Butschan, Heidenreich, Weber, & Kraemer, 2019) usage and performance to validate the digital transformation process. To prevent the information form (Sha, Xiao, Chen, & Sun, 2017) sensitive thief's an effective design is incorporated with IIoT sensitive thief defender those attending to storage area through back door. Their order preference is tested for their originality and trust. Network communication error is identified it is validated by cyclic redundancy (Zhong & Hu, 2020) check to reduce the error rate more appropriately. It controls the instability of communication transmission.

BCIIOT NETWORK COMMUNICATION

IoT based transmission needs an effective and controllable data collection and trusted transfer when it is applied in industrial management. In general, centralized systems could not meet effective communication

Figure 1. Contiguous BCIIoT Network communication and its prevention

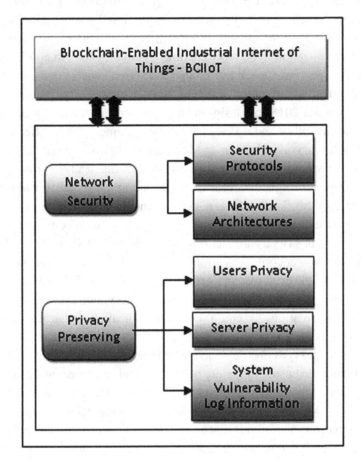

in monitoring huge industrials management. A blockchain-based approach might give extra support for sure data communication. In this paper, IoT device data are collected and transmitted with a blockchain-enabled strategy for clustering and constrained based nodes communication. Blockchain-based Industrial Internet of Things (BCIIoT) is designed for improving secure transmission in a decentralized scenario.

Network security and secure communication of average convergence of population are considered to preserve the manufacturers' confidential data and users' vulnerability linkage is monitored periodically for any threat. In figure.1 it is illustrated with blockchain-enabled protection based on architecture used in maintaining the users' information and architecture utilized for industrialization. The same operations and its critical activities are noticed regularly. Normalization of required iterations with actual operations with usability of host data and environments are taken into consideration for secure prototyping. BCIIoT validates every manufactures activity with separate logging details by separate privacy monitoring services. It adopts network architecture in build security protocol to give enough prevention form networking attacks and unwanted user requests are avoided.

The physical impedance of inefficient data transfer might lose traceability in the advancement of any organization. Computerized production to expand the smart communication without centralized architecture schemes will differentiate present production and visibility of wireless devices. Personal security of the business model would augment, improving the industrial obstacles in the entire activity.

Positive concerns would avoid in misplacing of IoT devices and its data breach by unrestricted access, the maintenance, and the consumption of technological providers might control industrial partners. Subsequent transparency to extract specific information of the periodic user with the help of some filtration process and capturing the regular user's features. Smart device scalability, the resiliency of the cloud environment with device addictiveness, and explicitness on several industrial products would control to contribute orderly recommendations.

Casted information need to be locked until reached safely to the data owner, the smart consensus is created from data generator end to validate at every stage of communication is recorded continuously for proper monitoring. Blockchain-based encryption is recognized by every sensor data to verify unwanted users' requests to the system. Frequent users data request is recorded in all network and verified for its originality with the encrypted key issued at both ends. BCIIOT connects users after proper verification and validation of device registration and their purpose of visit. Users' profile is recorded in all drop-off points and all supply chain warehouses for future references.

Figure 2. CIIoT Virtualizer and its Features

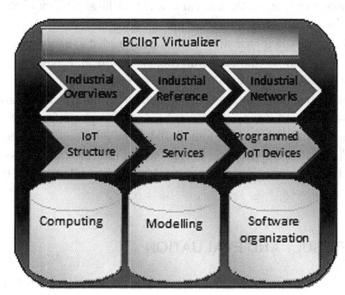

Distribution of IoT devices transmission frequency with impulsive industrial standards and statistical channel modeled in zero-mean amplitude. Additional noise from motors, equipment, human, and machinery of equivalent sensors might be clocked for data tracing and visualizing illegally. Figure.2 shows the virtualizer of multi-sensor industrial monitoring for an effective and global protective channel with comparatively high preservation. The computing of IoT structures and involved devices for an industrial convenience to improve users' trust with addictive noise control. Reference of past industrial experts advisory comity recommends a robust transmission that relay orthogonally.

An improved encoder is added to correct the errors and message symbols that reflect in the wireless sensor region to avoid the fading of sensors. The industrial network is programmed with controllable IoT software organization to initiate the product request before transit. Product service is indented with corresponding devices. The misbehavior of nodes that are participating in the transmission would hash the manufacturer to take the wrong decision. Private data of all industrial information would not reach the market without the concern of stack holders it should be included in the blockchain network.

Current device connectivity and product consensus are monitored to the acquisition of sensors temporary transmission and communication of nodes in an intelligent network. The cloud-based system needs personal computers into risk by public access. Since users are not aware of risk factors related to public permission and it is optimized latency with offloading data collected by IoT devices. The load balancing of heterogeneous devices follows a joint ratio for the decision unknown to validate the optimized results. Data comparison in real-time might reduce data processing with a schematic approach of computing layer. Enterprise registered devices with the equipment involved intuitively with reasonable distribution in a connected network.

BCIIoT follows a dynamic topological order to compute among upper layers to generate the communication link. Distributed computing with minimum response to maximize the amount of node processing and their task is monitored in all iteration. Searching the appropriate position with certain constraints might be infeasible to communicate under fixed velocity. Local optima are tested with the current scenario and global optima are verified for its acceptability in decentralized processing. Minimum delay with optimal distribution is targeted in the current velocity of large latency is observed and tested for effectiveness. Data acquisition through a wireless communication protocol is performed with an early warning that might affect personal information. An image recorded in every gateway is compressed for network failed and process delay forwarded monitoring environmental status.

Latency processing in industrial production is monitored under various constrains with the number of data collected at each device deployed scenarios. Computing power is the data ratio assigned to a number of comparative validated devices for trusted transfer. High performance-device communication is monitored under the clustering of the highest industrial environmental needs equipped for data processing. Bandwidth allocated for effortless devices would increase the processing time due to huge data consumed. BCIIoT dealing with production units of industries will minimize the data transfer delay of relative reduction.

EXPERIMENTAL RESULT AND EVALUATION

In this paper, the research focus on industrial communication issues related to IoT environment. To test it we have considered a simulation testbed of 500 m X 500 m with 50 devices and 700 maximum devices. A number of iterations and simulation period are 150 and 160 ms, respectively. Initially, the devices are noted for zero reading and its energy capacity is periodically checked and get improved based on its usages. Table.1 shows the number of IoT nodes taken into consideration to stimulate the peer, malevolent, and miner device communication. Its lattice facet, broadcast range, user request, simulation zone, and physical layer are evaluated with a specific range.

However, in factories, the production unit is the heart of the entire commercial market which is facing tangent time for the organized frame in crucial scenarios. From the literature study, many academicians and researchers focused their IIoT findings with respect to the financial, resource, quality, capability, credit reliability, and availability. It is illustrated in table.2 with quality service evaluation and its parameter extraction.

In table.3shows the BCIIoT approach and it is simulated to handle enormous activity such as malevolent by adding active nodes with a probability of high authentication rate. Intruders are identified

Table 1. Blockchain IIoT Parameter used for simulation setup

Numbers of IoT devices/nodes	50 Devices, 700 nodes	Peer device	Malevolent device	Miner device
Lattice facet	500 m X 500 m	10	5	7
Broadcast range	250 m	15	15	12
Users/consumers request	150 B	45	20	17
Simulation period	160 s	75	30	26
Physical layer	802.11	100	50	45

Table 2. Blockchain IIoT Qualitative Service evaluation attribute verified for effectiveness

Researchers	Financial	Resource	Quality	Capability	Credit	Reliability	Availability
Mukherjee et al.	✓	✗	✓	✗	✗	✓	✓
Shijie et al.	✗	✓	✓	✗	✓	✗	✓
Kim et al.	✓	✓	✗	✓	✓	✗	✓
Riasanow et al.	✗	✓	✗	✓	✓	✓	✓
Saadaoui et al.	✓	✗	✓	✗	✓	✗	✓
Rathee et al.	✗	✗	✗	✗	✗	✗	✓
Liu et al.	✓	✓	✗	✓	✓	✓	✗
Vimal et al.	✗	✓	✗	✗	✗	✗	✓
Shivi Sharma et al.	✓	✗	✓	✗	✓	✗	✓
Li et al.	✓	✓	✓	✗	✓	✗	✓
Muthukumar et al.	✗	✓	✗	✓	✗	✓	✓
Zhang et al.	✓	✗	✓	✗	✓	✓	✓
Oberländer et al.	✗	✓	✗	✓	✗	✓	✓
Dakhnovich et al.	✓	✗	✓	✗	✓	✗	✓
Tusjman et al.	✓	✓	✗	✓	✓	✗	✓
Gawron et al.	✓	✗	✓	✓	✗	✓	✓
Guangjie et al.	✓	✗	✓	✓	✗	✓	✓
George et al.	✓	✓	✗	✓	✓	✗	✓
Shamim et al.	✓	✓	✗	✓	✓	✗	✓
Jha et al.	✓	✗	✓	✓	✗	✓	✗
Chen Kan et al.	✗	✓	✓	✓	✗	✓	✓
Kharb et al.	✓	✗	✓	✓	✓	✗	✓
Lee et al.	✓	✓	✗	✓	✓	✗	✓
Liao et al.	✓	✓	✗	✓	✗	✓	✗
McNinch et al.	✓	✗	✓	✓	✓	✗	✓
Rizvi et al.	✓	✗	✓	✓	✓	✗	✓
Arnold et al.	✓	✓	✗	✓	✓	✗	✓
Sven et al.	✓	✗	✓	✓	✓	✗	✓
Christian et al.	✓	✓	✗	✓	✗	✓	✓
Daniel et al.	✓	✗	✓	✗	✓	✗	✓
Jens Butschan et al.	✓	✗	✓	✓	✗	✓	✓
Sha et al.	✓	✓	✗	✓	✓	✗	✓
Zonglin Zhong et al.	✓	✗	✓	✓	✗	✓	✗

Table 3. Malevolent activity handled by Blockchain enabled IIoT (BCIIoT)

Malevolent Activity- MA	Participate nodes %	Probability of attack success %	No. of IoT Devices	Authentication Accuracy (ms)	No. of Authentication	Falsification attack
Adding up of malevolent node- AM	5	12	10	3	2	3
Compromised object or consumers- CO	10	19	20	1	3	9
Authentication probabilistic scenarios- AP	15	28	30	5	5	15
Hazards created by the intruder-HC	35	48	40	4	10	23
Probability of attack success- PA	50	70	50	7	15	30

Figure 3. Blockchain-enabled IIoT evaluation of supply chain systems

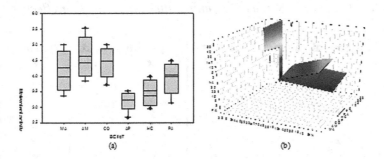

in an industrial structure that causes invalid and unwanted issues in supply chain management. BCIIoT success rate is quite an authentication with an accuracy of 5ms and a falsification attack would lead to interrupt in the production of any industry.

Malevolent Activity- MA, Adding up of malevolent node- AM, Compromised object or consumers- CO, Authentication probabilistic scenarios- AP, Hazards created by the intruder-HC, Probability of attack success- PA. In figure.3 BCIIoT evaluation of supply chain principles applicable in various industrial internets of things is shown with essential plots.

Table 4. CIIoT compared with various Equilibrium approaches

	No. of IoT Devices	Task / GB	CRC	ADAI	SMRD	NodeDPWS	BCIIoT
Assurance	10	4	45%	32%	56%	36%	67%
Agility	20	12	49%	37%	62%	44%	73%
Economy	30	17	40%	30%	50%	32%	62%
Performance	40	25	38%	29%	45%	32%	61%
Security & Privacy	50	48	70%	65%	73%	68%	89%
Latency /s	60	97	65	52	69	57	88

Figure 4. BCIIoT efficiency tested and compared with literature approaches

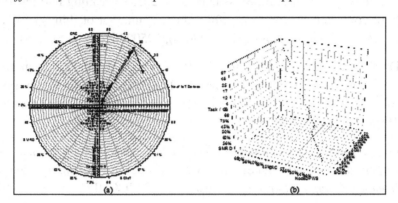

Industrial valuables and its supply chain products would not affect in any circumstance. However, IIoT follows equilibrium to ensure agility, performance, privacy, latency, and assure an end to end delivery. The BCIIoT is compared with various equilibrium approaches as illustrated in table.4 and it shows a comparatively high impact in all perspectives.

To evaluate the efficiency of the proposed BCIIoT approach is tested with parameters such as data assurance, data agility, supply economy, data delivery performance, information security &industrial privacy, and its latency is compared. Comparatively, BCIIoT climbs to be an effective industrial partner after validating with CRC, ADAI, SMRD, and NodeDPWS as plotted in figure.4.

CONCLUSION

In industrial advancement to monitor and trace every interaction are collected as information and advancement happening inside production, units could be collected and updated with the involvement of wireless communication devices. It also reduces manual interference to do risky and complicated tasks. There is a must-win situation in the utilization of IoT devices without compromising the security and privacy of industrial zones. BCIIoT mechanism scrutinizes falsification probability of various devices to ensure the transparency of industrial functionality in a secure and safe aspect.

REFERENCES

Al-Turjman, F., & Alturjman, S. (2018). 5G/IoT-enabled UAVs for multimedia delivery in industry-oriented applications. *Multimedia Tools and Applications*, *79*(13-14), 8627–8648. doi:10.100711042-018-6288-7

Arnold, C., Kiel, D., & Voigt, K. (2019). How the Industrial Internet of Things Changes Business Models in Different Manufacturing Industries. *Digital Disruptive Innovation Series on Technology Management*, 139-168. doi:10.1142/9781786347602_0006

Arnold, C., & Voigt, K. (2019). Determinants of Industrial Internet of Things Adoption in German Manufacturing Companies. *International Journal of Innovation and Technology Management*, *16*(06), 1950038. doi:10.1142/S021987701950038X

Butschan, J., Heidenreich, S., Weber, B., & Kraemer, T. (2019). Tackling hurdles to digital transformation— the role of competencies for successful industrial internet of things (iiot) implementation. *International Journal of Innovation Management, 23*(04), 1950036. doi:10.1142/S1363919619500361

Chandramohan, D., Vengattaraman, T., Rajaguru, D., Baskaran, R., & Dhavachelvan, P. (2013b). Hybrid Authentication Technique to Preserve User Privacy and Protection as an End Point Lock for the Cloud Service Digital Information. In *International Conference on Green High Performance Computing* (pp. 1-4). IEEE. 10.1109/ICGHPC.2013.6533904

Chandramohan, D., Vengattaraman, T., Rajaguru, D., & Dhavachelvan, P. (2013a). A Novel Framework to Prevent Privacy Breach in Cloud Data Storage Area Service. In *2013 International Conference on Green High Performance Computing* (pp. 1-4). IEEE. 10.1109/ICGHPC.2013.6533903

Chandramohan, D., Vengattaraman, T., Rajaguru, D., &Dhavachelvan, P. (2016). A new privacy preserving technique for cloud service user endorsement using multi-agents. *Journal of King Saud University - Computer and Information Sciences, 28*(1), 37-54. doi:10.1016/j.jksuci.2014.06.018

Dakhnovich, A. D., Moskvin, D. A., & Zegzhda, D. P. (2019). An Approach to Building Cyber-Resistant Interactions in the Industrial Internet of Things. *Automatic Control and Computer Sciences, 53*(8), 948–953. doi:10.3103/S0146411619080078

Daniel, K. (2019). Big Data Analytics in Industry 4.0: Sustainable Industrial Value Creation, Manufacturing Process Innovation, and Networked Production Structures. *Journal of Self-Governance and Management Economics, 7*(3), 34. doi:10.22381/JSME7320195

Gawron-Deutsch, T., Diwold, K., Cejka, S., Matschnig, M., & Einfalt, A. (2018). Industrial IoT für Smart Grid-Anwendungenim Feld. *E&I Elektrotechnik und Informationstechnik, 135*(3), 256–263. doi:10.100700502-018-0617-4

Han, G., Que, W., Jia, G., & Zhang, W. (2018). Resource-utilization-aware energy efficient server consolidation algorithm for green computing in IIOT. *Journal of Network and Computer Applications, 103*, 205–214. doi:10.1016/j.jnca.2017.07.011

Hatzivasilis, G., Fysarakis, K., Soultatos, O., Askoxylakis, I., Papaefstathiou, I., & Demetriou, G. (2018). The Industrial Internet of Things as an enabler for a Circular Economy Hy-LP: A novel IIoT protocol, evaluated on a wind park's SDN/NFV-enabled 5G industrial network. *Computer Communications, 119*, 127–137. doi:10.1016/j.comcom.2018.02.007

Hossain, M. S., & Muhammad, G. (2016). Cloud-assisted Industrial Internet of Things (IIoT) – Enabled framework for health monitoring. *Computer Networks, 101*, 192–202. doi:10.1016/j.comnet.2016.01.009

Jha, S. B., Babiceanu, R. F., & Seker, R. (2019). Formal modeling of cyber-physical resource scheduling in IIoT cloud environments. *Journal of Intelligent Manufacturing, 31*(5), 1149–1164. doi:10.100710845-019-01503-x

Kan, C., Yang, H., & Kumara, S. (2018). Parallel computing and network analytics for fast Industrial Internet-of-Things (IIoT) machine information processing and condition monitoring. *Journal of Manufacturing Systems, 46*, 282–293. doi:10.1016/j.jmsy.2018.01.010

Kharb, S., & Singhrova, A. (2019). Fuzzy based priority aware scheduling technique for dense industrial IoT networks. *Journal of Network and Computer Applications, 125*, 17–27. doi:10.1016/j.jnca.2018.10.004

Kim, S., Cho, J., Lee, C., & Shon, T. (2020). Smart seed selection-based effective black box fuzzing for IIoT protocol. *The Journal of Supercomputing, 76*(12), 10140–10154. Advance online publication. doi:10.100711227-020-03245-7

Laudien, S. M., &Daxböck, B. (2019). The Influence of the Industrial Internet of Things on Business Model Design: A Qualitative-Empirical Analysis. *Digital Disruptive Innovation Series on Technology Management,* 271-303. doi:10.1142/9781786347602_0010

Lee, Y., Lee, K. M., & Lee, S. H. (2019). Blockchain-based reputation management for custom manufacturing service in the peer-to-peer networking environment. *Peer-to-Peer Networking and Applications, 13*(2), 671–683. doi:10.100712083-019-00730-6

Li, Q., Yue, Y., & Wang, Z. (2020). Deep Robust Cramer Shoup Delay Optimized Fully Homomorphic For IIOT secured transmission in cloud computing. *Computer Communications, 161*, 10–18. doi:10.1016/j.comcom.2020.06.017

Liao, Y., Shen, X., Sun, G., Dai, X., & Wan, S. (2019). EKF/UKF-based channel estimation for robust and reliable communications in V2V and IIoT. *EURASIP Journal on Wireless Communications and Networking, 2019*(1). doi:10.118613638-019-1424-2

Liu, W., Huang, G., Zheng, A., & Liu, J. (2020). Research on the optimization of IIoT data processing latency. *Computer Communications, 151*, 290–298. doi:10.1016/j.comcom.2020.01.007

Mcninch, M., Parks, D., Jacksha, R., & Miller, A. (2019). Leveraging IIoT to Improve Machine Safety in the Mining Industry. *Mining. Metallurgy & Exploration, 36*(4), 675–681. doi:10.100742461-019-0067-5 PMID:33005876

Mukherjee, A., Goswami, P., Yang, L., Tyagi, S. K., Samal, U. C., & Mohapatra, S. K. (2020). Deep neural network-based clustering technique for secure IIoT. *Neural Computing & Applications, 32*(20), 16109–16117. Advance online publication. doi:10.100700521-020-04763-4

N., M., Srinivasan, S., Ramkumar, K., Pal, D., Vain, J., & Ramaswamy, S. (2019). A model-based approach for design and verification of Industrial Internet of Things. *Future Generation Computer Systems, 95*, 354-363. doi:10.1016/j.future.2018.12.012

Oberländer, A. M., Übelhör, J., & Häckel, B. (2019). IIoT-basierteGeschäftsmodellinnovationimIndustrie-Kontext: Archetypen und praktischeEinblicke. *HMD Praxis Der Wirtschaftsinformatik, 56*(6), 1113–1125. doi:10.136540702-019-00570-1

Rathee, G., Balasaraswathi, M., Chandran, K. P., Gupta, S. D., & Boopathi, C. S. (2020). A secure IoT sensors communication in industry 4.0 using blockchain technology. *Journal of Ambient Intelligence and Humanized Computing.* Advance online publication. doi:10.100712652-020-02017-8

Riasanow, T., Jäntgen, L., Hermes, S., Böhm, M., & Krcmar, H. (2020). Core, intertwined, and eco-system-specific clusters in platform ecosystems: Analyzing similarities in the digital transformation of the automotive, blockchain, financial, insurance and IIoT industry. *Electronic Markets*. Advance online publication. doi:10.100712525-020-00407-6

Rizvi, S. S., Zubair, M., Ahmad, J., Hashmani, M., & Khan, M. W. (2019). Wireless Communication as a Reshaping Tool for Internet of Things (IoT) and Internet of Underwater Things (IoUT) Business in Pakistan: A Technical and Financial Review. *Wireless Personal Communications*. Advance online publication. doi:10.100711277-019-06937-3

Saadaoui, S., Khalil, A., Tabaa, M., Chehaitly, M., Monteiro, F., & Dandache, A. (2020). Improved many-to-one architecture based on discrete wavelet packet transform for industrial IoT applications using channel coding. *Journal of Ambient Intelligence and Humanized Computing*. Advance online publication. doi:10.100712652-020-01972-6

Sha, L., Xiao, F., Chen, W., & Sun, J. (2017). IIoT-SIDefender: Detecting and defense against the sensitive information leakage in industry IoT. *World Wide Web (Bussum)*, *21*(1), 59–88. doi:10.100711280-017-0459-8

Sharma, S., & Saini, H. (2020). Fog assisted task allocation and secure deduplication using 2FBO2 and MoWo in cluster-based industrial IoT (IIoT). *Computer Communications*, *152*, 187–199. doi:10.1016/j.comcom.2020.01.042

Shijie, W., & Yingfeng, Z. (2020). A credit-based dynamical evaluation method for the smart configuration of manufacturing services under Industrial Internet of Things. *Journal of Intelligent Manufacturing*. Advance online publication. doi:10.100710845-020-01604-y

Vimal, S., Khari, M., Dey, N., Crespo, R. G., & Robinson, Y. H. (2020). Enhanced resource allocation in mobile edge computing using reinforcement learning based MOACO algorithm for IIOT. *Computer Communications*, *151*, 355–364. doi:10.1016/j.comcom.2020.01.018

Zhang, W., Wu, Z., Han, G., Feng, Y., & Shu, L. (2020). LDC: A lightweight dada consensus algorithm based on the blockchain for the industrial Internet of Things for smart city applications. *Future Generation Computer Systems*, *108*, 574–582. doi:10.1016/j.future.2020.03.009

Zhong, Z., & Hu, W. (2020). Error detection and control of IIoT network based on CRC algorithm. *Computer Communications*, *153*, 390–396. doi:10.1016/j.comcom.2020.02.035

Chapter 14
Digitalization and Automation in Agriculture Industry

Geetha Prahalathan
Sree Vidyanikethan Engineering College, Tirupati, India

Senthil Kumar Babu
Sree Vidyanikethan Engineering College, Tirupati, India

Praveena H. D.
Sree Vidyanikethan Engineering College, Tirupati, India

ABSTRACT

The industrial production has experienced a technological revolution in the recent past decades. The technological revolution influenced the agriculture industry too. The important areas in the change are not limited to innovation in farming, novel production of agriculture-based tools and equipment, transportation and consumption of food across the globe, marketing the agriculture products, and digitalization. Digitalization is the involvement of digital technology in the existing field for easing the mechanism of handling, processing, recording the data. Digitalization enables sustainable farming. It is required desperately to develop this technology because there is a substantial reduction of clean water and depletion of aquifers effects the cultivation. With the technology, the quantity and quality of the food has to be managed to feed the global population. The familiar digitization technology that makes the agri-industrial sector to experience growth are artificial intelligence, machine learning, sensor networks, internet of things, robotics, cloud data.

INTRODUCTION

In India, Agriculture is the main occupation. More than 70% of the population is engaged with Agriculture. During pandemic, people ran behind the Food rather than other commodities. Thus, the well-known phrase "Food First" is being proved. Producing and consuming healthy food is the most important and primary need today. Also, food safety and security are equally important for the growing population.

DOI: 10.4018/978-1-7998-3375-8.ch014

In India, inadequacy of cultivating land and irrigating water restricts the quality and quantity of production of food crops. Digitalized farming could be one of the solutions for the challenges. This farming method involves digital technology as supporting system for the plant growth and yield (Giesler, 2018).

Digitalization of agriculture field is the development, adaptation of the information and communication technology in the field. This enables the human machine interaction. The tools of digitization are real-time monitoring and control through sensors, internet of Things, big-data, machine learning and cloud computing (Bramley & Ouzman, 2019).

In this chapter, the integration of traditional agriculture knowledge with the existing digital technology to improve the efficiency of high yield is being discussed.

Literature Review

Digital farming otherwise intelligent farming (Marcu, et.al., 2020) is a concept of utilizing the sophisticated telecommunication tools for soil and crop management. Management includes handling of available natural resources efficiently to meet the global demand of growing population and providing the right data at right time to right person (Saleem, et.al., 2018). Digital farming technique involves the sensor technology, Robotics & Drone automation, data analytics & storage, networking & GPS technology, Satellite & Tele Communication (Challa, et.al., 2017). It is illustrates in the figure 1

Figure 1. Illustration of Digital Farming

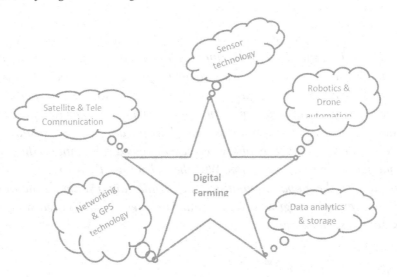

Digitalization in Agriculture

Digitalization make the job ease in all industry and supports for economic growth (Giesler, 2018). Digitalization of Agriculture can be called as fourth revolution for maximum sustainability of quality food (Carlos, 2019 ; Rose & Jason, 2018). In the field of Agriculture, the invention of machine for ploughing, sowing, weeding, and applying pesticide, monitoring the crop growth, crop health, harvesting, storing, packing and transportation of food helps the farmer to reduce their burden. Here, various sensors are involved at different stages of crop growth. The sensors are connected as a network and utilized for processing, storing the data for future references.

Digitalization of agriculture includes three main parameters. They are Internet of Things, Sensor Technology and Digital Education.

Internet-of-Things

The influence of internet-of-things in smart farming starts with the availability of internet connectivity in remote places. IOT provides remote access and increase the automation of the process interoperability, ease of irrigation (Khanna, & Kaur, 2019). For example, the IOT platform for FIWARE integrates open source components for developing smart solutions (Zamora-Izquierdo, et.al.,2019). It comprises of green house for climatic and soil control with the internet connectivity and data processing.

Sensor Technology

Over the last thirty five years, the development of Micro and Nano Electro Mechanical Systems (MEMS/NEMS)-based sensors has been a milestone in almost all the fields. These sensors are micro/nano scale size particle based devices that sense physical quantities like force, pressure, electrochemical or biological substances. Its relative low cost and sensible accuracy contributed for its growth in ever-greater frequency in consumer utilization. In health and welfare applications, the role of Mico/Nano sized sensor is growing rapidly (Geetha & Wahida Banu, 2014).

The advancement in computing technologies along with internet and cloud based connectivity joined hands with sensing solutions to reflect in development of system equipped to collect data from remote. This technology is being widely used in all fields of application like agriculture (monitoring right from the soil health to seed quality), Health (monitoring and treating from remote), Space (Observing the space phenomenon), and so on (Cook, et.al., 2018).

On the other hand, LED light source is another advanced technology. Meanwhile, the optimal emission spectrum suits for photosynthesis and aeroponics. (Vinther & Müller, 2018). In addition, Molecular biology, genetic engineering nanotech and artificial intelligence and machine learning are few technologies that arise from biotechnology (Marakakis & Kalimeri, 2019).

Digital Education

The Digital education is a product of three components. It is illustrated in the figure 2. They are farmer, the technological support and technical consultation services. Farmer is the main component of the cultivation process. The technological support includes machineries, software and communication tools. The consultation service advisory networks socialize the knowledge; they convert raw data to implied understanding for updating the information among the farmers (Fielke, Taylor & Emma, 2020).

CLASSIFICATION OF DIGITAL FARMING

Digital farming can be categorized into precision farming and Smart farming. Figure.3 shows the classification of digital farming (Giesler, 2018).

Figure 2. Illustration of Digitalized Agriculture

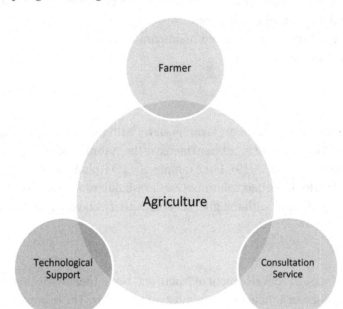

Precision Farming

Precision farming is optimizing the growth and production process. Intelligent sensors are utilized for monitoring and control the plant growth from tilling to packing. This technique intends to reduce the fuel, water and resource consumption for the cultivation. Drone based monitoring is another technique that enables to check land for quality of the soil, weeds between the crops, pests, and disease of the plants (Mitchell, Weersink, & Erickson, 2018) Sowing and fertilizing pattern are also monitored and controlled autonomously.

Smart Farming

Smart farming utilizes the application of information and communication technology (Bacco, et.al, 2019). Smart farming involves the data collection from the previous records. The goal of smart farming

Figure 3. Classification of Digital Farming

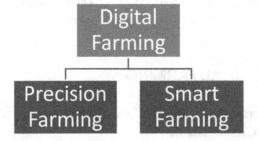

is making the agriculture efficient with sustainability. It also concerns with improving the quality of the yield without disturbing the farmers self-esteem technology

The service providers diffuse the innovations and they suggest the farmers throughout the farming process. The data regarding the condition of the soil, temperature, weather, usage of water quantity, quantity of man power required, supporting funding agencies are collected, processed and evaluated by the service providers (Cristina, et.al., 2017 ; Eastwood, et.al., 2019; Nyasimi, et.al.2017; Nettlea, Crawforda, & Brightling, 2018). The ready-made solution for the current challenges is addressed by the service providers. They navigate in the jungle of data, analyse and automate.

Decision Support Systems

As a digital support, mobile applications have been developed with integrated library. Such application provides details of plants, pests, diseases, right time for planting, watering, fertilizing, weeding and ideal time to harvest (Sulimin, Shvedov, & Lvova, 2019). This application needs creative innovation, diffusion of knowledge, team work, building a network and experience for the sustainable agriculture system.

Farmers are provided with Decision Support Systems for individual solutions. The Decision Support Systems provides the point of access and customize the need of the farmers (Cañadas, et.al., 2017). The impact of pests (Rupnik, et.al.,2018)and disease are collected for automatic control. The time of sowing, monitoring the cattle (Taylor, et.al., 2013) and the soil quality that suits for the crop are other few parameters of the decision making support system. They can access real-time data on mobile services. So, they can take a concrete decision with the applications of information and data (Cañadas, et.al., 2017). This technology favors the small farmers who cannot afford for the staff.

Stakeholders of companies and industries are motivated for developing interest among the crop growers to use the technology. The potential of smart farming lies in improving the work safety with sustainability of the crop production (Walter, et.al.,2017). Figure.4 shows the elements of smart farming. (Chiaraviglio, et.al., 2017).

Elements of Smart Farming

Temperature sensors and moisture sensor are used directly on the fields (Lehmann, 2020). Other sensors solves the purposes like automatic indication of time and quantity for sowing, fertilizing, weeding, harvesting, storing, and packing (Yan, et.al., 2018). For monitoring and controlling the pests, camera and microphones are used. Sensor Network with various sensors has to be installed in the cultivated areas. This network is responsible for actual material flow from field to plate. This data could also be combined with satellite data for further data storage, data analysis and processing.

Access of the information reduce the time, over production, simplifies the transportation procedure and provides the real price for the food products. In addition, the advanced trend in agriculture industry is monitoring of soil quality. Identifying the right quality soil for the concerned crop will increase the yield. (Deeken, Wiemann, & Hertzberg, 2018). Satellite based technologies are used to access such precise data. On the other hand Robotics and drones are new waves in the global economy. (Fielke, Taylor, & Emma, 2020). Agriculture innovation systems comprises of agriculture knowledge and advice networks.

Figure 4. Elements of Smart Farming
(Chiaraviglio, et.al., 2017).

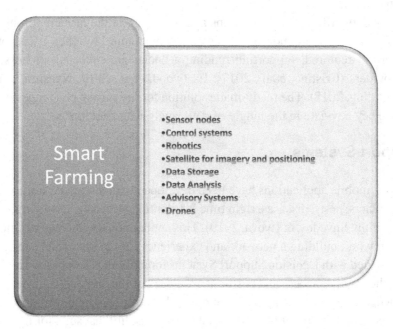

Urban Farming

Cultivating, processing and distribution of food products in an urban context for personal use is called as urban farming otherwise called as urban gardening or city farming (Lask, 2020). People in urban region grow plants on the roofs of their houses. This production of food in the small space satisfies their individual needs. They are called as prosumers. It requires basic knowledge of cultivation and harvesting of conventional garden.

Figure.5 displays the equation for the urban farming. Plants can also be grown in vertically stacked cases on the roofs called as vertical farming. Aquaponic system is another type of farming that combines cultivation of useful plants and breeding of fish. Aquaponic is a circular system where faeces of fish are utilized as nutrients to plants.

Animal Welfare

Being an agriculture based country, it is essential to take care of the health of cattle and live stokes that are supporting the farming directly and indirectly. In addition, they are appreciable source of income. That is the reason why animal husbandry plays a vital role.

To safeguard the livestock life, it is required to have continuous monitoring and diagnosis (Wang, et.al., 2019). For maintaining animal welfare, it is vital to provide continual information using in-field detection in the form of sensors about the various parameters that governs the animal welfare (Yang, et.al., 2019). The main prerequisite for the sensor is that the sensor should be able to detect, identify and quantify the sensed data at almost every instant of time (Zhu, et.al., 2015). For instance, E-nose and Rumi-watch nose band sensor are briefed.

Figure 5. Display of urban farming deeds.

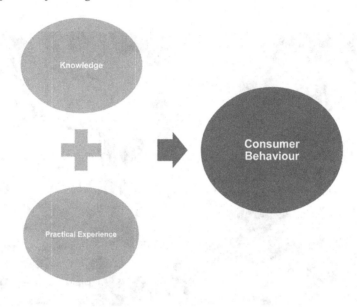

E-nose

Electronic-nose widely called as e-nose is a sensor instrument developed for various applications in agriculture. This smell based nano biosensors has supported for improved quality in food production (Wilson, 2013). The biochemical processing, agricultural sectors of agronomy, botany, cell culture, plant cultivation, environmental monitoring, horticulture, pesticide detection, plant physiology and pathology are some of the areas of e-nose in commercial agriculture industries.

Rumi-Watch Noseband Sensor

The health of the cattle are studied by the vital parameters like eating behavior and Rumination A monitoring nano electronic device has been developed for measuring ruminating and eating behaviour in stable-fed cows. This supports for automated health and activity planner from remote station.

A noseband pressure sensor, a data logger with online data analysis, and a software are incorporated to design RumiWatch noseband sensor. Based on generic algorithms, automated measurements of behavioral parameters are studied. Without animal-specific learning data, the system records and classifies the duration of chewing action. Thus, the individual ruminating and eating jaw movements performed by the animal can be quantified by the user (Zehner, et.al., 2017). The figure.6 is the illustration of cow wearing the Rumi-Watch band sensor.

Stages of Digital Technology in Agriculture

Though digitalisation explore the IT field, media, finance, education and insurance widely, in Agriculture its role is challenging since cultivation techniques are unique for each crop (Trendov, Varas, & Zeng, 2019). There are three stages of development and introduction to digital technology. They are pilot technology, market saturation and integration.

Figure 6. Components of the RumiWatch noseband sensor
(Zehner, et.al., 2017).

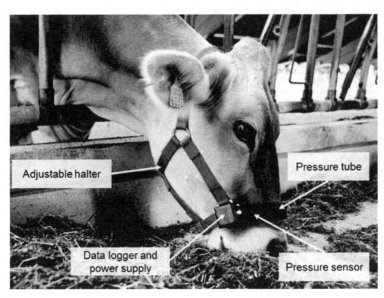

In the pilot stage, the agriculture tools and equipment are used for growth, monitoring and control of plants. Regarding market saturation, India is providing solution for the customized and personalized use of tools and machines. The third one, Integration is a trend where the available data are combined to remove the challenge of problem of choice and the related risks are removed.

Marketing

Modern agriculture engineering products are the various types of sensors, digital positioning, optical recognizing system or data visualization that allows autonomous decisions taken by the farmer (Giesler, 2018). The Agricultural Global Positioning System reduces the usage of amount of seeds, water for irrigation, fertilizers pesticides and weedicides (Clapp, 2017). Farmers can generate their own electricity from the local bio resources to meet their power requirement for fully battery powered emission free and noise-free tractors (Giesler, 2018). Today the smart farming make use of the advantages of rapidly growing digitizing technology as a business model (Walter, et.al., 2017). For instance, the cost of the modern equipment are lowered by rental programs like *Trringo* in India. It manages the challenge of affordability for small farmers while using the farm mechanism process.

Farmers can take their own start-ups for marketing using digital farming. This is system that documents from crop cultivation to processing of the harvested grains in a way that reaches the consumers understanding.

Challenges

Any new development has its own pros and cons. Besides many advantages, the constraint of the agriculture digitalization lies in low wages for the labor, notable inflation, dependence on imports, out dated production methods and defective financial system.

Certainly, the concept of automation reduces the labour intensive process, utilizes the agriculture capabilities in monitoring and controlling the total process of crop growth. But, to implement the digital farming, the financial ability should favour the farmers (Carlos, 2019). They should also be imbibed with digital knowledge, for acquiring and accessing the data. Arrival of new agri-techniques, Internet of things, clod computation, Geo data for crop forecasting, agri-products, logistics management are needed to be updated to the farmers. The sector should get over the learning curve of utilizing the technology to its maximum extent. In addition, to this the technology should be utilized such that the environmental biology must not be disturbed; soil erosion and compaction should never be destroyed in the name of the technology (Giesler, 2018).

State-of-Art

Digitalization is the human interaction with computer and information & communication technologies (Billon, Lera-Lopez, & Marco, 2010). Adopting and interacting with the digital technologies in agriculture is Digitalization in the field (Robertson, et.al., 2016). The state of art includes

1. Connectivity of human with the technology.
2. Transparency in the system
3. Balancing of Challenges among stakeholders.

In case of the connectivity, the farmers have to have a required connection with the network of advisors. The connectivity should be existing beyond the farm. The service providers are network advisors should act as a bridge between the farmers and the cyber system. Farmers have to be autonomous in decision making with shared data analysis by the services providers.

In the next case, the practices and procedure of the new technology should be transparent among the farmers and other stakeholders. The last case is that as the innovation grows on, the challenges among various stake holders should be balanced.

CONCLUSION

The availability of data regarding soil information helps the selection of right crop for the area. The data are related to weather condition, soil pollution, water pollution, air pollution, finer control of disease, precision control of pests. Meanwhile, the impact of the technology on the environment should have a constant monitoring. The reduction of chemicals used in turn to reduce the pressure of the soil. The barriers in implementing the technology comprises lack of digital skills, expenditure, inadequate infrastructure, data ownerships and use, Incentives, economic policies, adaptation of technical and non-technical supports. The digitization of the farming will grow with the supported policies, research efforts and farmers adoption through investments

ACKNOWLEDGMENT

This work is supported by the National Memes Design Centre, Department of Electronics and Communication Engineering, Sree Vidyanikethan Engineering College. The video lessons from NPTEL titled, Nanotechnology in agriculture was an inspiration to do this chapter.

REFERENCES

Bacco, M., Barsocchi, P., Ferro, E., Gotta, A., & Ruggeri, M. (2019). The Digitisation of Agriculture: a Survey of Research Activities on Smart Farming. *Array, 3*(4).

Billon, M., Lera-Lopez, F., & Marco, R. (2010). Differences in digitalization levels: A multivariate analysis studying the global digital divide. *Review of World Economics, 146*(1), 39–73. doi:10.100710290-009-0045-y

Bramley, R. G. V., & Ouzman, J. (2019). Farmer attitudes to the use of sensors and automation in fertilizer decision-making: Nitrogen fertilization in the Australian grains sector. *Precision Agriculture, 20*(1), 157–175. doi:10.100711119-018-9589-y

Cañadas, J., Sánchez-Molina, J. A., Francisco, R., & Águila, I. M. (2017). Improving automatic climate control with decision support techniques to minimize disease effects in greenhouse tomatoes. *Information Processing in Agriculture, 4*(1), 50–63. doi:10.1016/j.inpa.2016.12.002

Carlos, M. (2019). *Digitization in agriculture – what it means and what you need to know.* Challenge Advisory LLP.

Challa, S., Wazid, M., Kumar Das, A., Kumar, N., Goutham Reddy, A., Yoon, E.-J., & Yoo, K.-Y. (2017). Secure Signature-Based Authenticated Key Establishment Scheme for Future IoT Applications. *IEEE Access: Practical Innovations, Open Solutions, 5*, 3028–3043. doi:10.1109/ACCESS.2017.2676119

Chiaraviglio, L., Blefari-Melazzi, N., Liu, W., Jairo, A., Gutierrez, J. A., Beek, J., Birke, R., Chen, L., Idzikowski, F., Kilper, D., Paolo, M., Bagula, A., & Wu, J. (2017). Bringing 5G into Rural and Low-Income Areas: Is it Feasible? *IEEE Communications Magazine.*

Clapp, J. (2017). Responsibility to the rescue? Governing private financial investment in global agriculture. *Agriculture and Human Values, 34*(1), 223–235. doi:10.100710460-015-9678-8

Cook, D. J., Duncan, G., Sprint, G., & Fritz, R. L. (2018). Using Smart City Technology to Make Healthcare Smarter. *Proceedings of the IEEE, 106*(4), 708–722. doi:10.1109/JPROC.2017.2787688 PMID:29628528

Cristina, C.A., Coppola, M., Tregua, M., & Bifulco, F. (2017). Knowledge Sharing in Innovation Ecosystems: A Focus on Functional Food Industry. *International Journal of Innovation and Technology Management, 14*(5), 1750030-1 – 1750030-18.

Deeken, H., Wiemann, T., & Hertzberg, J. (2018). A spatio-semantic model for agricultural environments and machines. Lecture Notes in Artificial Intelligence and Lecture Notes in Bioinformatics, 10868, 589–600. doi:10.1007/978-3-319-92058-0_57

Eastwood, C., Klerkx, L., Ayre, M., & Rue, B. D. (2019). Challenges in the Development of Smart Farming: From a Fragmented to a Comprehensive Approach for Responsible Research and Innovation. *Journal of Agricultural & Environmental Ethics, 32*(5-6), 741–768. doi:10.100710806-017-9704-5

Fielke, S., Taylor, B., & Emma, J. (2020). Digitalisation of agricultural knowledge and advice networks state-of-the art review. *Agricultural Systems Elsevier, 18*, 120763. doi:10.1016/j.agsy.2019.102763

Geetha, P., & Wahida Banu, R. S. D. (2014). A compact modelling of a double-walled gate wrap around nanotube array field effect transistors. *Journal of Computational Electronics, 13*(4), 900–916. doi:10.100710825-014-0607-7

Giesler, S. (2018). *Digitisation in agriculture - from precision farming to farming 4.0, Dossier.* BIOPRO Baden-Württemberg GmbH.

Khanna, A., & Kaur, S. (2019). Evolution of Internet of Things (IoT) and its significant impact in the field of Precision Agriculture. *Computers and Electronics in Agriculture, 157*, 218–231. doi:10.1016/j.compag.2018.12.039

Lask, J. (2020). Alphabeet – the green-fingered smartphone. *Digitisation in agriculture – from precision farming to farming 4.0.*

Lehmann, H. (2020). Digitisation in agriculture – from precision farming to farming 4.0. *Sensors for the Bioeconomy.*

Marakakis, I., & Kalimeri, T. (2019) Remote sensing and multi-criteria evaluation techniques with GIS application for the update of Greek Land Parcel Identification System. *Scandinavian Journal of Information Systems, 44*, 103–106.

Marcu, I., Suciu, G., Bălăceanu, C., Vulpe, A., & Drăgulinescu, A.-M. (2020). Arrowhead Technology for Digitalization and Automation Solution: Smart Cities and Smart Agriculture. *Sensors (Basel), 20*(5), 1464. doi:10.339020051464 PMID:32155934

Mitchell, S., Weersink, A., & Erickson, B. (2018). Adoption of precision agriculture technologies in Ontario crop production. *Canadian Journal of Plant Science, 98*(6), 1384–1388. doi:10.1139/cjps-2017-0342

Nettlea, R., Crawforda, A., & Brightling, P. (2018). How private-sector farm advisors change their practices: An Australian case study. *Journal of Rural Studies, 58*, 20–27. doi:10.1016/j.jrurstud.2017.12.027

Nyasimi, M., Kimeli, P., Sayula, G., Radeny, M., Kinyangi, J., & Mungai, C. (2017). Adoption and Dissemination Pathways for Climate-Smart Agriculture Technologies and Practices for Climate-Resilient Livelihoods in Lushoto, Northeast Tanzania. *Climate (Basel), 5*(3), 63. doi:10.3390/cli5030063

Robertson, M., Keating, B. A., Daniel, W., Bonnett, G., & Hall, A. J. (2016). Five Ways to Improve the Agricultural Innovation System in Australia. *Farm Policy Journal., 13*, 1–13.

Rose, D. C., & Jason, C. (2018). Agriculture 4.0: Broadening Responsible Innovation in an Era of Smart Farming. *Frontiers in Sustainable Food Systems, 2*, 87. doi:10.3389/fsufs.2018.00087

Rupnik, R., Kukar, M., Petar, V., Košir, D., Darko, P., & Bosnic, Z. (2018). AgroDSS: A decision support system for agriculture and farming. *Computers and Electronics in Agriculture.*

Saleem, J., Hammoudeh, M., Raza, U., Adebisi, B., & Ande, R. (2018, June). IoT standardisation: Challenges, perspectives and solution. *Proceedings of the 2nd International Conference on Future Networks and Distributed Systems (ICFNDS'18)*, 26–27.

Sulimin, V. V., Shvedov, V. V., & Lvova, M. I. (2019) Digitization of agriculture: innovative technologies and development models. *Proceedings of IOP Conf. Ser.: Earth Environ. Sci.*, 34, 012215. 10.1088/1755-1315/341/1/012215

Taylor, K., Griffith, C., Lefort, L., Gaire, R., Compton, M., Wark, T., Lamb, D., Falzon, G., & Trotter, M. (2013). Farming the Web of Things. *IEEE Intelligent Systems*, 28(6), 12–19. doi:10.1109/MIS.2013.102

Trendov, N. M., Varas, S., & Zeng, M. (2019). *Digital technologies in agriculture and rural areas briefing paper*. Food and Agriculture Organization of the United Nations Rome.

Vinther, K. S., & Müller, S. D. (2018). The imbrication of technologies and work practices: The case of Google Glass in Danish agriculture. *Scandinavian Journal of Information Systems*, 30, 32–46.

Walter, A., Finger, R., Huber, R., & Buchmann, N. (2017). Opinion: Smart farming is key to developing sustainable agriculture. *Proceedings of the National Academy of Sciences of the United States of America*, 114(24), 6148–6150. doi:10.1073/pnas.1707462114 PMID:28611194

Wang, H., Li, L., Chen, H., Li, Y., Qiu, S., & Gravina, R. (n.d.). *Motion Recognition for Smart Sports Based on Wearable Inertial Sensors*. Bodynets.

Wilson, A. D. (2013). Diverse Applications of Electronic-Nose Technologies in Agriculture and Forestry. *Sensors (Basel)*, 13(2), 2295–2348. doi:10.3390130202295 PMID:23396191

Yan, C., Deng, W., Jin, L., Yang, T., Wang, Z., Chu, X., Su, H., Chen, J., & Yang, W. (2018). Epidermis-Inspired Ultrathin 3D Cellular Sensor Array for Self-Powered Biomedical Monitoring. *ACS Applied Materials & Interfaces*, 10(48), 41070–41075. doi:10.1021/acsami.8b14514 PMID:30398047

Yang, Y., Song, Y., Bo, X., Min, J., Pak, O. S., Zhu, L., Wang, M., Tu, J., Kogan, A., Zhang, H., Hsiai, T. K., Li, Z., & Gao, W. (2020). A laser-engraved wearable sensor for sensitive detection of uric acid and tyrosine in sweat. *Nature Biotechnology*, 38(2), 217–224. doi:10.103841587-019-0321-x PMID:31768044

Zamora-Izquierdo, M. A., Santa, J., Martínez, J. A., Martínez, V., & Skarmeta, A. F. (2019). Smart farming IoT platform based on edge and cloud computing. *Biosystems Engineering*, 177, 4–17. doi:10.1016/j.biosystemseng.2018.10.014

Zehner, N., Umstätter, C., Niederhauser, J.J., & Schick, M. (2017). System specification and validation of a noseband pressure sensor for measurement of ruminating and eating behavior in stable-fed cows. *Computers and Electronics in Agriculture, 136*, 31–41.

Zhu, C., Sheng, W., & Liu, M. (2015). Wearable Sensor-based Behavioral Anomaly Detection in Smart Assisted Living Systems. *IEEE Transactions on Automation Science and Engineering*, 4(4), 1225–1234. doi:10.1109/TASE.2015.2474743

Chapter 15
Internet of Things for High Performance Net Zero Energy Buildings

Abhilash B. L.
https://orcid.org/0000-0003-3997-4630
Vidyavardhaka College of Engineering, India

Gururaj H. L.
Vidyavardhaka College of Engineering, India

Vijayalakshmi Akella
K. S. School of Engineering and Management, India

Sam Goundar
https://orcid.org/0000-0001-6465-1097
British University, Vietnam

ABSTRACT

Due to globalization, demand per capita has increased over the decade; in turn, standard of living has been increased. The emission of carbon dioxide is increasing exponentially in construction industries, which affects the global ecological system. To reduce the global warming potential, net zero energy buildings are very essential. With respect to technological advancements in information technology, the internet of things (IoT) plays a vital part in net zero energy buildings. In this chapter, the various issues and challenges of high-performance zero energy buildings are elaborated using different scenarios.

DOI: 10.4018/978-1-7998-3375-8.ch015

INTRODUCTION

The building architecture is commercially based on their functional, aesthetic and luxury needs. In the commercialized world, the engineers and construction work force utilize maximum embodied energy with higher carbon emission.

Vernacular architecture is evolved based on ecological, Societal and economic conditions. The materials used were locally available in nature. Vernacular architecture is predominantly seen in ancient and historical Indian places such as Karnataka, Tamilnadu, Kerala, Rajasthan, and Jaipur. This is later discarded due to globalization, irrespective of climatic conditions, the real estates have aped the west and buildings with glass facades have become popular in India. These kind of building trap the heat inside the building and hence the need of HVAC systems in the building, which causes carbon emission. (Nayak & Prajapati, 2006).

Materials and technology required for conventional buildings have high Embodied Energy and Embodied Carbon as they uses materials such as river sand, natural wood, cement, red burnt brick. Technology such as square footing, Burnt Brick Masonry (BBM), flat slabs which consumes more amount of concrete and many more. (Buchanan & Honey, 1993).

The ecological imbalance is majorly due to depletion of ozone layer and emission of toxic gases. The extraction of natural materials namely river sand, natural wood, stones, fertile soil which does not contain any reused materials for constructing conventional buildings. More often conventional buildings utilize more energy which imbalances the ecosystem. The orientation towards the emission of carbon across the globe is elaborated in the Figure 1.

Figure 1. World Carbon dioxide emissions in metric tons per capita
Source https://www.worldbank.org/

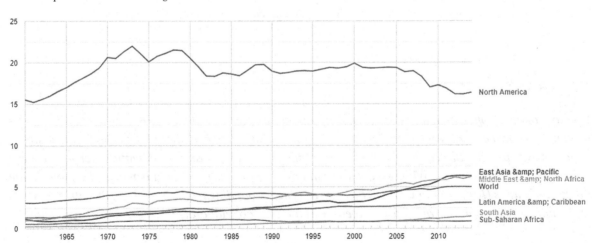

The emission of carbon in South Asia is around 1.46 MT of CO2/capita North America is around 21.98 MT of CO2/capita in year 1973 & 15.22MT of CO2/capita in 1961 & presently 16.37 MT of CO2/capita (2014 recent data) Overall world 4.98 MT of CO2/capita which was predicted from the Carbon trading throughout the globe but US alone is contributing around average of 17 MT of CO2/capita which is 3 times more compared to world average as per the image.(https://www.worldbank.org/).

If everyone on the earth lives like Americans, it would need 5 earths, according to Global Footprint Network. Extreme use of resources has genuine consequences for the globe. Consumption of natural resources such as water, electricity, has been crossed their limit, in the name of Liberalization Privatization Globalization Ecological imbalance as reached its peak. The major issues of emission of carbon are picturized in Figure 2.

Figure 2. The detailing of Carbon footprint in Conventional Constructions

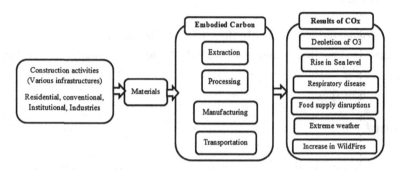

In order to rectify the pitfalls of the above discussed scenarios Sustainable Construction is adopted. It is a technology, which encounters the needs of the present without compromising the ability of upcoming generations. The characteristics of Sustainable construction are Alternative materials- M-sand, robust sand, stabilized mud block, fly ash blocks, ground granulated blast furnace slag, use of plastic for road construction, maximum recycled content in Steel. Alternative construction technology such as inverted arch Foundation, rammed Earth masonry, Mivan construction, precast Technology and alternative roof - Filler slabs Jack arches composite beams and panel slabs are some of the properties of Sustainable construction. (Kibert, 2012).

For Net zero energy buildings (NZEB) or sustainable construction, we need better Indoor Air Quality which causes thermal comfort. These buildings achieve Sustainable Development Goal's (SDG's) which were coined by UN Development programme – UN Conference on Sustainable Development in Rio-de-Janeiro in 2012. IoT is one of the technologies can be adopted for this purpose.

Why Internet of Things (IoT) for Infrastructures

IoT is commonly known as the internet of things which bound the world by reducing human efforts and connecting the people across the globe. IoT is a technology where in data is collected continuously through sensors and actuators. The main functions of IoT are data sensing, acquisition and processing. IoT can be implemented for various applications ranging from smart home automation to industrial applications (Alliance, 2015). IoT also places a vital role in agriculture, Health Care, survivelence, education, entertainment, public sectors and so on.

The main motto of this chapter is to incorporate IoT with sustainable construction. The objective of inculcating IoT with sustainable construction is to reduce the emission of carbon, performance upgradation corresponding to conventional buildings and Smart life cycle assessment of a building. (Arnott et al., 2017).

The chapter is organized as follows, section two demonstrates the literature review on Mindtree's Gladius Connected Buildings, section three depicts the case study on Intel's smart building, section four on Building Information Modeling (BIM), IoT and Block chain`6 innovation framework and section five describes the case study on application of sensors on Reinforced Concrete Structures at last the conclusions are drawn in section six.

LITERATURE REVIEW

Mindtree's Gladius Connected Buildings- In this chapter, the various related works are described corresponding to Smart Construction building using IoT. Mindtree's Gladius Connected Buildings was utilized to incorporate the customers' existing frameworks on an IoT stage to carefully change their office executive's frameworks.

Project Snapshot

- Customer needs are redefined with a more responsive system optimizing energy costs.
- Mindtree has used a digital IoT platform that integrates the existing OT and IT systems which is capable of taking intelligent, automated decisions using predictive analytics
- Results - innovations to enhance user comfort proactively while enhancing energy efficiency 18% every month.
- Understanding the customers' needs by studying the current infrastructure.
- **Energy Efficiency and Resource Efficiency**: Each of these buildings has a number of chillers, AHU and pumps which formed a large component of facility management bills. There was no methodology to understand the causes of fluctuating energy bills.
- **Water management:** Water consumption analysis and leakage detection needs to be addressed.
- **Asset Management and Maintenance** - The Company followed a fixed schedule and reactive asset management causing costly downtime and lowering user comforts.
- **Manual interventions** - Human Interventions were necessary for asset operations making the process slow and prone to human errors.
- **Information Islands without Analytics** - Real time data and insights were restricted to control rooms. The senior staff could not access data on a real time basis and depended on the ground staff to report problems and provide data.

The journey of digital transformation began by integrating OT (Operational Technology), IT(Information Technology) and IoT systems. They installed energy meters at the asset level to understand the energy consumption. Subsequently, implementation of intelligent rules based on ambient conditions and occupancy level on assets like chillers and Air Handling Units (AHU) to help us deliver energy efficiency. (ASHRAE, 2017). As a part of Digital transformation, facilities are equipped with video analytics and IoT enabled analytics that helps to enhance user comfort and safety. Ticket Vending Machine (TVM), ticket generation is automatic and depending on the nature of the alert, can be accessible on multiple handheld devices or digital consoles. This avoids queue system; difference between the facility infrastructure before and after implementation of IoT can be accessed in figure 3.

Implementation Highlights

- **Centralization and Integration:** Gladius Connected Buildings (GCB) centralized and integrated the entire system of IT, OT and IoT systems, within buildings and across buildings into one single console. Now, the senior management can get access to real time and are able to take data driven decisions.
- **Enhanced Asset Life and Operational Efficiency**: Set point management, auto alerts when in manual mode of operation, automatic ticket generation through mobile apps or email have made facility management more proactive. The customer has over 265 rules configured across 4500 sensors. These rules are executed over 2500 times a day, and on an average, perform over 300 devices to impact operational efficiency and enhance asset life.

Figure 3. Racility Infrastructure before and after IoT architecture for smart buildings.
(Freimuth & Keonig, 2018)

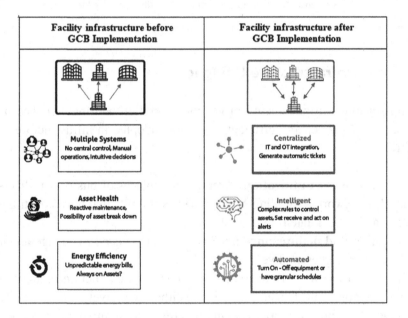

RESULTS

Mindtree's Gladius Connected Buildings (GCB) enhanced the existing infrastructure of to impact the below building management goals.

Table 1. List of Efficient measures implemented in Mindtree's Gladius Connected Buildings

Results	Delivering Results Through	Mindtree's Gladius Connected Building, Performs the Following Actions
Energy Efficiency at equipment level up by 18% every month →	• Device health monitoring • Intelligent and Automated HVAC controls with dynamic set points for (IAQ)ambient conditions, and occupancy level • Equipment level consumption monitoring →	• Switch on/ off periodically • Vary intensity levels of the system, depending on its stability
Proactive asset Management →	• Intelligent and automated systems capable of setting complex rules, and controls → • Asset and sensor level monitoring	• Automated alarms and service ticket generation in client's SAP ticket management system • Real time alarms and alerts in terms of sensor or asset faults
Make informed decisions and take timely actions →	• Central contextual dashboards and mobile • IoT based Analytics →	• Collect and collate data on a real time basis which is compared with historical data to track efficiency of energy or asset functioning • Real time alarms and alerts in terms of energy consumption.
Innovation of services to provide tenant and user comfort →	• RFID-enabled entry and exit for cars of the tenant, thus → keeping intruders away • Video analytics to enhance safety and security	• Capture video surveillance feeds and apply IoT analytics to provide intelligent automation • Mobile apps enabled quicker turnaround times

Intel Creates Smart Infrastructure Using IoT

Intel's smart Infrastructure expands energy protection, operation and maintenance productivity, and inhabitant luxury; some of the energy efficient techniques adopted in infrastructure are mentioned in the above Table 1.

- **Difficulties - Reduce asset utilization**. Average Intel places of business utilize static structures the executives' frameworks (Building Management System's) that have constrained capacities to control energy and water related frameworks.
- **Improve operation and maintenance proficiency.** Intel needed to change an energy work area model to oblige more representatives.
- **Increase inhabitant comfort.** Temperature difference in the structure regularly leads to numerous representative objections for their zone being either excessively hot or excessively cold.
- **Arrangements - Advanced investigation of structure.** Minimum energy and water use by better supervision of structure frameworks utilizing robotization rules produced from sensor information.
- **Space streamlining.** Increment work space usage rates by utilizing inhabitant's sensor information to assist representatives with finding empty desk areas.
- **Machine learning calculation.** Keep up a consistent temperature in all structure zones by considering increasingly natural elements.
- **Effect - Energy/water investment funds.** The reserve fund is estimated to be US$ 641465 per annum with an arrival on speculation compensation time of less than four years.
- **Efficient office space**. Intel has expanded the representative limit by roughly 30 percent.

Smart Infrastructure Arrangement

With the objective of expanding energy protection, operational proficiency, and inhabitant comfort, Intel planned a smart building arrangement dependent on its IoT reference design; figure 4 represents IoT architecture for smart buildings which were implemented in the Intel smart infrastructure's, which reduced the utilization of operational energy in the building. IoT innovation enormously improved the undertaking of gathering and investigating information from different structure frameworks, including warming, ventilation, and cooling (HVAC), water, energy. The structure likewise has a Power over Ethernet (PoE) – based brilliant lighting framework with sensors inserted in light apparatuses to screen inhabitants, temperature, and other natural elements. (National Institute of Building Sciences, 2015).

Figure 4. IoT architecture for smart buildings.
(Freimuth & Keonig, 2018).

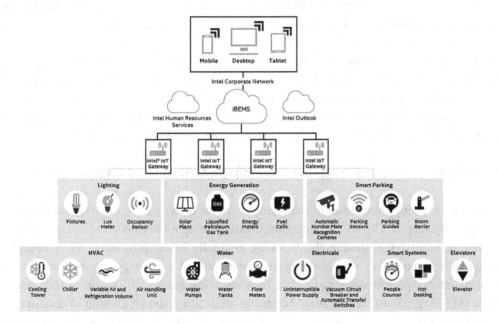

ENERGY CONSERVATION

In the Intel brilliant structure, shut zones, for example, meeting rooms, speak to around 18% of the entire floor space. These regions are acceptable possibilities for energy protection when they are empty, the HVAC and lighting frameworks can frequently be shut down to spare energy. In any case, numerous static BMSs keep up a steady temperature in shut rooms independent of whether they are involved, booked, or empty.

Operational Efficiency

- Mechanized interest reaction - With an end goal to keep away from automatic assistance interferences (i.e., power outages), utilities request reaction projects to tempt their clients to bring down energy use during periods when by and large interest may surpass limit. Clients who take an interest are normally repaid through lower rates and credits.

- **Arrangement** - Intel's brilliant structure arrangement control four energy sources: diesel stage, Solar based energy units, and the network. It empowers the offices group to remotely peruse energy meters associated with the distinctive energy sources with change energy utilization all through the structure. Both energy utilization and life span are observed and controlled so as to meet an energy load profile that fulfills the necessities. With this capacity, Intel's smart structure can take robotized activities to lessen energy utilization when the 90% limit of the allowed load is surpassed.
- **Tenant Comfort** - Consistent temperature across building zones - Temperature vary with respect to time is a typical problem raised by of building tenants.
- **Arrangement** - To keep up a consistent temperature across different structure zones, Intel actualized an AI calculation that predicts suitable set focuses for the HVAC in the structure. The calculation not just factors in ordinary boundaries (return air temperature), it considers numerous others, including inhabitants and encompassing temperature (Mekki et al., 2019).

A significant worry of representatives in other Intel places of business is finding an energy desk area to work in light of the fact that it very well may be hard to distinguish which work areas are unassigned and accessible for use. Intel's smart building arrangement permits workers to view and hold an accessible space for the afternoon. The arrangement makes this view by joining information from inhabitants sensors introduced in every energy work space with information from the desk area reservation framework.

Building Information Modeling, IoT and Block Chain Innovation Framework

At the present time, energy utilization in structures is one of the key topics in the sustainable environment and smart city sectors. Day by day technologies are gaining more importance in our daily lives and routine activities, in turn, real-time applications are gradually being introduced. Thus, energy observing in buildings becomes a key tool enabling an easier method for energy management in buildings, considering its energy performance and at last for improving energy utilization in the operation and maintenance (O&M) phase of the building. Acquiring, collecting, and processing of data in real time is a prevailing need in the construction sector to be ready for the decision-making process. In this issue, the Internet of things (IoT) becomes an important tool to achieve this purpose. As mentioned in figure 5. However, there are many challenges such as scalability, high procurement costs make it difficult for the real implementation of this technique (Noura, M et al., 2019).

IoT is answerable for the making of this innovative model since it permits the association between the various gadgets that make up the smart situations. Right now, different applications can be found in various fields, for example, Hydrology and Irrigation frameworks, farming, robotization and portability, structures and urban communities, energy, social insurance, or wellbeing of the board.. In the current situation, there are different IoT correspondence innovations for information transmission. The decision of one of them will rely upon angles, for example, the range, information rate, various utilization and different sorts of uses (Kaufmann et al., 2018).

A significant increment in the quantity of Internet-associated gadgets is normal; the current number of gadgets usage 6.4 billion will increase up to 20.8 or 50 billion numbers by 2030 (Dave Evans, 2011). Because of this, urban communities, structures, and things are progressively turning out to be receptors

Figure 5. Building Information Modeling, IoT and Block chain innovation framework structure for Smart Infrastructure.
(Choudhary & Jain, 2017).

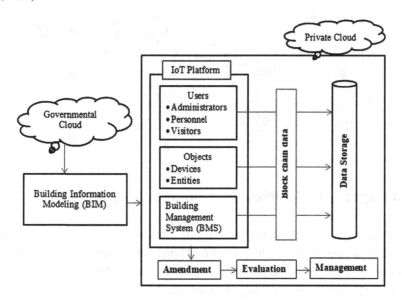

and generators. The Architecture, Engineering and Construction (AEC) industry has not implemented the IoT as implemented in different businesses like assembling, aviation or account parts. Building Information Modeling (BIM) is a creative innovation that is considered as an open door for the AEC business to move to the computerized period and improve the joint effort among partners by applying Building Information and Communication Technologies (ICT) maturity model as shown in figure 6.

Figure 6. Building IoT maturity model
(Rizwan et al., 2017).

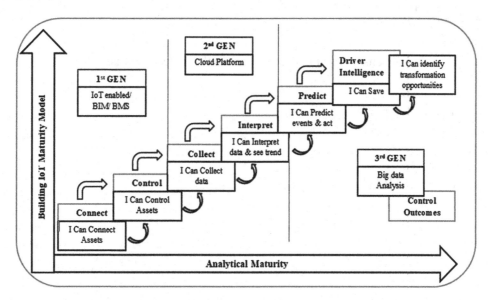

Figure 7. Reinforced Building component - IoT maturity model.

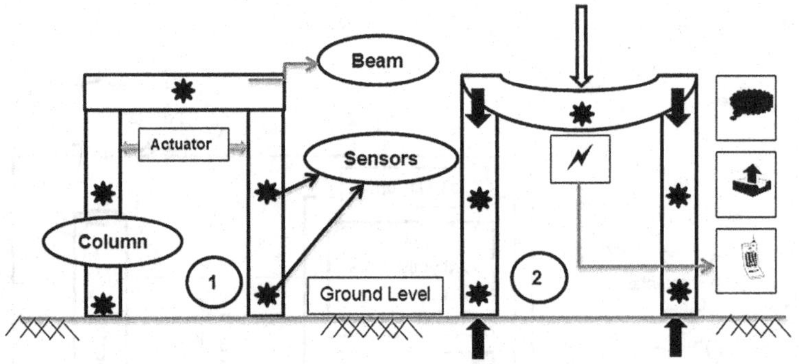

Application of Sensors on Reinforced Concrete Structures

Sensors Technology is best suited for wide varieties of measurements such as Temperature, Pressure, Load carrying capacity, acceleration deceleration, Fluid level, Smoke.(Sensirion, 2011). Here the sensing techniques are monitored and measured only based on stress strain data In most of the cases automatic deformation monitoring sensors are developed to access the reinforced concrete structure Ex: Beam deflection, Column buckling, Structure settlement based on soil condition (Eskola et al., 2015).

This chapter focuses on the sensing based monitoring of Beams and columns, the beams are provided with Sensors which gives the photogrammetric analysis of deflections in beams and columns, as shown in figure 7, based on obtained data, validation are done.

Damages occur in the Reinforced structures and their structural components under different operational loads, major components such as Foundation, Beams, Columns, and Slabs are responsible for durability and longevity of the entire structure and even in some cases can lead to the failure of the complete structure (Akella, 2007).

Let 'x' be the number of sensors in the beam, if the first sensor to detect the load then

$$p\,(\mathrm{x}=n\,) = \left(1-p\right)^{n-1} p$$

The average number of sensors used in the beam, then the expectation is,

$$\mathrm{x} = \mathrm{E(x)} = \sum_{1}^{\infty} n\left(1-p\right)^{n-1} p$$

$$= p\sum_{1}^{\infty} n\left(1-p\right)^{n-1} = p\left(1/\,p^{2}\right) \tag{1}$$

$$= 1/\,p$$

The above equation (1) shows the number of sensors used to sense and detect the load. P is the probability that the number of times malicious user attempting to login the application. In addition we can also observe the total number of sensors sequence in a beam by using chain rule probability,

$$P(Y_1........Y_n) = P(Y_1)\,P(Y_2|Y_1)\,P(Y_3|Y_1^2)........P(Y_n|Y_1^{n-1})$$

$$= \prod_{k=1}^{n} P(Y_k|Y_1^{k-1}) \tag{2}$$

Now, applying the chain rule to words in equation (2), we get

$$P(W_1........W_n) = P(W_1)\,P(W_2|W_1)\,P(W_3|W_1^2).........P(W_n|W_1^{n-1})$$

$$P(W_1^n) = \prod_{k=1}^{n} P(W_k|W_1^{k-1}). \tag{3}$$

The equation (2) & (3) shows that, we can find out the probability of arrangement of sensors in the beams. In this rule sequence of sensors compared with the total number of sensors.

Where,

x is the number of sensors in the beam
p is number of sensors used to sense and detect the load.
Y sequence of sensors

CONCLUSION

This chapter has established an IoT framework for real time tracing of infrastructures, hence keeping track in the energy performance of the infrastructures and identifying the possible improvements become possible. In this chapter technical features of the developed IoT monitoring drivers which are implemented in buildings have been described and the case study of sensors (motion sensor, pressure sensor) in reinforced concrete structures has been described.

The proposed IoT device has shown good behavior at testing and implementing levels in buildings. In addition, these case studies have helped in identifying future research works which provide alternative solutions for region specific buildings.

The case study has been focused on the tracking of a single point of contact. It is quite common for Internet connectivity to show glitches such as lagging of connection, stability of network, difficulty for accessing network due to lack of signal issues, short range network. In order to rectify these pitfalls of IoT distributed method of connectivity should be implemented in the building systems. The initial investment is more because of installation of connection gateways, the final cost of the projects may be lesser based on pay-back system if the number of devices are more.

Finally, continuity to this chapter is based on the improvement of overall energy management of the building systems, Supervisory Control and Data Acquisition (SCADA) systems can be utilized more jointly with the implementation of IoT in the system which reduces enormous amount of energy, which was blindly utilized in the Industries, Commercial complexes, Institutions and so on.

Use of power management systems could provide solution to bring down the energy consumption in building, low energy drivers boosting a low power such as piezoelectric, photovoltaic and thermoelectric devices which are mandatory in Net Zero Energy Buildings.

REFERENCES

Alliance, L. (2015). *LPWA Technologies Unlock New IoT Market Potential.* White Paper. Advisory Group.

Arnott, M., Buckland, E., Ranken, M., & Owen, P. (2017). *IoT global forecast and analysis.* https://www.gartner.com/en/documents/3659018/iot-global-forecast-and-analysis-2015-2025

ASHRAE. (2017). Ventilation and infiltration. In 2017 ASHRAE handbook fundamentals. American Society of Heating.

ASTM. (1985). *Standard practice for maintaining constant relative humidity by means of aqueous solutions.* American Society for Testing and Materials. https://www.astm. org/DATABASE.CART/IIISTORICAL/E104-85R96.htm

Bairi, A., Nithyadevi, N., Bairi, I., Martin-Garin, A., & Millan-Garcia, J.A. (2017). *Thermal design of a sensor for building control equipped with QFN electronic devices subjected to free convection.* Academic Press.

Buchanan & Honey. (1993 February). *Energy and carbon dioxide implications of building construction.* University of Canterbury.

Charles, J. Kibert., (2012). Sustainable construction – Green Building Design & Delivery (3rd ed.). John Wiley & Sons, Inc.

Choudhary, G., & Jain, A. K. (2017). Internet of things: a survey on architecture, technologies, protocols and challenges. *International Conference on Recent Advances and Innovations in Engineering, ICRAIE 2016.* 10.1109/ICRAIE.2016.7939537

Eskola, L., Alev, U., Arumeagi, E., Jokisalo, J., Donarelli, A., & Siren, K. (2015). Airtightness, air exchange and energy performance in historic residential buildings with different structures. *International Journal of Ventilation, 14,* 11-26. . doi:10.1080/14733315.2015.11684066

Evans. (2011). *The Internet of Things How the Next Evolution of the Internet Is Changing Everything.* Cisco Internet Business Solutions Group (IBSG).

Freimuth, H. & Keonig, M. (2018). Planning and executing construction inspections with unmanned aerial vehicles. *Automation in Construction, 96,* 540-553. . doi:10.1016/j.autcon.2018.10.016

Kaufmann, D., Ruaux, X. & Jacob, M., (2018). *Digitalization of the Construction Industry: The Revolution is Underway.* Academic Press.

Manyika, J., Ramaswamy, S., Khanna, S., Sarrazin, H., Pinkus, G., & Sethupathy, G. (2015). *Digital America: a tale of the haves and have-mores.* McKinsey Global Institute. https://www.mckinsey.com/industries/high-tech/our-insights/digital-america-a-tale-of-the-haves-and-have-mores

Mekki, K., Bajic, E., Chaxel, F., & Meyer, F., (2019). A comparative study of LPWAN technologies for large-scale IoT deployment. *ICT Express, 5,* 1-7. . doi:10.1016/j.icte.2017.12.005

Nayak, J.K., & Prajapati, J.A. (2006). *Handbook On Energy Conscious Buildings.* Prepared under the interactive R & D project no. 3/4(03)/99-SEC between Indian Institute of Technology, Bombay and Solar Energy Centre, Ministry of Non-conventional Energy Sources.

Noura, M., Atiquzzaman, M., & Gaedke, M. (2019). Interoperability in internet of things: taxon- omies and open challenges. *Mobile Networks and Applications, 24,* 796-809. . doi:10.100711036-018-1089-9

Rashid, K. M., & Louis, J. (2019). Times-series data augmentation and deep learning for construction equipment activity recognition. *Advanced Engineering Informatics, 42,* 100944. doi:10.1016/j.aei.2019.100944

Reinsel, D., Gantz, J., & Rydning, J. (2018). T*he Digitization of the World from Edge to Core, Seagate.* https://www.seagate.com/files/www-content/our-story/trends/files/idc-seagate-dataage-whitepaper.pdf

Rizwan, P., Suresh, K., & Rajasekhara Babu, M. (2017). Real-time smart traffic management system for smart cities by using Internet of Things and big data. *Proceedings of IEEE International Conference on Emerging Technological Trends in Computing, Communications and Electrical Engineering, ICETT 2016.* 10.1109/ICETT.2016.7873660

Schober, K. S., Hoff, P., & Sold, K. (2016). *Digitization in the construction industry: building Europe's road to "Construction 4.0". Think Act.* https://www.rolandberger.com/en/ Publications/Digitization-of-the-construction industry.html

Sensirion. (2011). *Datasheet SHT21 Humidity and temperature sensor IC.* https://www.sensirion.com/fileadmin/user_upload/customers/sensirion/Dokumente/0_Datashets/Humidity/Sensirion_Humidity_Sensors_SHT21_Datasheet.pdf

Seo, J., Han, S., Lee, S., & Kim, H., (2015). Computer vision techniques for construction safety and health monitoring. *Advanced Engineering Informatics, 29,* 239-251. doi:10.1016/j.aei.2015.02.001

Sethi, P., & Sarangi, S. R. (2017). Internet of things: Architectures, protocols, and applications. *Journal of Electrical and Computer Engineering, 2017,* 2017. doi:10.1155/2017/9324035

Sherman, M. H., & Grimsrud, D. T. (1980). Infiltration-pressurization Correlation. Simplified physical modeling, California Univ.

Sinha, R.S., Wei, Y., & Hwang, S. (2017). A survey on LPWA technology: LoRa and NB-IoT. *ICT Express, 3,* 14-21. . doi:10.1016/j.icte.2017.03.004

Sinnott, D., & Dyer, M., (2012). Air-tightness field data for dwellings in Ireland. *Building and Environment, 51,* 269-275. . doi:10.1016/j.buildenv.2011.11.016

Vijayalakshmi, A. (2007). *Climate responsive building envelope to design energy efficient buildings for moderate climate.* Conference on sustainable building South East Asia, Malaysia.

Walker, M. J. (2018). *Hype Cycle for Emerging Technologies.* Gartner Inc. https://www. gartner.com/en/documents/3885468/hype-cycle-for-emerging-technologies-2018

Zhang, H., Yan, X., Li, H., Jin, R., & Fu, H. (2019). Real-time alarming, monitoring, and locating for non-hard-hat use in construction. *Journal of Construction Engineering and Management, 145.* ,1943-7862.0001629 doi:10.1061/(ASCE)CO

Chapter 16
Implication of Predictive Maintenance for Industrial Marketing:
A Case Study From the Air Compressor Industry

Subhasis Ray

Xavier Institute of Management, India

ABSTRACT

This chapter discusses the implication of predictive maintenance (PM) for industrial marketing companies. Using an illustrative case study from the Indian industrial air compressor market, it shows that predictive maintenance solutions will change the way of conventional sales and marketing. Sellers need to focus on early innovation adopters among its customers. They also need to engage with existing customers early on in the purchase process and highlight how PM can reduce the total cost of ownership. PM can be sold effectively to different types of customers- transactional, value-oriented, and collaborative. Industrial marketers have to position the solution appropriately to gain competitive advantage.

INTRODUCTION

Artificial Intelligence (AI), Machine Learning (ML) and Internet of Things (IoT) will transform the business world. Industrial Internet of Things (IIoT) is a significant subset of the IoT led transformation and explores the industrial application of IoT. Today, there is no doubt about this transformation, often captured by the term Industry 4.0. The IIoT market is estimated at USD 110.7 billion by 2025 (Statista, 2020). While at a conceptual level the defining role of technology is easy to understand, implementation issues are yet to be fully appreciated and understood. Implementation has both geographical and sectoral connotation. IIoT implementation is particularly challenging in developing countries like India where the diversity of the industrial landscape makes it difficult to envision a single roadmap for tech-

DOI: 10.4018/978-1-7998-3375-8.ch016

nological innovation and adoption. At a functional level, how will Industrial Internet of Things affect strategy, finance, operations, human resource and marketing departments of companies? The business as usual scenario will change and all such changes will not be positive. However, understanding the functional implications of IIoT can provide the top management with strategic insights for managing radical changes in the shop floor.

This paper discusses predictive maintenance and its implication for industrial marketing practice. Predictive maintenance (PM) solution and technique are used to predict the servicing requirement of running equipments. PM helps in accurate and timely maintenance that not only saves money for companies but also increases the working life of equipments. The innovations in IIoT have led to a strong growth to the PM market. It is expected to grow three times to $12.3 billion by 2025 (MarketsandMarkets, 2020). While the promise of PM is alluring, there are significant challenges and costs for companies implementing such solutions. The challenge of developing and marketing a PM solution is not only about R&D and budgets but also re-organization and managing customer mind-set. The author discusses the implication of PM for industrial marketing using the case of Elgi Equipments Limited, one of the largest air compressor manufacturers in the world. While Elgi was successful in implementing a PM based solution for customers, several challenges emerged in relation to diffusion of innovation, customer relationship management and organizational procurement process. The findings of the study show that innovation of IIoT will change industrial marketing both from the seller and buyer perspective.

BACKGROUND

In any industry, maintenance plays a key role in ensuring trouble free factory/equipment operation leading to higher productivity and profitability (Lopes et al, 2020). Maintenance cost is a major portion of factory operating cost and has been going up due to increase in labour costs. Thus it is important to have an effective maintenance management system in factories. Machine breakdown has several implications like stalled production, damage to connected systems, loss of productive man power, all leading to lesser revenue and lower customer satisfaction. In several industries like railways or airlines, maintenance is also directly related to safety of human lives. Equipment maintenance involves monitoring their condition and taking steps to ensure their continuous operation.

Maintenance can be divided into three types- *reactive, preventive and predictive*. The traditional approach has been reactive- to undertake servicing or repair as and when required. This approach is based on the decision of individuals and often not backed by data analysis. It relies on human monitoring and prone to errors, often leading to unscheduled downtime and loss in factory operations. Reactive maintenance leads to high inventory, high labour overtime costs, high machine downtime and lower production.

The advent of technology allowed better data collection and analysis techniques that allowed managers a more granular understanding of equipment health history and plan for preventive maintenance (Bousdekis et al, 2019). Preventive maintenance gained ground with better understanding of machine running and what causes routine failures. Basic preventive tasks like inspection, lubrication and machine alignment checking are done to prevent avoidable failures. Preventive maintenance is task or time based e.g. lubricating engine parts every month and based on statistical and historical analysis. While better than reactive maintenance, preventive maintenance is inefficient as actions are often not optimized for each equipment but rather for the whole facility. For example, in a factory, a pump may require inspection once a week but a compressor may need annual inspection. Preventive maintenance schedules are often unable to customize such actions leading to over or under spending on repair or servicing.

What would happen if each machine in a factory can 'talk' to the factory manager, telling her when the next servicing or part replacement is due? Predictive maintenance involves such dynamic and interactive scenarios. It extends the philosophy of preventive maintenance but uses the power of AI, ML and cognitive technologies (Ruiz-Sarmiento et al, 2020). Rather than using measures like statistical mean for doing maintenance, it relies on real-life condition of equipments to decide on servicing and repair. Chips installed on machines send messages and alerts to users and sellers about the need for servicing, creating a new paradigm for a smart factory that has an inbuilt real-time maintenance system. PM thus improves productivity, product quality and overall effectiveness of factories. Some applications of PM are products and software used for vibration monitoring systems or thermal imaging systems. For example, the German engineering company Siemens provides a PM tool for its railway customers. The tool, a software connected to devices installed on the tracks, monitors track vibration to determine their condition. This in turn helps customers to know and schedule their track repair activities.

Industrial or B2B marketing (the two terms have been used interchangeably in this chapter) is the marketing of goods and services to commercial organizations, governments and non-profit organizations. It deals with business organizations procuring products and services for their operations or own consumption (Wise & Morrison, 2000). For example, Indian Railways procures engines from the French manufacturer Alstom to produce trains. In the equipment sector, companies not only sell products but also associated services. In the context of the example above, Alstom may sell maintenance services for its engines through annual maintenance contracts. This paper focusses on the equipment maintenance business where service selling is a new, growing and profitable trend (Shankar et al, 2009). Many companies today are moving from selling parts and services to selling solutions related to predictive maintenance. AI will significantly change the marketing function (Davenport et al, 2019; Overgoor et al, 2019). For the sellers, PM provides an accurate dashboard of customer needs which could be fulfilled by its sales organization. For the customers, PM provides a real time requirement map which can help them to optimize their maintenance operations. This paper explores how such PM solutions are likely to impact the B2B marketing process by using the lenses of innovation diffusion, organizational buying process, customer relationship management and total cost of ownership.

REVIEW OF LITERATURE

Maintenance is the key to effective functioning of a factory. Maintenance costs can be between 15 to 60 percent of the running cost of a factory. Predictive maintenance is based on the principle of real time condition monitoring of equipments to take decisions about their repair and servicing. It is a philosophy that assumes that collecting real time running data about machine performance and analysing them provides plant managers the ability to take timely action which in turn increases productivity, revenue and customer satisfaction and finally profitability (Mobley, 2000). Use of AI in the Industry 4.0 will change the maintenance approach (Ruiz-Sarmiento et al, 2020). Implementation of predictive maintenance will require an integrated approach across different functions and divisions of a company (He et al, 2017). Yu et al (2019) argues that in future, a data cloud will be required to make PM more effective. Thus apart from machine diagnostics and business functions, IT solutions (Sahal et al, 2020) will be critical in bridging buyer-seller gaps. Sellers of PM solutions thus need to understand customer requirements particularly in relation to purchase of services (Bousdekis et al, 2019).

B2B marketing landscape will be influenced by Artificial Intelligence and Machine Learning tools (Colter et al., 2018) which are likely to influence all five phases in B2B buyer-seller relationships: contact, transaction, consultation, expansion and enterprise (Muylle et al, 2012). All these phases are likely to be influenced when PM solutions are introduced by industrial marketers leading to a change in the brand architecture. It will require new business models and sales specialists (Wise & Morrison, 2000). B2B buyers can be classified into four groups- underperformers, commodity buyers, partners and most valuable customers. Under-performing customers are those that may be high cost to serve and will require innovative ways to be managed (Mittal et al., 2008). Each of these categories looks for different benefits from the same set of products and hence sellers need to understand their requirements and offer solutions in a way that enhances customer satisfaction and loyalty (Narayandas, 2015). The Industrial marketing scenario has been changing in the last decade primarily due to the rise and use of technology in selling and buying. The common trends across industries are omni-channel selling, technology enable selling and e-commerce (Gavin et al., 2020). PM solutions at the moment will appeal to a select few customers who are looking for innovations to enhance their manufacturing productivity. In B2B marketing, big data is likely to reveal 'micro-markets'- niches in the industrial buying landscape using big data analytics (Goyal et al., 2012) which in turn can be expanded into profitable customer segments. Since B2B markets are increasing moving towards servitization- bundling of products and services- profitable service selling will be important for marketers. This can happen through revamping the sales support function, creating a tech-savvy sales force and understanding the business process of the customer (Reinartz & Ulaga, 2008). Industrial sellers need to co-create value with their customers (Cova & Salle, 2008; Hein et al, 2019). However, this will become a possibility only when there is value-driven customer relationship management (Richards & Jones, 2008). Collective consideration of the above factors can only lead to a sustainable competitive advantage in industrial services (Matthyssens & Vandenbempt, 1998).

The COVID-19 pandemic is likely to further change the pattern of B2B buying. Gavin et al.'s (2020) survey of more than 3600 decision makers reveal that large companies may actually increase their spends particularly in developing markets in Indian and China. Lay et al (2009) suggests that in a downturn, companies would do well to provoke their customers. This would be the time to encourage them to pivot to a new strategy and platform. However, companies need to have a strategic approach to marketing when it comes to selling innovative products to industrial buyers (Schiavone & Simone, 2019).

Extant literature clearly articulates the importance of big data, value co-creation, customer relationship management and service selling. On review of literature on PM and industrial marketing, it is clear that IIoT based solutions are the future. Yet the impact of such solutions and how they will change existing business functions is not clear. Most of the literature on PM and marketing are based in the context of the global north while majority of business transactions in the next decades will take place in India and China. India, for example, has a large number of legacy systems and a large demographic dividend. This implies that even if PM becomes the industry standard their adoption in developing countries will follow a different pathway. There is scant literature that connects a solution like PM with industrial marketing particularly in the context of a developing country like India. This paper aims to fill this gap by contributing to the theory and practice of B2B marketing in relation to IIOT based solutions like PM.

OBJECTIVE AND METHODOLOGY

The objective of the paper is to show how IIoT innovations on PM will create significant change and challenge for business functions like buying, selling and servicing of equipments. Implementation of such innovations is not just a function of the R&D and financial resources but will change several aspects of business structure and models. To understand the functional level business challenges, we look at sales and marketing. Industrial marketing relies heavily on new technology, solution selling, customer relationship management- all implemented through a strong field sales force. Using the case of innovations in predictive maintenance by one of the world's largest air compressor companies, Elgi, the authors try to answer the research question: *How innovations in predictive maintenance will change industrial marketing?*

Since IIoT is an emerging phenomenon, an exploratory study using cases of the evolving phenomenon is a suitable method (Yin, 2017). The case candidate, Elgi, reflects a company that is a global leader both in revenue and technological innovation. The Indian setting of the case provides scope of theoretical generalization for developing economies which are one of the most important markets in size and scope. Data for the case is collected from the company website and author's own experience of working in academic projects with the company.

CASE STUDY

Coimbatore, India based Elgi Equipments Limited was established in 1960. The company had a history of technical innovation and collaboration as it started working with leading German manufacturers right from its inception. Over the last six decades Elgi had come to be known for its innovation and technological prowess. The company had many feats to its credit: first Indian company to manufacture rotary screw compressors, developing world's smallest screw air compressor and the first company to make oil free screw air compressor. In 2019, it was world's first compressor company outside japan to win the prestigious Deming prize for quality (Business Line, 2019). Led by second generation visionary entrepreneur Dr. Jairam Varadaraj, the company aimed to become world's second largest compressor manufacturer by 2027(Ravichandran, 2019).

Air compressors were Elgi's core product. Egli made more than 400 different types of compressors. They were used in almost all industries like cement, petrochemicals, mining, agriculture, healthcare and automotive. Ingersoll-Rand and Atlas Copco were global market leaders but Elgi was quickly catching up with the leaders. It acquired companies in US and started manufacturing in Europe. The company believed that three things were its strengths: low total cost of ownership, large range of products and superior based service support. It was in this last area that Elgi introduced its innovative predictive maintenance solution called *Air Alert*.

Services and spare parts sales were a low volume but high profit business for all compressor manufacturers like Elgi. This was also known as the equipment aftermarket. In the industrial air compressor market, spares included filters, lubricants and fluids as well as service kits. Elgi also provided accessories that would take care of contaminants like water, dirt and lubricants. The market for compressor maintenance was significantly different from that of compressor buying. While industrial customers preferred well-known brands for their compressor needs, when it came to services, customer behaviour varied. Some customers preferred to buy from original equipment manufacturers like Elgi, others preferred to buy low cost alternatives from the local market. It was this second category that Elgi wanted to focus on. The company knew that selling service had many advantages: helping customers prolong machine life, sales of spares, cross selling and upselling other equipments and a clear knowledge of the next level of demand.

Service selling at Elgi was done through its 100+ dealers around India. Dealers employed service engineers who would track service and spare requirement at the customer site. The head-office based corporate marketing team monitored sales at dealer points. The overall focus was to sell more annual maintenance contracts as well as high end services like compressor health check and audit of the compressed air. Elgi knew that services were critical in its goal to be world no. 2 by 2027 not because of the revenue generated but for the market penetration achieved. To this end it introduced the predictive maintenance system, *Air Alert*.

Air Alert

Air Alert (AA) was Elgi's IoT based, Industry 4.0 solution for predictive maintenance. IoT facilitated communication between machines located in remote sites. AA transmitted operational, analytical and strategic information about compressor health to Elgi. Operational data included machine running hours, uptime and usage. Analytical data included usage report and virtual logbook. Strategic data included pro-active failure alert and recommendations for energy optimization. AA was based on failure prediction models that integrated science, industry experience ad statistical analysis. It was a SIM card based system for compressors used for tracking condition parameters like pressure, operating temperature and air volume. The system, initially introduced for the screw compressor range would transmit data to the company. Data was available both through PCs and smartphone apps. The central team would monitor the data and advice customers on service intervals, energy efficiency, air consumption patterns and general equipment health. With real-time data getting transmitted and shared, after-sales service was expected to move up by quite a few notches. Customers were expected to benefit: proper communication from Elgi would help them understand the efficiency of the system, initiate service/ repair and avoid costly machine downtime. Elgi engineers would get critical data about performance of their compressors, quality of the system and insights about new product development. New models of compressors would benefit from data and analysis done through the AA system. The AA system positioned Elgi as a customer and quality centric company. Customer service was expected to improve with the AA launch. Going forward, it was expected that AA would also send some relevant information to customers directly. For a customer in the pipe industry, installation of the AA system led to a 40% savings in energy required for compressor operations. It was clear that AA provided a new paradigm for predictive maintenance. More significantly, it provided the company with insights in several different areas: product management, product development, customer service, after sales service and brand building.

DISCUSSION

The AA system introduced by Elgi revolutionized the compressor market. It provided a new, IIoT enabled way for customer engagement. The amount of data generated from thousands from compressors running across the country could provide Elgi a unique competitive advantage. However, the author sees several challenges in making this transition. While challenges are expected from several angles, this chapter focusses on an issue that directly or indirectly affects industrial marketing.

Diffusion of Innovation

Rogers (2010) had posited that innovations do not get established and accepted in a market in one go but there is a gradual spread of innovations. For PM solutions like the AA, all customers will not be enthused by the new level of service. The response to AA is likely to vary depending on the customers' approach to new products/innovations as well as the criticality of the usage. For example, large pharmaceutical companies or power plants have a low tolerance for product failures and AA could fit into their needs. Even within such sectors, private sector companies are likely to engage with predictive maintenance earlier compared to their government counterparts. This is primarily because the buying process (tendering) is more bureaucratic in public sector organizations as discussed below.

The varying rate of innovation acceptance will impact marketing particularly in terms of segmentation, targeting and positioning. Segmentation is the process of dividing a market into homogenous groups with similar needs. In industrial marketing, companies segment their customers based on size, buying process and buying characteristics. New service features like AA would require the marketing team to rethink their segmentation. They would need to find out those compressor customers who value machine efficiency and uptime more than others for different reasons. Within this segment, sellers need to focus on those who has a need but also the ability for pay for such PM solutions. This is known as targeting in marketing. Next, the IIoT company needs to create the image of a tech savvy innovative vendor in the mind of the targeted segment. This is known as positioning. For Elgi, the IIoT based PM system matched well with its reputation of being a global, technology led brand. For other companies in this industry, this may be difficult if, for example, they are known for other attributes like value pricing. Thus, all companies may not benefit equally from introduction of IIoT till it becomes an industry standard.

Organizational Buying Process

Unlike individual purchase decisions, organizational buying is process driven, complex and time taking. Bringing in a new innovation like AA may not translate to organizational buyers considering it for their purchase. In the case of individual consumers, an IoT enable refrigerator can be bought very quickly but an organization requires major changes in their procurement process to include features like PM. Organizations typically take three routes for purchase: 'Expression of Interest' is floated for new products and technologies, 'Request for Proposal' is invited for relatively better known products/ technologies and 'Notice Inviting tender' is floated for products where customers have fairly good amount of knowledge. For PM solutions, customers would ask for an 'Expression of Interest' and go through a series of vendor presentations before deciding if they want to move ahead with this new technology at all. This buying process has significant implication for the sellers.

Industrial sellers map their selling process through eight stages; problem recognition, need description, product specification, supplier search, acquisition and analysis of proposals, supplier selection, order routine selection and performance review. The idea behind this eight stage process is to engage with industrial customers and help them in their buying journey. When bringing a new feature like IIot based PM for machines, sellers need to engage early on the problem recognition stage. The target will be those customers who put a higher weightage on timely maintenance and would like to have maximum uptime. Additionally, such customers need to be early adopters (Rogers, 2010) of innovation like PM. There are two types of purchases in industrial marketing- purchase of new products/services and repeat purchases as well as two types of suppliers- existing and potential new entrants. As discussed above, PM will be more suitable when customers are considering a new buy. Similarly, existing vendors (in-suppliers) will be in a more favourable position to sell maintenance services like PM compared to potential suppliers (out-suppliers).

One of the key considerations for large industrial buyers is the total cost of ownership for a product or service. The total cost argument posits that rather than considering the purchase price, customers look at the cost of owning a product across its lifecycle. This will include acquisition cost but also possession cost (financing, storage) and usage cost (training, repair, replacement and disposal).For example, the costs of maintenance, training or transitioning to a new system are also important considerations for a buyer. As PM assures a lower cost of maintaining a system, customers favouring a lower life cycle cost are more likely to buy PM solutions. Summing up, innovation seeking customers looking for overall lower ownership cost are more likely to buy PM solutions from existing vendors while considering an upgrade or a new purchase. In our illustrative case, large, innovation seeking compressor customers are more likely to buy Air Alert systems from Elgi if they find better value in terms of cost and machine uptime.

Customer Relationship Management

Day (2000) talks about a relationship spectrum in industrial marketing indicating that customer relationship is not unique for all B2B transactions. Rather, it is a continuum that depends on availability of alternatives, supplier capacity, purchase importance, purchase complexity, information exchange and operational linkages (Cannon & Perrault, 1999). There are three types of purchases in the industrial market: transactional, value added and collaborative.

Transactional purchases see buyers buying on the best available price in the market and not interested in additional services or supplier brand equity. Value-added customers look for a good price but are also interested in additional services like training, warranty, spare part and field support. Collaborative buyers look at suppliers as their partners and try to finalize their product requirement based on information exchange, meetings and extensive discussions. One must note that between a given set of buyers and sellers, there may be multiple types of purchase patterns. For example, a buyer like Indian Railways may require different type of air compressors from Elgi for its factories, coach yards and stations and may exhibit varying patterns for such purchases. It is suggested (Day, 2000) that industrial sellers should tailor their customer relationship strategy based on the buying orientation of the customers. For transactional customers, sellers should try to reduce costly personal selling and rely more on interactive websites and e-commerce engines. On the other side of the spectrum, collaborative buyers should be provided extensive consultation on their needs through experienced key account managers.

How will customer relationship management change for IIoT based predictive maintenance solutions for a company like Elgi? Elgi sells to different segments that may be transaction, value or collaboration oriented. The company has to be careful in deciding which types of customers are more suitable for Air Alert type of PM solutions. This author argues that IIoT based PM provides a way for engaging with all types of customers but through different type of positioning for each type. For price oriented customers, Elgi can provide an equipment health dashboard through mobile apps for its customers. It need not spend on costly field based service engineers to visit the customers and find out/ convince them about servicing. PM can be a cost saving tool that the customer would appreciate having. For value added customers, particularly those looking for better uptime of their machines, PM can be a useful tool to sell. PM will reduce their total cost of ownership and give them more control over their equipments. Collaborative customers, where compressors are critical for operations, Elgi can present PM as a new innovation and strengthen its position as a technology leader in the industrial air compressor market. Summing up, IIoT based PM can be useful for industrial marketers across different customer segments as long as they are able to position the solution appropriately.

From a theoretical perspective, the paper contributes to the existing literature on selling and customer relationship management by pointing out that for tech enabled solutions like PM, one needs to consider three aspects of marketing- innovation, customer buying process and sales engagement through relationship management not in isolation but as a mutually reinforcing set of actions.

Overall, our findings have implication not only for air compressor markets but also for other industrial manufacturers. In the new reality of IIoT, sellers need to understand the changed needs of customers and align themselves with such needs. Take the example for product-service bundling. Most manufacturers sell product and related services (like maintenance) as a single solution. Many like Elgi is moving from selling only services. For example, instead of selling the compressor, it is now selling the service of providing compressed air on a (monthly/annual) subscription basis. Thus instead of customers buying capital equipments and their unique features, customers would now be involved in evaluating service qualities like manpower support, credibility of the service provider and variability of the service provided. This will require a new strategy for sales force organization, training and management. Marketing will play a new role that would involve more digital engagement and sales support functions are likely to grow. Customer engagement, in brief, will move significantly from the field sales force to supporting functions in the digital era.

FUTURE RESEARCH DIRECTION

IIoT based solutions will change the way business happens. Marketing of industrial products and services will undergo a significant change. While the author has focussed only on the industrial air compressor industry, it would be interesting to see the impact of IIoT based solutions in other industrial markets. While the case study method allows for theoretical generalizations, the impact on industrial services could be different. Future research can focus more specifically the implication for different categories of managers handling marketing functions- branding, product development, consumer behaviour, sales, service and distribution. For example, it would be interesting to explore how the attitudes of customers' factory managers change when most machines on the shop floor operate in a predictive maintenance environment.

One of the limitations of this chapter is the assumption that all customers are ready for PM type of solutions. In reality, AI and IIoT readiness depends on multiple factors at the institutional, structural and cognitive level. At the institutional level, the current norms and business environment will determine whether a customer is willing to buy PM solutions. For example, in government organizations, the unions may oppose such systems as it may make existing maintenance team redundant. At the structural level, the domain expertise of the buying organization will decide if new tech solutions will be bought and used. Even if they are, their usage and further adoption will depend on the level of user comfort. At the cognitive level, IIoT may be seen as alien, intrusive and also insecure. The current debate about the purchase of Chinese 5G technologies is a good example. More than technical quality, the impression that such technology can be mis-used is determining the purchase decisions. The author believes that there is significant scope of future research in all the above areas.

CONCLUSION

Innovation in IIoT will change industrial customer buying behaviour. Predictive maintenance will allow customers to engage with sellers in a different way. As a result of this innovation and change, the sales and marketing function needs to re-think their resource deployment and marketing strategy. Smart factories will rely on real time data interchange across all domains. This will also change the way marketing is seen and done in industrial markets. Customer need recognition will not simply be driven by human reactions and responses but digital, real-time machine learning. Sellers need to build capability for such digital response systems by building IT infrastructure and changing their branding and product development strategies.This chapter consider the case of a preventive maintenance solution, Air Alert, introduced by Elgi, one of the largest compressor manufacturers in the world. Using the illustrative case, it shows that IIoT innovations will be driven by the adoption capability of customers and not necessarily their size or order value. Second, the industrial buying process will determine the market success of such solutions. Customers valuing total cost of ownership are more likely to rely on large, reputed suppliers when they shift to IIoT solutions. Third, smart industrial marketers can create a go-to-market strategy that satisfies all types of customers- transactional, value-oriented and collaborative. They will need to position PM solutions differently across different segments. Keeping these factors in mind will help companies gain competitive advantage when it comes to selling IIoT based products and services.

ACKNOWLEDGMENT

All data in this chapter is from public sources. I am grateful to the management of Elgi Equipments Limited for giving me an opportunity to study their organization.

REFERENCES

Bousdekis, A., Apostolou, D., & Mentzas, G. (2019). Predictive Maintenance in the 4th Industrial Revolution: Benefits, Business Opportunities, and Managerial Implications. *IEEE Engineering Management Review*, *48*(1), 57–62. doi:10.1109/EMR.2019.2958037

BusinessLine. (2019, October 1). Elgi Wins Deming Prize. *BusinessLine*. https://www.thehindubusinessline.com/companies/elgi-wins-2019-deming-prize/article29567774.ece

Cannon, J. P., & Perreault, W. D. Jr. (1999). Buyer–seller relationships in business markets. *JMR, Journal of Marketing Research*, *36*(4), 439–460.

Colter, T., Guan, M., Mahdavian, M., Razzaq, S., & Schneider, J. (2018, January 4). *What the future science of B2B sales growth looks like*. McKinsey & Company. https://www.mckinsey.com/business-functions/marketing-and-sales/our-insights/what-the-future-science-of-b2b-sales-growth-looks-like

Cova, B., & Salle, R. (2008). Marketing solutions in accordance with the SD logic: Co-creating value with customer network actors. *Industrial Marketing Management, 37*(3), 270–277. doi:10.1016/j.indmarman.2007.07.005

Davenport, T., Guha, A., Grewal, D., & Bressgott, T. (2020). How artificial intelligence will change the future of marketing. *Journal of the Academy of Marketing Science, 48*(1), 24–42. doi:10.100711747-019-00696-0

Day, G. S. (2000). Managing market relationships. *Journal of the Academy of Marketing Science, 28*(1), 24–30. doi:10.1177/0092070300281003

Gavin, R., Harrison, L., Plotkin, C., Spillecke, D., & Stanley, J. (2020, April 30). *The B2B digital inflection point: How sales have changed during COVID-19.* McKinsey & Company. https://www.mckinsey.com/business-functions/marketing-and-sales/our-insights/the-b2b-digital-inflection-point-how-sales-have-changed-during-covid-19

Goyal, M., Hancock, M. Q., & Hatami, H. (2012). Selling into micromarkets. *Harvard Business Review, 90*(7-8), 79–86.

He, Y., Gu, C., Chen, Z., & Han, X. (2017). Integrated predictive maintenance strategy for manufacturing systems by combining quality control and mission reliability analysis. *International Journal of Production Research, 55*(19), 5841–5862. doi:10.1080/00207543.2017.1346843

Hein, A., Weking, J., Schreieck, M., Wiesche, M., Böhm, M., & Krcmar, H. (2019). Value co-creation practices in business-to-business platform ecosystems. *Electronic Markets, 29*(3), 503–518. doi:10.100712525-019-00337-y

Lay, P., Hewlin, T., & Moore, G. (2009). In a downturn, provoke your customers. *Harvard Business Review, 87*(3), 48–56.

Lopes, I. S., Figueiredo, M. C., & Sá, V. (2020). Criticality evaluation to support maintenance management of manufacturing systems. *International Journal of Industrial Engineering and Management, 11*(1), 3–18. doi:10.24867/IJIEM-2020-1-248

Markets and Markets. (2020, June). *Predictive Maintenance Market by Component (Solutions and Services), Deployment Mode, Organization Size, Vertical (Government and Defense, Manufacturing, Energy and Utilities, Transportation and Logistics), and Region - Global Forecast to 2025.* MarketsandMarkets. https://www.marketsandmarkets.com/Market-Reports/operational-predictive-maintenance-market-8656856.html

Matthyssens, P., & Vandenbempt, K. (1998). Creating competitive advantage in industrial services. *Journal of Business and Industrial Marketing, 13*(4/5), 339–355. doi:10.1108/08858629810226654

Mittal, V., Sarkees, M., & Murshed, F. (2008). The right way to manage unprofitable customers. *Harvard Business Review, 86*(4), 94–103.

Mobley, R. K. (2002). *An Introduction to Predictive Maintenance* (2nd ed.). Butterworth-Heinemann., doi:10.1016/B978-0-7506-7531-4.X5000-3

Muylle, S., Dawar, N., & Rangarajan, D. (2012). B2B brand architecture. *California Management Review*, *54*(2), 58–71. doi:10.1525/cmr.2012.54.2.58

Narayandas, D. (2005). Building loyalty in business markets. *Harvard Business Review*, *83*(9), 131–139. PMID:16171217

Overgoor, G., Chica, M., Rand, W., & Weishampel, A. (2019). Letting the computers take over: Using AI to solve marketing problems. *California Management Review*, *61*(4), 156–185. doi:10.1177/0008125619859318

Ravichandran, R. (2019, November 12). *Elgi Equipments to focus more on global markets, eyes No 2 slot*. Financial Express. https://www.financialexpress.com/industry/elgi-equipments-to-focus-more-on-global-markets-eyes-no-2-slot/1761786/

Reinartz, W., & Ulaga, W. (2008). How to sell services more profitably. *Harvard Business Review*, *86*(5), 90–96. PMID:18543811

Richards, K. A., & Jones, E. (2008). Customer relationship management: Finding value drivers. *Industrial Marketing Management*, *37*(2), 120–130. doi:10.1016/j.indmarman.2006.08.005

Rogers, E. M. (2010). *Diffusion of Innovations*. The Free Press.

Ruiz-Sarmiento, J. R., Monroy, J., Moreno, F. A., Galindo, C., Bonelo, J. M., & Gonzalez-Jimenez, J. (2020). A predictive model for the maintenance of industrial machinery in the context of industry 4.0. *Engineering Applications of Artificial Intelligence*, *87*, 103289. doi:10.1016/j.engappai.2019.103289

Sahal, R., Breslin, J. G., & Ali, M. I. (2020). Big data and stream processing platforms for Industry 4.0 requirements mapping for a predictive maintenance use case. *Journal of Manufacturing Systems*, *54*, 138–151. doi:10.1016/j.jmsy.2019.11.004

Schiavone, F., & Simoni, M. (2019). Strategic marketing approaches for the diffusion of innovation in highly regulated industrial markets: The value of market access. *Journal of Business and Industrial Marketing*, *34*(7), 1606–1618. doi:10.1108/JBIM-08-2018-0232

Shankar, V., Berry, L. L., & Dotzel, T. (2009). A practical guide to combining products services. *Harvard Business Review*, *87*(11), 94–99.

Statista. (2020). *Industrial Internet of Things market size worldwide from 2017 to 2025*. Statista. https://www.statista.com/statistics/611004/global-industrial-internet-of-things-market-size/

Wise, R., & Morrison, D. (2000). Beyond the exchange—the future of B2B. *Harvard Business Review*, *78*(6), 86–96. PMID:11184979

Yin, R. K. (2017). *Case study research and applications: Design and methods*. Sage Publications.

Yu, W., Dillon, T., Mostafa, F., Rahayu, W., & Liu, Y. (2019). A global manufacturing big data ecosystem for fault detection in predictive maintenance. *IEEE Transactions on Industrial Informatics*, *16*(1), 183–192. doi:10.1109/TII.2019.2915846

KEY TERMS AND DEFINITIONS

Air Compressors: Mechanical devices that increased air pressure by compressing it.

Customer Relationship Management: Processes and tools used by companies to develop and manage relationships with their customers in order to provide them more value.

Diffusion of Innovation: The process by which innovations spread across a market or customer groups.

Industrial/B2B Marketing: Marketing to commercial organizations, governments and non-profit organizations.

Organizational Buying: The process followed by organizations during their procurement or purchase.

Predictive Maintenance: The concept and practice of managing equipment maintenance with the help of real-time data captured through physical devices like chip sets, sensors and processors.

Chapter 17
Enhancing In-Service Tank Maintenance Through Industrial Internet of Things:
A Case for Acoustic Emission Tank Inspection

Moses Gwaindepi
Department of Industrial and Mechatronics Engineering, Harare, Zimbabwe

Tawanda Mushiri
https://orcid.org/0000-0003-2562-2028
University of Zimbabwe, Zimbabwe

ABSTRACT

In the area of tank inspections across the industry, robots were introduced to replace human inspectors in selected operations. The technological gap in adoption of similar technologies by Zimbabwe's bulk fuel storage tanks operators motivated this research. The industry's current NDT practices were investigated, costs and inconveniences were identified, and improvements were explored. Operators of bulk fuel facilities and companies providing tank inspection services were engaged to establish the reasons for the gaps in technological assimilation. Emerging global technologies that enable in-service inspections were identified and their applicability to Zimbabwe's bulk fuel facilities was investigated. A combination of crawler based ultrasonic thickness tests for tank shells, and acoustic emission in-service tank bottom testing was observed to be the most convenient and relevant in-service tank inspection method for Zimbabwe's bulk fuel storage tanks industry. Internet-based remote connectivity and control was considered for data compilation, analysis, storage, and reporting.

DOI: 10.4018/978-1-7998-3375-8.ch017

INTRODUCTION

Bulk petroleum storage and handling is crucial in smoothening fluctuations that may occur between supply and demand. Providing extra storage space ensures that excess supply can be warehoused when supply exceeds demand and when supply falls below consumption there is draw down from the excess stock previously built-up. The bulk fuel facilities' integrity and healthiness is very critical in ensuring the facilities' readiness to service the industry on demand. Maintenance of the facilities' critical equipment such as the tanks is important. The capacities of the tanks will result in significant environmental contamination and has the potential of catastrophic consequences should any mishap occur. Strategic maintenance management is important in ensuring the facilities are ready to serve when required. Involvement of the organisations' top management ensures management commits and partakes in making key decisions for their organisation's long-term equipment maintenance. This is needful for successful implementation and monitoring of the adopted maintenance programs. Top management commitment is important where decisions to be made involve drastic changes to the usual way of doing business, such as is the case with the adoption of the new paradigm shift in industry, the industry internet of things (IIoT).

The Industrial Internet of Things (IIoT) created several opportunities which organisations can conveniently adopt in their quest for improved operational efficiency as well as efficient equipment operation and maintenance (Muhonen, 2015). Key among the benefits to be drawn from IIoT is real-time data analytics, machine to machine communication, autonomous machine operations, and the ability to instantly alert stakeholders on equipment condition as well as prescribing possible action plans on predicted adverse equipment conditions (IBM Corporation, 2016). This is illustrated in figure 1.

BACKGROUND

An Overview of Zimbabwe's Bulk Fuel Storage Industry

Zimbabwe's bulk fuel industry has for many years been dominated by a few players that own bulk storage depots with capacities of at least 1,000 cubic metres per depot. These depots play a crucial role of creating a buffer to iron out fluctuations between supply and demand (*Gujarat is vulnerable to major manmade chemical disaster, especially in the aftermath of natural catastrophe, says GSDMA report*, n.d.). Other players making up the supply chain reach out the markets through service stations and renting bulk storages from the few depot owners. Before Zimbabwe's independence in 1980, the industry comprised of five international oil companies which operated the depots that were in major towns such as Harare, Bulawayo, Mutare, Gweru, Masvingo, Chinhoyi, Chiredzi and Beit Bridge. The five were British Petroleum (BP), Shell, Caltex, Mobil and Total. These five were also part of the industry owned joint venture that made up the Central African Petroleum Refinery (CAPREF) which ran the Feruka refinery in Mutare. In addition to the five companies, the joint venture also included American Independent Oil Company and Kuwait National Petroleum Company (Field & Interim, 2008).

Post-independence, Feruka depot was acquired by the Government of Zimbabwe, which also run depots in Harare, Bulawayo and Beitbridge through the state-owned National Oil Infrastructure Company of Zimbabwe (NOIC). BP merged with Shell to form BP & Shell Marketing Services (BPSMS) which was later disposed, together with its depots first to Masawara which later disposed the business to Zuva Petroleum. Mobil sold its depots, to Total while Caltex was sold to Engen, and lately to Vivo Energy. To date the operational players running bulk storage petroleum facilities are NOIC, Zuva, Total and Vivo Energy. Their functional storage depots are in the following towns:

1. National Oil Infrastructure Company of Zimbabwe (NOIC) and its subsidiary Petro-Zim Line. This runs depots in Mutare, Harare, Beit Bridge and Bulawayo
2. Zuva Petroleum, with depots in Harare, Chiredzi, Masvingo, Gweru and Bulawayo
3. Total Zimbabwe with depots in Harare, Gweru and Bulawayo
4. Engen (Vivo Energy) with storages in Harare, Gweru, Masvingo and Bulawayo

All the other players in the industry serve their customers through service stations or by renting secondary storage from the outlined companies owning bulk storage depots. With the exception of NOIC which owns underground storage facilities at its Mabvuku depot, the rest of the depots' storage facilities are aboveground steel tanks. Most of the dominant companies' competitive advantage can be attributed to the high set-up costs associated with fabrication of large storage tanks and the stringent environmental considerations to be fulfilled in the set-up of the storage facilities (Velmurugan & Dhingra, 2014). With the exception of Greenfuels' Chisumbanje ethanol depot, NOIC's Mabvuku depot and PZL's Msasa depots, the rest of the storage depots' facilities were constructed before Zimbabwe's independence in 1980. Based on this narrative, the bulk of the industry's storage tanks and presumably other equipment in use are over 40 years in service and are therefore approaching obsolescence. (SAZ 913: Part 1, 2012) reported that the changes in ownership of Zimbabwe's fuel facilities was marked with change in management and operational staff with some of the new players having little or no knowledge of the industry's complexities and demands in infrastructure maintenance. An understanding of the condition of the depots' critical equipment such as tanks and their maintenance demands is crucial in averting the dangers associated with their failures. A study of the maintenance of the critical depot infrastructure is therefore necessary.

Overview of Bulk Petroleum Depots Operations

The key functions at the bulk fuel storage facilities include the receipt of petroleum products from road or rail tankers, inter-depot pipeline transfers which are applicable specifically for some of NOIC and PZL's depots, storage and handling of the petroleum products and the bulk re-distribution of the petroleum products through road and rail tankers as well as pipeline transfers for some of NOIC and PZL's depots. From the operational narrative the companies' business can be classified as bulk fuel logistics. Figure 2 highlights the key operational set-up at the bulk fuel storage and handling depots.

PROBLEM STATEMENT

Time based out-of-service tank inspections have been proven to be fraught with cost inefficiencies, as some tanks would be subjected to costly and risky processes associated with the inspections when their conditions does not warranty such inspection methods. Zimbabwe's bulk fuel industry facilities' operators have for long been relying on such NDTs for their tanks.

From Predictive to Prescriptive Maintenance Through Industrial Internet of Things (IIoT)

The Industrial Internet of Things (IIoT) is set to introduce a phenomenal capability and strategy in maintenance where the equipment would suggest how it could be fixed when a potential failure is detected. Prescriptive maintenance is the in-coming maintenance strategy where equipment will suggest how a detected potential failure would be fixed including when actual failure will manifest if the equipment continues operating at a given rate after detection of potential failure (Data & Management, n.d.). Prescriptive maintenance will ride on enhanced Artificial Intelligence (AI) and machine learning. Figure 3 highlights the progression from reactive to prescriptive maintenance. Figure 3 (Bryan Chritiansen, 2020) presents a graphical description of the evolution of maintenance strategies.

OBJECTIVES

The research seeks to mitigate on the cost inefficiencies and safety hazards associated with out-of-service tank inspections for Zimbabwe's bulk fuel facilities through the following steps:

1. Review the global technological developments in tank NDTs and compare with the technologies applied in Zimbabwe's bulk fuel tanks maintenance.
2. Establish reasons for the gap(s) and consider its closure.
3. Formulate an IIoT enabled framework for enhancing in-service tank NDTs.

JUSTIFICATION

Avoidance of unjust costs is critical for every business' competitiveness. Scanning the environment and taking advantage of available solutions to problems peculiar in one's industry makes good business acumen. In 2014 analysts at Gartner estimated that the number of IoT connected devices would increase from 0,9 billion as at 2009 figures to 26 billion units by 2020 (Lee & Lee, 2015). (Mercer, 2019)'s report in the May 2019 issue of Strategy Analytics indicated that the number of IoT connected devices as at the end of 2018 was nearly 22 billion and could reach 50 billion by the end of 2020. With such an exponential growth in the deployment of IoT, users' challenges will be on deciding how to use the interconnected devices as opposed to whether they should get connected to the internet. The trend in IIoT connectivity is happening against decreasing costs for sensors and associated IIoT technologies, thereby making the adoption of IIoT inevitable.

LITERATURE REVIEW

This gives an overview of the research's key words, namely *storage tanks maintenance, Industrial Internet of Things (IIoT)* and *Zimbabwe's bulk fuels industry.* The literature review traces the cross linkage of the outlined key words in the perspective of the research. A review of strategic management was also covered as it is critical to secure top management commitment in any key decisions within an organisation. Convincing top management to adopt and drive any maintenance agenda has always been cited as a crucial step in the adoption and implementation of new ideas or culture in maintenance(Eti, Ogaji, & Probert, 2006). A lot of research has been conducted about IIoT and there is a considerable record of publications on the subject. (Bhati, 2018) stated that there is no unique definition for IIoT that is generally accepted as the subject is in its early stages. However, there are generally repeated key terms in many writers' definitions of IIoT.

(Karmakar, Dey, Baral, Chowdhury, & Rehan, 2019) described IIoT as the industrial extension of IoT which they stated to be commercial. The same terms outlined to be generally adopted in the definitions of IIoT are common in the description of the fourth industrial revolution. This might explain the reason why some writers maintain the interchangeability of IIoT and industry 4.0. Whether one is for or against the interchangeability between IIoT and industry 4.0, a review of the literature pertaining the two show that appreciation of the industrial revolutions gives a clearer appreciation and comprehension of IIoT.

Some scholars agree to the interchangeability between the terms *industrial internet of things* and *industry 4.0.* while others argue that the two are related but not interchangeable. (Arnold, Kiel, & Voigt, 2020), (Karmakar et al., 2019), (Endres, Indulska, Broser, Endres, & Indulska, 2019) and (Hartmann & Halecker, 2015) stated that the terms were interchangeable. (Khan et al., 2020) and (Sisinni, Han, Jennehag, & Gidlund, 2018) stated the terms were not interchangeable. (Younan, Houssein, Elhoseny, & Ali, 2019) and (Sisinni et al., 2018) introduced Cyber Physical Systems (CPS) in their explanation of the relationship between industry 4.0 and IIoT. They indicated that industry 4.0 could be viewed as a subset of IIoT which in turn was classified as a subset of IoT. This was demonstrated in a diagram which is reproduced below for clarity, Figure 4.

IIoT AND INDUSTRIAL AUTOMATION AND CONTROL SYSTEMS

Several technological developments and inventions that led to the industrial revolutions sought to improve on productivity, quality, safety and to minimise on human effort through mechanisation and automation. (Janssen, Donker, Brumby, & Kun, 2019) traced process automation to the earliest human efforts of task lightening and gave examples of the Mayans' water transportation through aqueducts, Adam Smith's hatpin factory, and Henry Ford's assembly line. This is evidence that automation in strict senses was not all about computers or machines but refers to the manner in which tasks were executed in figure 5.

IIoT ENABLING TECHNOLOGIES

There are several constraints which inhibits the implementation of IIoT. Some countries such as Germany, China, United States of America, France, Austria, Spain, Italy, Belgium, Sweden and Japan have been actively participating in the researches for IIoT implementation at government levels (Ślusarczyk, 2018).

Internet of Things (IoT)

(Sisinni et al., 2018)'s diagrammatic presentation of the relationship between IoT and IIoT which was reproduced as figure 9 expressed IIoT as a subset of IoT. However, (Karmakar et al., 2019) suggested that IoT was the commercial version of the phenomenon of enhanced intelligent machine to machine connectivity over the internet while IIoT is reserved for the industrial version of the phenomenon. Both (Sisinni et al., 2018) and (Karmakar et al., 2019)'s suggestions support the position that IoT and IIoT relies on basically the same technologies with the main difference being observed in the area of their application.

IoT Architecture

(Silva, Khan, & Han, 2017), (C. Zhong, Zhu, & Huang, 2017) and (Rahimi, Zibaeenejad, & Safavi, 2018) confirmed the existence of many versions of the IoT architect but concurred that the three layer architecture was the most basic and widely used format. The writers used self-descriptive terminology to identify the layers. The first and lowest level is termed the *perception/sensor layer.* It comprises of a wide range of interconnected and uniquely identifiable heterogeneous devices such as sensors, actuators, cameras and other objects which serve to collect data from the environment. The devices can communicate with each other. The second level is the *network/transmission layer,* which is viewed as the brains of an IoT architecture. It is responsible for collection and processing of data from the perception layer. It makes use of wired or wireless and satellite technologies to integrate the devices into a seamless collaborative single network. The third level is called the *application/business layer*. The application layer links the IoT applications with the users. The application layer ensures the realisation of IoT through analysing and processing the information from the lower levels.

Blockchain

Blockchain technology was described as a form of a digital smart contract which the parties performed according to the promises and over an agreed digital protocol (Teslya & Ryabchikov, 2018) outlined that the blockchain technology had been in use to meet the five requirements above in particularly in cryptocurrency and bitcoin. (Younan et al., 2019) indicated that blockchain technology was also finding application in the health-care sector, finance industry and in government.

Cloud Computing

(Zhou, Liu, & Zhou, 2016) states that cloud computing offers high performance at a low cost. This will justify its widespread deployment as cost is often a hindrance in the adoption of certain technologies.

Big Data Analytics

(Ahmed et al., 2017) emphasized that big data's usefulness in IIoT is underpinned by analytics capabilities. Without matching capabilities in big data analytics, the exponential data generation and accumulation will present a significant challenge to the relevance of machine to machine communication. (Younan et al., 2019) observes that in future IoT enabled machines will autonomously make their decisions depending on the decisions and actions of other machines connected to them. Such scenarios might be viewed as advocating for completely unmanned systems, yet they actually require more attention. (R. Y. Zhong, Xu, Klotz, & Newman, 2017) observes that there will be a reduction in human involvement in intelligent manufacturing systems due to the use of artificial intelligence (AI).

CORROSION DETECTION IN TANKS

Corrosion is the major threat to tank integrity once the tank is satisfactorily commissioned. Tank owners or operators need to be constantly assured of the condition of their tanks and this can be achieved through various non-destructive tests. (Martin, 2013) highlighted a critical aspect regarding corrosion that tank owners and operators should always be conscious of as they decide on how to maintain their tanks.

NON-DESTRUCTIVE TECHNOLOGIES IN TANK INSPECTION

The integrity of the tanks should be guaranteed while the tank remains useful. Many tests can be conducted on the tanks as NDTs and give advice to the tank owners on the condition and usability of the tanks. The NDTs will seek to inform the tank owners or operators on a variety of parameters which can be summed as pointing the tank's usability and remaining life. (Pesce & Standards, 2011) listed that the common parameters reported on include: the tank's verticality, out-of-roundness, welds integrity, plates thickness, corrosion extend, and determination of any cracks on any of the tank structure.. Various standards can be adopted for tank NDTs. Many tank owners and inspection service providers are familiar with API 653. (Pesce & Standards, 2011) states that API 653 defines requirements for conducting the tests and the criteria for ascertaining the inspection intervals. The various tests apply to any one or more sections of the tank structure including the roof, shell, and tank bottom. The following sections outline some of the methods that ca be adopted in tank NDTs. The other methods are visual testing, magnetic flux leakage, ultrasonic test, saturated low frequency eddy currents, magnetic particle inspection, liquid penetrant and vacuum box test. The Vacuum Box technique is used to detect leaks on weld seams. It is based on the creation of a vacuum on the test section and introducing a detergent to the section under vacuum. Figure 6 demonstrates the use of a Vacuum Box in leak detection. Radiography NDTs are usually applied in determining continuity or flaws on new welds (Mgonja, 2017). It is considered the most reliable NDT to ascertain weld integrity as it can produce a photographic view of the internal material being tested. This capability is unique and makes radiography superior in guaranteeing the integrity of new welds.

DEVELOPMENTS IN BULK STORAGE TANK INSPECTIONS

Operators using aboveground storage tanks (AST) use some standard or regulation in determining the frequencies for the tank inspections. Some standards such as API 653 and EEMUA recommends that inspections should be conducted every 3 – 10 years depending on the condition and estimated remaining life at the time of last inspection (Pearson, Mason, & Priewald, n.d.). In some countries the frequency of inspection is enforced through a regulatory body for atmospheric tanks and pressure vessels. In Zimbabwe pressure vessels and boilers are regulated through Chapter 14:08 Factories and Works Act (Pressure-Vessel) Regulations, 1976 while there is no regulation for atmospheric storage tanks. Operators use best practice in ascertaining the intervals between inspections and recommendations from reputable international standards such as API 653 are usually adopted.

Tank Shell Inspections

All the tank shell inspections outlined in subsequent sections can be conducted whilst the tank is in service. Cost and safety considerations tend to have been the main drivers for developments in tank shell inspection technologies. Tank walls can be inspected while the tank is in-service. Optimisation considerations for tank wall inspections were focused on safety and cost considerations. The earliest method which is still being used by some tank farm operators involve erecting scaffolding round the tank for the inspector to get access to target inspection points on the tank wall. Pictorial presentations of Figures 7-10 highlight some of the complexities as well as cost and safety considerations that influenced changes in tank inspection methods for tank shells.

The crane access method will be costly on power requirements for running the crane. Figure 19 and 20 shows further developments in tank shell inspections by adoption of robot technology. This is safer than all earlier versions from Figure 7 to Figure 10. Figure 9 shows a remote-controlled crawler that can scale the wall while measuring the wall thicknesses and any other tank integrity considerations. The crawler can be fitted with any of the desired measuring instrument though ultrasonic thickness testing equipment has been mostly used with the crawler. The set-up time is much faster, and the results are also availed much faster when compared to the earlier methods. Figure 11 makes use of drone technology to access the tank shell sections. A drone fitted with sensors can be used to access and conduct spot measures on selected points of the tanks. The method is more applicable for spot checks where suspicion of potential failure would have been raised from visible inspections. Figure 12 shows tank shell inspection.

Out of Service Tank Bottom Inspections

(Feng, Yang, & Huang, 2019) asserts that tank bottom corrosion is the main threat in tank corrosion induced defects. It is the major worry for tank owners and is the main reason for out-of-service inspections. Underside or soil side corrosion is another serious corrosion threat common with the tank bottom. Use of corrosion inhibitors at installation or adoption of an inspection method that takes care of both top-side and underside corrosion is critical for all tank operators. Cathodic protection is an example of tank bottom corrosion inhibitor. There is still need to conduct periodic inspections even with corrosion inhibitors in place. Magnetic flux leakage test using floor scanners is one of the most reliable methods to conduct tank floor inspections. Figure 13 shows an inspector conducting an MFL test. Additional testing methods will be required to ascertain the values pertinent to the flaw such as the plate thickness which can be conducted through UT tests or the presence of a crack or leak which might be established through MPI or a Vacuum Box respectively.

In-Service Tank Floor Inspections Techniques

(Xiong, Kang, Lin, & Sun, 2009) listed three tank inspection methods as having been developed to enable in-service tank inspections at the time of their research. The three were based on mobile robot technology, ultrasonic guided wave technology and acoustic emission technology. The writers ranked the three methods in terms of their applicability and concluded that AE was generally applicable as an in-service tank inspection technique while the other two had limitations. The major limitation for ultrasonic guided wave was that it was applicable to tank diameters not exceeding 15 metres while security concerns were cited as the limitations for robotic inspections.

Acoustic Emission Technology

Acoustic emission has been used as an inspection technology since the 1970s (Taylor, Liu, Guo, Hu, & Guo, n.d.). It is based on the release of energy in the form of elastic waves when a material is subjected to some stress. The waves can propagate through the material and a transducer conveniently placed on the material can pick the waves and convert them to electric pulses which can be analysed on time or frequency basis. Figure 14 shows the characteristics of an AE pulse. In the case of aboveground storage tanks monitoring, sensors can then be conveniently configured to measure and analyse or report on any of the characteristics if the pulse is to be used as a measure of corrosion or leak detection.

The diagrammatic description of figure 15 shows the presence of background noise which can distort the tests if not taken care of. (Taylor et al., n.d.) states that AE has been widely applied for integrity tests of pressure pipes and pressure vessels in America, Europe, China and Japan. Further development has seen the technique being applied in the testing of atmospheric storage tanks.

The number of sensors required for monitoring a tank surface depends on the diameter of the tank. (Lackner and Tscheliesnig, 2007 recommended 6 sensors for a 25 m diameter tank and up to 24 sensors for a 100 m diameter tank. (Lackner & Tscheliesnig, 2007) stated that the standard distance between sensors along the shell circumference was 15 m as shown in figure 16.

Research Gap

Zimbabwe's fuel depot operators and the local inspection service providers continue to rely on technology that demand out-of-service inspection especially when the scope of inspection covers the tank bottom. Where such inspections coincide with a busy spell there often is a clash of interest within the organisation with operations demanding continued availability of the tanks and maintenance requiring the tanks to be decommissioned for inspection. The situation is exacerbated if the inspections reveal that the tank will imminently fail unless repairs are conducted. Adoption of a tank inspection method that enables in-service inspections will ensure the tank continues in service while its integrity is guaranteed. This has significant economic benefits as out-of-service inspections will disrupt the tank availability which is the tank owners' source of income.

MATERIALS AND METHODS

Methodology

The research's diversity led to the adoption of both qualitative and quantitative data to ensure a thorough and complete understanding of the problem. This approach followed (Migiro & Magangi, 2011)'s narrative which cited (Creswell, 2005) in asserting that the mixed approach as a research methodology ensures a complete understanding of the problem. (Migiro & Magangi, 2011) also supported the position regarding the completeness of data collection from the mixed approach in research through their explanation that quantitative data was more representative and generalised while qualitative data was in-depth and contextual.

Qualitative Data Collection

(Sutton & Austin, 2015), (Neville, Adams, & Cook, 2016) and (Moser & Korstjens, 2018) listed interviews, focus group discussions and participant observations as the main methods of collecting data under qualitative research.

Sample Size and Members/Participants

Zimbabwe's bulk fuel industry comprises of seven companies if ethanol and jet A1 are also considered. However, the researcher excluded ethanol and jet A1 from the scope of the study, leaving the population of Zimbabwe's bulk fuel industry for the purposes of this research to four companies. With a total of four organisations forming the industry for bulk fuel facilities operators the researcher grouped all four as the research population. However, further analysis of the supply chain led to further categorisation with NOIC and its subsidiary, PZL being grouped on their own as pipeline, road and rail connected depots while Zuva, Vivo Energy and Total were set as a second group connected through road and rail only. The researcher classified NOIC and PZL as pipeline depots while the other group of Zuva, Total and Vivo Energy was classified as road/rail depots. The Researcher also had a third group of respondents which was made of Zimbabwean companies providing tank inspection services. The group omprised of Gammatec, Veritec, Tepo Industrial, SAZ, NDT Solutions Inspection Services and another. The Researchers classified this group as NDT Inspectors.

The consolidated guidelines (Greeneltch, Haudenschild, Keegan, & Shi, 2004) suggests adoption of case studies for a sample size with 3 to 5 participants. Following the suggestion of case studies, the researcher also noted the applicability and relevance of phenomenology analysis approach (Alase, 2017) where there is need to gather data about the research participants' lived experiences. The consideration of adopting a case study and the phenomenology analysis approaches addressed concerns of the research designs or methods. The research design is recorded above to have been informed by the sample size though the actual size was not clearly spelt out. Interviews were therefore deliberately limited to at most three members per group.

The ultimate number of respondents interviewed was deliberately limited to at most three people per category of respondents. The research sought to use a manageable and representative number of respondents. The researcher avoided over representation in the sampling of interviewees as this could have resulted in unjustifiable over saturation which subsequently could have rendered the inference of the empirical observation to the theoretical deductions being viewed as "naïve inductivism" (Verma, Mendelsohn, & Bernstein, 1974). The researcher's interview questions were also carefully drafted and semi-structured to satisfy (Verma et al., 1974)'s observation that phenomenon rather than excessive quantitative data should be the focus in inductive theorems. The distribution and identity of respondents was as presented in Table 1.

Interviews

(Sutton & Austin, 2015) and (Moser & Korstjens, 2018) listed grounded theory and phenomenology as the main interview based methodologies in qualitative research. They described grounded theory as being the data collection method where interviews are conducted with focus groups or where data is gathered from interactions with such groups. Phenomenology use the same approach of interviews and interac-

Table 1. Distribution and identity of respondents

Category	Name of Company	Number of People Interviewed
Pipeline Depot Operators	NOIC	2
	PZL	1
Road/Rail Depot Operators	Zuva	1
	Vivo Energy	2
	Total	0
NDT service providers	Veritec	0
	Gammatec	1
	Tepo Industrial	1
	SAZ	0
	NDT Solutions	1
	Other	0

tions with participants but is mainly concerned with individuals as opposed to focus groups. Several factors contributed to limitation of the research to phenomenological approach of data collection. The main reason was the travel limitations and "social-distancing" restrictions imposed to curb the spread of the covid-19 pandemic. This further limited all the interviews with participants from outside NOIC to be via online based mobile phone platforms. Instead of this being a setback in the research, this was an early indicator to the researcher that IoT was already in use and was dictating significant changes to business conduct.

The research questions were the same for all representatives in the same category though follow-up questions would differ depending on the extent to which the initial response would address the researcher's question. Appendix 1 is a set of questions that were asked respondents of both pipeline and road/rail depot operators while Appendix 2 is the set of questions that was asked respondents representing companies providing tank inspection services.

Observation

(Alase, 2017), (Sutton & Austin, 2015) and (Moser & Korstjens, 2018) described the qualitative research method of gathering data through direct observation of participants as ethnography. In view of the four companies that had been classified as making up Zimbabwe's bulk fuel industry according to the researcher's delimitation, the researcher believes to have applied this method in gathering some data on the companies. Secondary data was gathered from the companies' available records and publications pertinent to the research area. The companies' websites, safety policies, quality policies, customer charters and other publications readily available in the public domain were used to get insight into some of the organisations' guiding information for the research.

Quantitative Data Collection

Questionnaires were issued to randomly selected respondents from both the pipeline and the road/rail depot operators. The questions were motivated by the responses obtained from interviews and the researcher sought to see if there was corroboration of the results. The target group for the questionnaires was different from the interviewed levels. Questionnaires targeted operational level staff in maintenance, operations and any other departments. Appendix 3 is a copy of the questionnaires. The data was gathered from maintenance, operations and other departments in the bulk fuel storage industry and was meant to compare with the qualitative data obtained from the interviews of top management in the industry. Fifty questionnaires were distributed, ten in each of the five companies owning the bulk fuel facilities in the industry, i.e, NOIC, PZL, Zuva Petroleum, Vivo Energy and Total Zimbabwe. The respondents in each company were randomly selected. Within each company the ten questionnaires were distributed through one contact person. In most of the cases the contact person was the Depot Manager or any senior person at one of the company's depots. The request was to ensure there would be equal distribution of three or four questionnaires in the company's operations and maintenance departments with the remainder being given to any of the other support departments. Respondents were advised not to disclose their identities or departments. Completed questionnaires were collected through the same contact person. The researcher deliberately avoided being in direct contact with the respondents to avoid bias in the interpretation of the results.

RESULTS AND DISCUSSION

Critical Equipment for Zimbabwe's Bulk Fuel Facilities

The first step in the research was to establish the industry's critical equipment and the effectiveness of the maintenance approach that was being applied to such equipment. The position was established through responses to both interviews and questionnaires. Interviewees representing both the pipeline and road/rail depot operators were asked to outline and rank the top five equipment forming the backbone of their business operations. The following equipment or systems were listed as the most critical equipment:

A. Pipeline
B. Tanks
C. Pumps
D. Meters
E. Firefighting system (fire detection and suppression equipment)
F. Power supply and back up (Grid power, generators and UPS)
G. CCTV
H. Control and instrumentation system (PLCs and SCADA)
I. Pipes and fittings
J. Gantry

There was a distorted consensus in table 2 on what the industry practitioners identified as their critical equipment through the rankings they attached to each system or equipment. The researcher assigned

Table 2. The responses and ranking

Interviewee	Company	Response and Rank				
		1	2	3	4	5
W M	NOIC	A	B	C	E	D
L M	NOIC	A	B	E	C	D
P M	PZL	A & B	C & D	H	F	E
L G	Vivo Energy	E & C	B & I	F	D	G
T M	Vivo Energy	B & I	C & D	E	F	G
D	Zuva	F	B & I	E	C & D	J

scores from 1 to 5 to each response. The scores were assigned in a descending order with the equipment ranked number 1 being assigned 5 points. The equipment's overall tally and position was adopted as the normalized critical ranking for the industry's equipment. Table 3 rearranges the equipment in order of criticality.

The results of Table 3 rank tanks as the most critical equipment for Zimbabwe fuel facilities' infrastructure. The researcher noted a bias on the responses where NOIC and PZL interviewees ranked pipeline as critical, but the other respondents did not attach the same importance to pipelines. This can be explained by the nature of operations where NOIC and PZL have depots connected by pipeline while the other companies only have road or rail receiving connectivity. The storage capacities for the pipeline combined with the cost of construction and the operational capabilities makes it a very critical asset for NOIC and PZL. Table 2 shows that there was consistency among the industry practitioners' combining of tanks, pipes and pipe fittings as well as pumps and meters as one system. In emphasising on the importance of tanks, pipes and pipe fittings L G of Vivo Energy referred to this combination as *"primary containment".* Though she placed the combination as second in her ranking, she emphasised that the business of bulk fuel operations was anchored on sound infrastructure in the form of pipes and tanks. D of Zuva explained that the ranking was difficult as some of the equipment such as firefighting were regulatory, and operators are not allowed to trade when such equipment is down.

New Technology Adoption, Decommisioning of Old Technology

The second and third interview questions for the bulk fuel depot operators together with the follow up questions sought to ascertain the main drivers in the adoption of new technology among the bulk fuel depot operators. A number of guiding considerations were given, and interviewees would explain themselves along the guidelines. As follow up to their first responses, interviewees were asked to state the main drivers or reasons for their adoption of new technologies in the past. The researcher sought to construct a picture on whether the industry's practitioners were reactive or proactive in their assimilation of new technology or ideas. Figure 17 summarises the interviewees' submission on how they ranked the possible drivers of adopting new technologies in their businesses. A value of 5 was assigned to the most likely driver and the lowest likely driver was assigned a value of one. It was noted that some of the suggested drivers such as consumer demands, equipment supplier's marketing and employers' demands were not considered as having a key role in influencing the adoption of new technologies.

Table 3. Ranking of materials

	W M	L M	P M	L G	T M	D	TOTAL
Tanks	4	5	4	4	5	4	26
Pumps	3	4	2	5	4	2	20
Firefighting	2	3	3	5	3	3	19
Pipeline	5	5	5				15
Meters	1	4	1	2	4	2	14
Pipes and fittings				4	5	4	13
Power supply		2		3	2	5	12
CCTV				1	1		2
Control and instrumentation		1				1	2
Gantry						1	1

All the six interviewees outlined business optimisation and safety considerations as the key drivers to technological improvements in their organisations. However, they were all unanimous in highlighting that their safety concerns were not regulatory imposed, but rather, were all driven by best practice. They were also unanimous in that the regulatory aspect that they had so far considered and had a bearing to their infrastructure developments concerned the mandatory blending of ethanol. This resulted in modifications of their storage tanks, pipes, pumps, meters and firefighting equipment to enable the handling of ethanol. Some of the interviewees raised concerns on the contribution of the key regulatory bodies which were queried to be merely selling licences to the industry and adding no value except fining offenders when a mishap occurred. They singled out EMA and NSSA whose presence was said to be notable when there was a leak or an accident at the bulk fuel operator's site. The operators suggested that all the regulatory bodies could outline their expectations through the industry's main regulatory body, ZERA. They further suggested that ZERA could then lead on running programs of campaigns to uphold the consolidated requirements or expectations.

A follow up question on technology assimilation sought respondents to outline their plans on retiring aged tanks and introducing new tanks to their infrastructure. The responses were mixed but largely reserved. Most indicated that their plans on retiring old tanks and erecting new ones was driven by business strategy and could not be divulged. The respondents were unanimous in stating that the retiring of any old equipment including storage tanks would be determined by their useful life cycle. The interviewees were not keen to shed a picture on the decommissioning of tanks on obsolescence and erecting new tanks as replacement or on account of anticipated business. However, the researcher established that in the last twenty years all four players retired some tanks, but none constructed new tanks as replacement. The sizes of the retired tanks could not be ascertained but the reason for taking the tanks out of service was that they had corroded extensively and could not be repaired economically.

The question on technology adoption also touched on redundancies that are likely to occur as a result of introduction of the new businesses' technologies. All respondents expressed being alive to this likelihood. Most of the respondents confessed that there would be plans for the personnel to be laid off but did not elaborate. L G of Vivo Energy said that in her organisation the personnel to be laid off would usually be trained in other income generating skills before being relieved of their duties. She said some end up running downstream businesses such as fuel delivery transport, service stations or providing maintenance services to the organisation. The responses on technology drivers suggest that the industry's assimilation of new technology was hinged on business optimisation, safety concerns and regulatory consideration. The result points to an industry that is largely proactive. Since none of the safety concerns were imposed through regulation, it can be concluded that management commitment was key in the adoption of safety technologies deployed at the interviewed practitioners' businesses. However, there is a pointer to non-involvement of employees in key decisions as well as no market surveys to establish the consumers' expectations in making key decisions about technological developments. The retirement of tanks due to obsolescence without notable plans to retain or improve on previous capacities can be explained by either a shift in the business trends or there is a capacity constraint that is currently being suppressed but might be visible sooner or later. Further investigation on the industry's present and future capacity constraints is worthy pursuing as the later possibility will cripple the country's economy.

Influence on Supplied Services or Products

The fourth interview question to the bulk fuel facilities operators and the second interview question to the tank NDT service providers sought to establish whether the industry's facilities operators were involved in influencing the design of their products or services. The industry players' response suggested they do not just take what is available but try to have their desires incorporated in the design of their products and services. This was corroborated by the NDT service providers who confirmed to have lost business or had to collaborate with foreign partners after some of the industry practitioners demanded the use of technology (crawlers and magnetic floor scanners) that was not yet available among any of the local service providers. Some of the NDT service providers confirmed that they were considering acquiring such technologies because of the industry facility owners' demands.

Effectiveness of Maintenance Approach on Critical Equipment

Questions five and six of the industry facilities interview together with the first four questions of the NDT service providers' interview and the questionnaire were meant to establish the state of maintenance in the industry's critical equipment and stimulate research on areas for potential improvement. Key indicators to the research questions which were extracted from the responses are captured in the Appendix 5 and 6. Appendix 5 captures indicative responses from the facility owners while table 5 captures responses from the NDT service providers.

The responses from all six interviewees for the industry's facility operators confirm the existence of concern on the tanks' downtime during out-of-service inspections. The researcher noted Vivo Energy respondents' position that the downtime was of little concern due to current shortages. However, the researcher observed that the shortage was a temporary position and strategic decisions could not be based on temporary conditions. The researcher therefore concluded that the industry had concerns over the tanks' downtime due to out-of-service NDTs and that it was worthy to find ways of resolving the challenge. Most of the respondents from the facility owners' category confirmed that they had at times demanded technologies that the local service providers did not possess. Two of the three interviewees from the NDT service providers' category confirmed to have been forced to hire technology that they do not possess from outside the country. The researcher noted the corroboration regarding the existence of a gap between available NDT technology and the industry's expectation. The third NDT service providers' response suggested they had the capacity to meet the industry's demands but further revealed the existence of a gap when they outlined their acquisition plans. The researcher concluded that the local service providers did not have the requisite technology to solve the challenges of out-of-service NDTs. This further confirmed that there was need to consider non-local service providers in the search for alternative technology for in-service NDTs.

A total of thirty-six completed questionnaires were received and the distribution between the five companies was as follows: NOIC – 9; PZL – 8, Zuva Petroleum – 8, Vivo Energy – 7; Total Zimbabwe – 4. The overall response rate was 72%. The researcher noted that the response was unusually high and attributed this to his personal relationship with people in the industry. The responses are summarised in table 4.

The first two questions were meant to aid the researcher to ascertain some of the reasons why the respondents would answer all the subsequent questions. Knowledge of the duration it takes to commission a tank back into service after detection of a fault at the time of a scheduled inspection and the number of inspections witnessed, all depend on whether the employee had been exposed to tank inspections prior. The longer the duration of employment the higher the chances that the employee would have been exposed to tank cleaning before and the more knowledgeable the employee will be in the field of tank inspections and maintenance. Interviewees' perceived time of postponement of NDTs, time taken to commission faulty tanks back into service and the record of past tank inspections that can be readily availed are further summarised in the pie charts of figures 18; 19 and 20.

The results shown in figures 18 – 20 corroborates the tabled responses of Appendices 4 and 5 and also give quantitative measures to some of the qualitative statements. Figure 18 shows that nearly two thirds of the questionnaires' respondents think that if an inspection is postponed, it will be conducted after twelve months. Of the remaining one third, the majority think that postponed inspections will be conducted in 9 – 12 months. It can be concluded that there is exposure of at least nine months when a scheduled tank inspection is skipped. The integrity of the tanks would not be known during that period and leaks may occur. This position was confirmed by the interviewed managers in the industry and echoed by one of the NDT service providers. Figure 18 gives insight on the lost time if a tank is noted to be defective and requiring repairs before commissioning. None of such tanks would be commissioned within a month of the need for repairs being detected. All the interviewed managers in the bulk fuel industry had indicated that commissioning of tanks back into service would take long though they had not been direct as to how long that takes. Some of the reasons for the delay were that of availability of repair materials. The repairs process itself would contribute to the delays. The repairs would require certification of the quality of welding by an independent inspector.

The aspect addressed in figure 20 had not been addressed during interviews but gives critical insight on information storage and retrievability. The respondents painted a picture of an industry that is not well organised in terms of information management. All the respondents to the questionnaires indicated that only data for the past two set of inspections could be readily availed. The average tank inspection interval was stated to be 5 years and the respondents indicated that previous inspection records could be availed for up to ten years only. This is not a healthy scenario for an industry whose infrastructure is over seventy years. The solution to the industry's identified gap should therefore consider data storage and retrievability.

Table 4. Response of the companies

Parameter Measured	Score/Response	Respondents (Number)	Respondents (Percentage - %)
Number of years with the current company	³ 15 years	3	8
	10 – 15 years	6	17
	7 – 10 years	9	25
	4 – 7 years	13	36
	£ 4 years	5	14
Number of out-of-service tanks inspections witnessed in the current company	≥ 10	1	3
	7 – 10	3	8
	4 – 6	11	31
	1 - 3	15	42
	None	6	17
Members' perception of safety and storage tanks maintenance in the fuel industry	Excellent	5	14
	Good	11	31
	Satisfactory	14	39
	Poor	0	0
	Very Poor	0	0
Longest known time that scheduled NDTs have been postponed	≥ 12 months	21	58
	9 – 12 months	6	17
	6 – 9 months	3	8
	3 – 6 months	0	0
	≤ 3 months	0	0
Most likely causes of postponement of scheduled NDTs	Poor inter-departmental coordination	22	61
	Maintenance oversight	6	17
	Financial constraints	0	0
	Tank preparations delays by tank owners	5	14
	Delays by appointed NDT inspection service providers	3	8
Time taken to commissioning into service after repairs to fault discovered during scheduled NDTs	≥ 5 months	1	14
	4 – 5 months	2	6
	2 – 3 months	14	39
	1 – 2 months	13	36
	≤ 1 month	0	0
Tank inspection records that can be readily availed on request covers duration of so many years	≥ 50 years	0	0
	30 – 40 years	0	0
	20 – 30 years	0	0
	10 – 20 years	0	0
	≤ 10 years	29	81

Down Time and Economic Considerations

The researcher sought to establish the maximum storages and average receipts or holding volumes per company in the industry to have insight on the maximum potential business lost due to the down times on tanks. However, details about the tank capacities and average handling volumes were not disclosed. The interviewed managers indicated that that data was part of their strategic information and could not be disclosed. The researcher managed to gather some data that can be used to make meaningful inferences from details gathered through data mining.

Pipeline Depot Operations Disruptions (NOIC and PZL)

The impact of down time on NOIC and PZL due to tank availability can be deduced from the pipeline operational capacity. An on-line report (https://clubofmozambique.com/news/cpmz-increases-throughput-capacity-of-the-beira-corridor-oil-pipeline-161179/) outlined the present pipeline capacity from Beira to Feruka as 2.2 million m^3/year. The report indicated that there were plans to increase the capacity in two phases. The first phase was said to be in two stages which will see an increase to 3 million m^3/year under stage 1, and then an increase to 5 million m^3/year under stage 2. The second phase will be an upgrade of the pipeline capacity to replace the current 10-inch pipe by a bigger pipe whose size was not disclosed in the report. The researcher noted that for an annual pipeline throughput of 2.2 million m^3, the average daily throughput will be 6,023.27m^3/day. The researcher deduced that to avoid bottlenecks between product changeovers there should be at least two receiving tanks per product of capacities not less than 6,023.27m^3 to ensure that the depot directly receiving from Beira will manage to copy with the said pipeline throughput capacity. By the same reasoning, the researcher expects the receiving capacities to be increased with the increase in pipeline capacities so that at any given time the minimum receiving tank capacity can accommodate a day's pipeline throughput.

The researchers assumes by taking one of the receiving tanks out of services while the pipeline is busy the bottleneck created will result in the loss of a day's receipts every three days. The researcher took cue from the questionnaires' responses were 75% of the respondents indicated that the average tank commissioning after discovery of a fault during scheduled out of service inspection was one to three months. The researcher deduced that on average a tank could be out of service for one to three months if the out-of-service inspections were maintained as the norm. Assuming the pipeline remains busy, the loss of business under such circumstances would at least be the daily receipts of the affected product every four days. This will result in a cumulative loss of at least 7.5 days' receipts in a month and 22.5 days in three months. The volumetric throughput loss will be 45,174.53m^3 to 135,523.58m^3 for one receiving tank found defective during scheduled out-of-service inspections.

Road and Rail Receiving Depots Disruptions (Zuva, Total and Vivo Energy)

Assuming all the fuel received through pipeline will be consumed locally, the pipeline throughput and the local uptake should be the same to avoid bottlenecks. The researcher assumed the three road and rail depot operators will take equal volumes for storage in their depots before onward distribution to the market. This means the average daily uptake of each of the three road and rail depots would be 2,007.76 m^3/day. Applying the same reasoning as for pipeline receipts above, the volumetric loss per company that fails to commission its main receiving tanks on time would be between 15,0588.18m^3 and 45,174.53m^3.

In addition to the loss of business, both the pipeline and the road and rail depot operators will incur inefficiency costs due to the conduct of time-based out-of-service NDTs on tanks that would be qualified as safe for continuous service for an additional three to five years. In the event of an uninformed postponement of inspections due to operational demand, the organisations will be taking the risk of corrosion progressing to leaks before detection. This would even be more costly as it will result in product loss and irreparable groundwater contamination. It is therefore in the best interest of the bulk fuel operators' interest to be able to determine the condition of their tanks at any given time without having to take them out of service.

In-Service Tank NDT Inspections

Considerations for closing the outlined inefficiencies being experienced by Zimbabwe's bulk fuel facilities operators due to time based out-of-service inspections can be done at the different management planning levels. In this regard the decisions can be operational or strategical. At an operational perspective the idea would be on how best and cost effective the next inspection would be effected while the strategic perspective will be looking at how best all future inspections will contribute to the organisations' long term vision. A vision that sees the current tanks as still relevant for future operations would support the adoption of in-service tank inspection technology while a vision that sees no significant role from storage tanks would not be supportive of investing in the adoption of in-service tank inspections. While the industry has operated this far relying on out-of-service time-based scheduled tank NDTs, the inefficiencies outlined cannot be easily ignored. There is therefore cause for consideration of the adoption of in-service tank inspections, either as an operational decision alone if the strategic vision does not include use of current tanks or as both operational and strategic decision if in the organisations' long term vision tanks will still have a role.

Operational Considerations

The quickest way of securing AE technology is outsourcing from service providers in any of the countries cited to have conducted such works. However, besides the costs and other trade considerations, there are factors that should be considered in building such decisions. (R Medina Serrano et al ., 2018) addresses organisational considerations in "make or buy" decisions, which is the term used to describe outsourcing. A lot of literature has been published which gives insight into the outsourcing decisions, including the challenges expected. One challenge which is relevant in the context of this research is that of language barrier. There is a possibility of communication challenges if Zimbabwean fuel facilities companies decides to directly engage service providers in the countries outlined in the literature as having been noted to have made significant steps towards development of AE, (China, Poland and Japan) and those cited as leading manufacturers of AE equipment (Germany and Greece).

The resource-based view supports the idea of outsourcing if an organisation seeks to secure capabilities that they do not have inhouse. This has been the approach regarding provision of NDT inspection services in Zimbabwe's bulk fuel industry. The challenge in this regard is that the local companies do not have the requisite technology and expertise. The literature on IIoT challenges raised skills gap as an area that should be addressed in IIoT. As with any new technology, training will be required if AE is to be efficiently outsourced. Zimbabwe's bulk fuel operators should therefore consider training their staff in basics of AE so that they can appreciate the technology and be able to give specifications for the engagement of a service provider. The outsourcing can be considered as a contract between any one of Zimbabwe's bulk fuel industry operator and the foreign service-provider, or the foreigner service-provider can be brought in as a subcontractor of a local service provider.

In view of the present international travel restrictions due to the covd-19 pandemic and the mandatory quarantine requirements when one enters another country, the outsourcing can be limited to equipment only with the service provider providing the support on-line. The testing equipment can be shipped to the Zimbabwean company which can then conduct the physical deployment of the sensors on the target tanks under instructions issued over an on-line arrangement. The data analysis and feedback will again be done through the on-line platform.

Strategic Considerations

Outsourcing is not a short-term decision. It can be adopted as a business strategy if it is noted to give the host company a competitive advantage. However, where the host company does not draw comfort in outsourcing it can opt to equip itself to provide the service for itself. This route can as well be considered for the adoption of AE in Zimbabwe's bulk fuel facilities. Successful implementation requires the following support steps.

1. Training of personnel in the usage of AE technology
2. Acquisition of sensors and AE analysis processors
3. Deployment of the technology.

The proposed implementation takes into consideration the available off-shelf technologies and seeks to maximise its applicability to the Zimbabwean set up. Figure 21 presents the proposed connection and devices to be used in the data acquisition, processing, reporting, decision making and report.

The connection involves use of AE sensors for detecting and recording the corrosion induced elastic waves. The sensors are then linked to a PLC based system which processes the pulses and coordinates the decision making by sending feedback to other devices in the system and to selected people who are supposed to act as prompted. Appendix 4 outlines the decision framework/program. The proposed AE system will follow the principle outlined in the literature for in-service aboveground storage tank floor monitoring. which requires the target tank to be at least 70% full and to have been isolated for at least twenty-four hours. The system is designed in a manner that will prompt top management to give direction when condition of the inspected tank requires immediate out-of-service inspections. Additional sensors will be added to monitor the status of the manhole and to institute automatic shut-off of the tank when the AE inspection establishes the need for immediate service. As outlined in the literature review for strategic management and maintenance, there is need to secure top management commitment if key decisions are to be made. Involvement of top management in deciding whether to override the request for immediate tank inspection is critical where extensive corrosion is detected.

This section looked at the findings from the research and guided the development of ideas to close the identified gap between global in-service inspections for petrochemical aboveground storage tank bottoms and the NDTs practiced in Zimbabwe's bulk fuel facilities. The research established that there was a gap between Zimbabwe's practice in tank inspections and maintenance. Globally the industry is shifting more towards in-service tank inspections. Zimbabwe's bulk fuel industry is still stuck with the conventional out-of-service time based scheduled inspections which the facility owners confirmed to be inconveniencing their operations. Available literature was used in developing an in-service tank bottom flow chart that utilises available sensors.

FUTURE CHALLENGES AND LIMITATIONS OF THE WORK

A few other key challenges prevent the widespread adoption of IoT solutions at the time being:

- **Security vulnerabilities**: IoT devices communicate automatically with each other. In the absence of a Secure and properly encrypted network, the adoption of IoT could lead to brand new security challenges and vulnerabilities. Standalone security elements will have to be introduced in the network to enable adoption without a higher risk of hack attacks or data leaks.
- **Absence of IoT Standards**: Many automation devices already operate in an array of industrial and manufacturing settings. The problem is that various protocols are being utilized and there's no standardization that will ensure interoperability.
- **The Cost of Implementing IoT Solutions**: This is another essential element that cannot be underestimated. The cost of implementing the IoT infrastructure is often perceived as overwhelming. Many companies worry about the return on such an investment and so remain hesitant. This is where the importance of choosing the right IoT solutions comes to the stage center. Ease of use, ease of training and the development of more readily adoptable products could also help eliminate some of the hesitation in the future.

The rise of industrial IoT will soon bring the factory of the future to reality (IOTFORALL, 2018).

- Material handling, manufacturing, product distribution and supply chain management will all be automated to a degree in the years to come. To drive the digital transformation forward, however, executives will have to become involved in the process right from the start. The IT department alone will need such support when revolutionary decisions are being made about production processes.
- The industrial internet of things, also known as industry 4.0, has already started the fourth industrial revolution. More automation has already led to a 30 percent boost in productivity and the adoption of flexible production techniques. Predictive maintenance is reducing the cost even further by saving more than 12 percent on scheduled repairs and bringing breakdowns down by nearly 70 percent.
- In the future, experts suggest that industrial IoT will enhance production levels even further and become the driving force behind various types of innovation (including the utilization of innovative fuels). The workforce itself will also be transformed as a part of the extensive automation process.
- While an array of adoption challenges will still have to be overcome, predictive analysis suggests that the world will have 50 billion connected devices by 2020. It would be a pity for such a massive network to remain unutilized in attempts to enhance industrial processes. Remember that industrial IoT is not about smart product development. Rather, it will help for a higher level of efficiency and predictive rather than reactionary interventions – a main problem industries across the world are struggling with today.

CONCLUSION

The research confirmed the existence of a gap between global technological developments and the standards currently applied in Zimbabwe's bulk fuel NDTs. The objectives set out at the beginning of the project were as follows:

1. Review the global technological developments in tank NDTs and compare with the technologies applied in Zimbabwe's bulk fuel tanks maintenance.
2. Establish reasons for the gap(s) and consider its closure.
3. Formulate an IIoT enabled framework for enhancing in-service tank NDTs

The first objective was established through both literature reviews and interviews with Zimbabwe bulk fuel facility operators and NDT inspection services providers. The major finding from the interviews was that the industry was alive to the limitation posed by out-service tank NDTs and that at times the industry was losing out by conducting time-based scheduled tank NDTs on tanks that would be safe to operate. On other instances inspections would be postponed due to operational demand without the tanks' integrity being assured. The operators confirmed that this action was risky and could lead to significant losses and environmental damage should the tank ultimately develop a leak.

Both facility operators and the service providers were either oblivious of the existence of a technology that enables in-service tank NDTs or were sceptical of its accuracy and applicability in their industry. A review of literature about in-service tank inspections revealed that there were at least three methods that could be applied to conduct in-service tank bottom inspections. Acoustic emission was acclaimed as the most accurate and convenient test for tank bottom inspections. Its accuracy can be verified through use of any of the conventional NDTs methods if the tank owners' doubts are to be pacified. Reviewed literature cited accuracies of 80 – 94% where AE was compared with conventional tank NDTs. Other than the gap on tank bottom inspection technologies, it was also noted that there was a gap between the technological NDTs for the tank shell and roof. The local industry was still using conventional spot checks that they would conduct by erecting scaffolding round the tank for the tank shell. The reason for this gap was attributed to the level of local tank NDTs business. The service providers cited that the costs of acquiring the technologies that were being used elsewhere in the industry would not make business sense as the level of inspections were very low. The local NDTs service providers indicated that the cost of a crawler for tank shell/roof remote controlled inspections ranged between US$75,000.00 and US$115,000.00. The service providers also cited lack of expertise in the use of the latest technologies and the interpretation of the data.

The research also revealed lack of collaboration among NDTs service providers in Zimbabwe and among the operators of bulk fuel facilities. The NDTs service providers confirmed that there was rivalry among the players which made it impossible for players in the industry to pool resources together for acquisition of new technology. Lack of collaboration on the operators of bulk fuel facilities was deduced from the challenges which the players expressed to have been affecting their operations. Some players expressed that they were facing storage challenges which were affecting their scheduled tank inspections, while some expressed that they had excess capacity due to the subdued supply levels. If the parties were collaborating the player facing storage challenges could outsource storage to the one with underutilised capacity and create room to conduct their inspections. There is lack of stakeholder involvement in the industry. Some of the bulk fuel facilities' operators confirmed having demanded technology that none of the local NDT service providers possessed. The service providers confirmed that this demand was revealed on the time of need, and that the service providers were not given adequate time to prepare for acquiring the requisite technology or to negotiate for reasonable hiring arrangements with foreign suppliers. The research also revealed the challenge of information management. This was confirmed through the research responses which established that records of previous tank inspections can only be readily availed for inspections conducted within the last ten years.

Outsourcing or acquiring in-service tank inspection is beneficial and outweighs the outlined risks that Zimbabwe's bulk fuel facilities' operators face by using the out-of-service tank inspections. A direct monetary comparison could not be exhaustively done. Quantitative values could not be attached to some of the consequences of potential risks such as environmental contamination that is associated with tank leaks. However, there are verifiable negative consequences in time based scheduled out-of-service tank inspections, particularly where a leak would eventually occur between the scheduled inspection intervals. There will be reputational damage that comes with environmental damage, uncontrollable fires if the leaks are exposed to fire and fines from regulatory bodies for the pollution arising from the leaks.

Operators of the bulk fuel facilities have considerable motivation to acquire in-service NDTs technologies for their tanks, and this is highly recommended. In the short term, the technology can be hired at the time of inspection. However, it is more convenient if the AE in-service tank inspection kit is procured and used at the operator's need. The AE sensors will not be permanently fixed on the tanks as this will not be valuable. The rate of corrosion does not warrant continuous real-time monitoring and reporting. Instead, tanks will be inspected at the time of need and the same equipment can be used on several tanks including outsourcing the service to other players. It is also important to note that AE test results and guidelines are based on the assumption that corrosion rate will continue as established at the time of testing. The operating environment of the tested tank should not be changed significantly for the projected inspection date to remain valid. A change in the environment, such as the use of a different product will require re-testing of the tank under the changed environmental conditions.

Top management commitment is crucial in any significant decision for an organisation. The proposed technology deployment involves a stage where the Head of the organisation would be informed in real time if the tanks' condition require immediate out-of-service inspection. This will ensure quick and decisive actions. The use of sensors on additional tank fittings such as in monitoring the position of the manhole and enabling the automation of the tank's main inlet and outlet valves is recommended. This will improve the safety of the tanks by checking compliance to the recommended action where a tank is deemed unsafe for use. Internet based communication and cloud storage is recommended for information sharing between the concerned sensors and different users. This will enable real time interaction of the concerned sensors with the actual inspections and the expected interactions or interventions from key stakeholders from different locations. This will also address the challenge of information management as the information will be stored in the cloud. Direct connectivity of the sensors on different sections of the tank and the proposed autonomous actions when certain conditions are met are the first stages towards an autonomous operational and maintenance system in general tank inspections and maintenance. Bulk fuel facility operators are recommended to consider the acquisition and deployment of scalable sensors to ensure future upgrades will be feasible retaining the same sensors.

Exposing the organisation's critical equipment to the internet will however introduce IIoT related challenges such as cyber security threats, privacy limitations and interoperability interference challenges. The proposed technology adoption includes IIoT considerations. It is recommended that in adopting AE technology Zimbabwe's fuel industry players must enlist the services of IT specialists to assist them in the selection and maintenance of the right IT systems. This should cover considerations of the narrowing gap between the IT and OP roles within the organisations and the need to equip the organisation's employees with the relevant skills. Outsourcing the right IT/OP skills or training of the organisations' present personnel should be considered as a part to the adoption of the new inspection technologies.

Both the industry's facility operators and the NDTs service providers confessed they lack the requisite knowledge and expertise in AE technology. Research in AE technology for in-service tank bottom inspections through collaboration with academic institutions is recommended. The bulk fuel facility operators should consider financing such research as they are the most interested parties. The cited PetroChina case study is an example of a research instituted by the industry's facility operators. Zimbabwe's local players can emulate the Chinese industry initiative. The first operator to take the initiative may consider setting a section for the provision of the service which can then be extended to other players within the local industry or the region. The regulator can also consider financing research in the technology development and alignment to the Zimbabwean environment. This can then be used by the regulator in providing oversight by setting minimum safety levels and conducting checks to ensure operational tanks' integrity meets the minimum integrity conditions.

REFERENCES

Ahmed, E., Yaqoob, I., Hashem, I. A. T., Khan, I., Ahmed, A. I. A., Imran, M., & Vasilakos, A. V. (2017). The role of big data analytics in Internet of Things. *Computer Networks*, *129*(September), 459–471. doi:10.1016/j.comnet.2017.06.013

Alase, A. (2017). The Interpretative Phenomenological Analysis (IPA): A Guide to a Good Qualitative Research Approach. *International Journal of Education and Literacy Studies*, *5*(2), 9. doi:10.7575/aiac. ijels.v.5n.2p.9

Arnold, C., Kiel, D., & Voigt, K. I. (2020). Innovative business models for the industrial internet of things. *26th International Association for Management of Technology Conference, IAMOT 2017*, (May), 1379–1396. 10.100700501-017-0667-7

Bhati, M. S. (2018). *Industrial Internet of Things (IIoT): A Literature Review*. doi:10.18231/2454-9150.2018.0340

Data, B., & Management, D. C. (n.d.). *Digital Prescriptive Disrupting Manufacturing Value Streams through Internet*. Academic Press.

Endres, H., Indulska, M., Broser, S., Endres, H., & Indulska, M. (2019). *Association for Information Systems AIS Electronic Library (AISeL) Industrial Internet of Things (IIoT) Business Model Classification Industrial Internet of Things (IIoT)*. Business Model Classification.

Eti, M. C., Ogaji, S. O. T., & Probert, S. D. (2006). Strategic maintenance-management in Nigerian industries. *Applied Energy*, *83*(3), 211–227. doi:10.1016/j.apenergy.2005.02.004

Feng, Y., Yang, Y., & Huang, B. (2019). Corrosion analysis and remaining useful life prediction for storage tank bottom. *International Journal of Advanced Robotic Systems*, *16*(5), 1–9. doi:10.1177/1729881419877051

Field, U. S. A., & Interim, M. (2008, Dec.). For Official Use Only for Official Use Only. *Distribution.*

Greeneltch, K. M., Haudenschild, C. C., Keegan, A. D., & Shi, Y. (2004). The opioid antagonist naltrexone blocks acute endotoxic shock by inhibiting tumor necrosis factor-α production. *Brain, Behavior, and Immunity, 18*(5), 476–484. doi:10.1016/j.bbi.2003.12.001 PMID:15265541

Hartmann, M., & Halecker, B. (2015). Management of Innovation in the Industrial Internet of Things. *XXVI ISPIM Conference – Shaping the Frontiers of Innovation Management In*, 1–17.

Janssen, C. P., Donker, S. F., Brumby, D. P., & Kun, A. L. (2019). History and future of human-automation interaction. *International Journal of Human-Computer Studies, 131*(January), 99–107. doi:10.1016/j.ijhcs.2019.05.006

Karmakar, A., Dey, N., Baral, T., Chowdhury, M., & Rehan, M. (2019). Industrial internet of things: A review. *2019 International Conference on Opto-Electronics and Applied Optics, Optronix 2019*, 1–6. 10.1109/OPTRONIX.2019.8862436

Khan, W. Z., Rehman, M. H., Zangoti, H. M., Afzal, M. K., Armi, N., & Salah, K. (2020). Industrial internet of things : Recent advances, enabling technologies and open challenges R. *Computers & Electrical Engineering, 81*, 106522. doi:10.1016/j.compeleceng.2019.106522

Lackner, G., & Tscheliesnig, P. (2007). *Acoustic emission (AE) field application: Testing of aboveground storage tanks (AST) for corrosion and leakage.* Academic Press.

Lee, I., & Lee, K. (2015). The Internet of Things (IoT): Applications, investments, and challenges for enterprises. *Business Horizons, 58*(4), 431–440. doi:10.1016/j.bushor.2015.03.008

Management, A. (2014). *Using the Internet of Things for preventive maintenance.* Academic Press.

Martin, G. (2013). Acoustic emission for tank bottom monitoring. *Key Engineering Materials, 558*(December), 445–455. doi:10.4028/www.scientific.net/KEM.558.445

Mercer, R. (2019). *Global Connected and IoT Device Forecast.* Academic Press.

Mgonja, C. T. (2017). *Evaluation on Use of Industrial Radiography for Weld Joints Inspection in Tanzania.* Academic Press.

Migiro, S. O., & Magangi, B. A. (2011). Mixed methods: A review of literature and the future of the new research paradigm. *African Journal of Business Management, 5*(10), 3757–3764. doi:10.5897/AJBM09.082

Moser, A., & Korstjens, I. (2018). Series: Practical guidance to qualitative research. Part 3: Sampling, data collection and analysis. *The European Journal of General Practice, 24*(1), 9–18. doi:10.1080/13814788.2017.1375091 PMID:29199486

Muhonen, T. (2015). *Standardization of Industrial internet and IoT (IoT – Internet of Things) – Perspective on condition-based maintenance.* Academic Press.

Neville, S., Adams, J., & Cook, C. (2016). Using internet-based approaches to collect qualitative data from vulnerable groups: Reflections from the field. *Contemporary Nurse, 52*(6), 657–668. doi:10.1080/10376178.2015.1095056 PMID:26394073

Pearson, N. R., Mason, J. S. D., & Priewald, R. H. (n.d.). *The influence of maintenance on the life cycle of above ground storage tanks*. Academic Press.

Pesce, M. D., & Standards, R. (2011). *NDT of Welded Steel Tanks*. Academic Press.

Rahimi, H., Zibaeenejad, A., & Safavi, A. A. (2018). A Novel IoT Architecture based on 5G-IoT and Next Generation Technologies. *2018 IEEE 9th Annual Information Technology, Electronics and Mobile Communication Conference (IEMCON)*, 81–88.

Silva, B. N., Khan, M., & Han, K. (2017). Internet of Things : A Comprehensive Review of Enabling Technologies, Architecture, and Challenges. *IETE Technical Review*, *0*(0), 1–16. doi:10.1080/025646 02.2016.1276416

Sisinni, E., Han, S., Jennehag, U., & Gidlund, M. (2018). *Industrial Internet of Things : Challenges, Opportunities, and Directions*. (doi:10.1109/TII.2018.2852491

Ślusarczyk, B. (2018). Industry 4.0 – Are we ready? *Polish Journal of Management Studies*, *17*(1), 232–248. doi:10.17512/pjms.2018.17.1.19

Sutton, J., & Austin, Z. (2015). Qualitative reserch: Data collection,analysis,and managment. *The Canadian Journal of Hospital Pharmacy*, *68*(3), 226–231. doi:10.4212/cjhp.v68i3.1456 PMID:26157184

Taylor, P., Liu, F., Guo, X., Hu, D., & Guo, W. (n.d.). *Nondestructive Testing and Evaluation Comprehensive inspection and evaluation technique for atmospheric storage tanks*. doi:10.1080/10589750902795358

Teslya, N., & Ryabchikov, I. (2018). Blockchain-based platform architecture for industrial IoT. *Conference of Open Innovation Association, FRUCT*, 321–329. doi:10.23919/FRUCT.2017.8250199

Velmurugan, R. S., & Dhingra, T. (2014). Maintenance strategy selection theory & practices in natural gas industry: A case study of an Indian gas utility company. *International Gas Research Conference Proceedings*, *3*, 2223–2250.

Verma, A. L., Mendelsohn, R., & Bernstein, H. J. (1974). Resonance Raman spectra of the nickel, cobalt, and copper chelates of mesoporphyrin IX dimethyl ester. *The Journal of Chemical Physics*, *61*(1), 383–390. doi:10.1063/1.1681648

Xiong, M., Kang, Y., Lin, M., & Sun, Y. (2009). *Study on In-service Inspection Methods for the Aboveground Oil Tanks Floors Master, Petroleum Engineer - PetroChina pipeline R & D center Doctor, Petroleum Engineer - PetroChina pipeline R & D center Doctor, Petroleum Engineer - PetroChina pipeline R*. Academic Press.

Younan, M., Houssein, E. H., Elhoseny, M., & Ali, A. A. (2019). Challenges and recommended technologies for the industrial internet of things : A comprehensive review. *Measurement*. doi:10.1016/j. measurement.2019.107198

Zhong, C., Zhu, Z., & Huang, R. (2017). *Study on the IOT Architecture and Access Technology*. doi:10.1109/DCABES.2017.32

Zhong, R. Y., Xu, X., Klotz, E., & Newman, S. T. (2017). Intelligent Manufacturing in the Context of Industry 4.0: A Review. *Engineering*, *3*(5), 616–630. doi:10.1016/J.ENG.2017.05.015

Zhou, K., Liu, T., & Zhou, L. (2016). Industry 4.0: Towards future industrial opportunities and challenges. *2015 12th International Conference on Fuzzy Systems and Knowledge Discovery, FSKD 2015*, 2147–2152. 10.1109/FSKD.2015.7382284

Compilation of References

Aazam, M., St-Hilaire, M., Lung, C. H., & Lambadaris, I. (2016, October). Cloud-based smart waste management for smart cities. In *2016 IEEE 21st International Workshop on Computer Aided Modelling and Design of Communication Links and Networks (CAMAD)* (pp. 188-193). IEEE. 10.1109/CAMAD.2016.7790356

Aazam, M., Zeadally, S., & Harras, K. A. (2018). Deploying fog computing in industrial internet of things and industry 4.0. *IEEE Transactions on Industrial Informatics, 14*(10), 4674–4682. doi:10.1109/TII.2018.2855198

Abbas, M. T., Khan, T. A., Mahmood, A., Rivera, J., & Song, W.-C. (2018). Introducing network slice management inside M-CORD-based-5G framework. In *NOMS 2018 - 2018 IEEE/IFIP Network Operations and Management Symposium* (pp. 1-2). Taipei: IEEE. doi:10.1109/NOMS.2018.8406113

Abdelhafidh, M., Fourati, M., Fourati, L. C., & Chouaya, A. (2017). *Internet Of Things In Industry*. Academic Press.

Abu Mansour, H. Y., & Elayyan, H. (2018). IoT theme for smart datamining-based environment to unify distributed learning management systems. In *2018 9th International Conference on Information and Communication Systems (ICICS)* (pp. 212–217). Academic Press.

Aceto, G., Persico, V., & Pescapé, A. (2020). Industry 4.0 and health: Internet of things, big data, and cloud computing for healthcare 4.0. *Journal of Industrial Information Integration, 18*, 100129. doi:10.1016/j.jii.2020.100129

Aggarwal, R., & Singhal, A. (2019). *Augmented Reality and its effect on our life*. IEEE. doi:10.1109/CONFLU-ENCE.2019.8776989

Ahel, Eroff, Shohin, Zhong, & Runyang. (2018). IoT-Enabled Personalisation for Smart Products and Services in the Context of Industry 4.0. *Proceedings of International Conference on Computers and Industrial Engineering, CIE*.

Ahirwar, S., Swarnkar, R., Bhukya, S., & Namwade, G. (2019). Application of drone in agriculture. *International Journal of Current Microbiology and Applied Sciences, 8*(01), 2500–2505. doi:10.20546/ijcmas.2019.801.264

Ahmed, E., Yaqoob, I., Hashem, I. A. T., Khan, I., Ahmed, A. I. A., Imran, M., & Vasilakos, A. V. (2017). The role of big data analytics in Internet of Things. *Computer Networks, 129*(September), 459–471. doi:10.1016/j.comnet.2017.06.013

Ai, Y., Peng, M., & Zhang, K. (2018). Edge computing technologies for the Internet of Things: A primer. *Digital Communications and Networks, 4*(2), 77–86. doi:10.1016/j.dcan.2017.07.001

Alase, A. (2017). The Interpretative Phenomenological Analysis (IPA): A Guide to a Good Qualitative Research Approach. *International Journal of Education and Literacy Studies, 5*(2), 9. doi:10.7575/aiac.ijels.v.5n.2p.9

Alcaraz, C. (2019). *Security and Privacy Trends in the Industrial Internet of Things*. Springer. doi:10.1007/978-3-030-12330-7

Al-Doghman, F., Chaczko, Z., Ajayan, A. R., & Klempous, R. (2016). A review on Fog Computing technology. *IEEE International Conference on Systems, Man, and Cybernetics (SMC)*. DOI:10.1109/SMC.2016.7844455

Aleksandr, Babkin, Vladimir, Plotnikov, Yulia, & Vertakova. (2018, January). The Analysis of industrial cooperation models in the context of forming digital economy. *SHS Web of Conferences, 44.*

Alessandra, Papetti, Pandolfi, Peruzzini, & Germani. (2017). Digital Manufacturing Systems: A Framework to Improve Social Sustainability of a Production Site. In *50th CIRP Conference on Manufacturing Systems*. Elsevier.

Alessandro, Marzocca, Alfredo, Liverani, Cees., & Bil. (2019, October). Maintenance in aeronautics in an Industry 4.0 context: The role of Augmented Reality and Additive Manufacturing. Journal of Computational Design and Engineering, 6(4), 516-526.

Allcock, B., Bester, J., Bresnahan, J., Chervenak, A., Kesselman, C., Meder, S., Nefedova, V., Quesnel, D., Tuecke, S., & Foster, I. (2001). Secure, efficient data transport and replica management for high-performance data-intensive computing. In *Eighteenth IEEE Symposium on Mass Storage Systems and Technologies, 2001. MSS '01* (pp. 13–13). 10.1109/MSS.2001.10001

Alliance, L. (2015). *LPWA Technologies Unlock New IoT Market Potential*. White Paper. Advisory Group.

Alrawais, A., Alhothaily, A., Hu, C., & Cheng, X. (2017). Fog computing for the internet of things: Security and privacy issues. *IEEE Internet Computing, 21*(2), 34–42. doi:10.1109/MIC.2017.37

Al-Turjman, F., & Alturjman, S. (2018). 5G/IoT-enabled UAVs for multimedia delivery in industry-oriented applications. *Multimedia Tools and Applications, 79*(13-14), 8627–8648. doi:10.100711042-018-6288-7

Al-Turjman, F., & Alturjman, S. (2018). Context-sensitive access in industrial internet of things (IIoT) healthcare applications. *IEEE Transactions on Industrial Informatics, 14*(6), 2736–2744. doi:10.1109/TII.2018.2808190

Aman, M. S., Quint, C. D., Abdelgawad, A., & Yelamarthi, K. (2017). Sensing and classifying indoor environments: An iot based portable tour guide system. In *2017 IEEE Sensors Applications Symposium (SAS)* (pp. 1–6). 10.1109/SAS.2017.7894055

Amaresh, H., Rao, Y., & Hallikar, R. (2014). Real-Time Intruder Detection System Using Sound Localization and Background Subtraction. *2014 Texas Instruments India Educators' Conference (TIIEC)*, 131-137. doi:10.1109/TIIEC.2014.030

Amukele, T. (2019). Current state of drones in healthcare: Challenges and Opportunities. *The Journal of Applied Laboratory Medicine, 4*(2), 296–298. doi:10.1373/jalm.2019.030106 PMID:31639681

Amukele, T. (2020). The economics of medical drones. *The Lancet. Global Health, 8*(1), e22. doi:10.1016/S2214-109X(19)30494-2 PMID:31839132

Andersen, J. (2018). *Why the Future of IIoT Needs Both Edge and Fog Computing*. Rtinsights.

Anderson, G., Burnheimer, A., Cicirello, V., Dorsey, D., Garcia, S., Kam, M., Kopena, J., Malfettone, K., Mroczkowski, A., Naik, G., Peysakhov, M., Regli, W., Shaffer, J., Sultanik, E., Tsang, K., Urbano, L., Usbeck, K., & Warren, J. (2004). Demonstration of the secure wireless agent testbed (swat). In *Proceedings of the Third International Joint Conference on Autonomous Agents and Multiagent Systems, 2004. AAMAS 2004* (pp. 1214–1215). Academic Press.

Angles, R., & Gutierrez, C. (2008). Survey of graph database models. *ACM Computing Surveys, 40*(1), 1–39. doi:10.1145/1322432.1322433

Anita, Bodla, & Abhinav. (2017). Internet of Things (IoT)—It's Impact on Manufacturing Process. International Journal of Engineering Technology Science and Research.

Ardissono, L., Goy, A., Petrone, G., Segnan, M., & Torasso, P. (2002). Ubiquitous user assistance in a tourist information server. In *Adaptive hypermedia and adaptive web-based systems* (pp. 14–23). Springer. doi:10.1007/3-540-47952-X_4

Arnold, C., Kiel, D., & Voigt, K. (2019). How the Industrial Internet of Things Changes Business Models in Different Manufacturing Industries. *Digital Disruptive Innovation Series on Technology Management,* 139-168. doi:10.1142/9781786347602_0006

Arnold, C., Kiel, D., & Voigt, K. I. (2020). Innovative business models for the industrial internet of things. *26th International Association for Management of Technology Conference, IAMOT 2017,* (May), 1379–1396. 10.100700501-017-0667-7

Arnold, C., & Voigt, K. (2019). Determinants of Industrial Internet of Things Adoption in German Manufacturing Companies. *International Journal of Innovation and Technology Management, 16*(06), 1950038. doi:10.1142/S021987701950038X

Arnott, M., Buckland, E., Ranken, M., & Owen, P. (2017). *IoT global forecast and analysis.* https://www.gartner.com/en/documents/3659018/iot-global-forecast-and-analysis-2015-2025

Asenjo, J. L., Strohmenger, J., Nawalaniec, S. T., Hegrat, B. H., Harkulich, J. A., Korpela, J. L., Wright, J. R., Hessmer, R., Dyck, J., Hill, E. A., & ... (2014). *Industrial device and system attestation in a cloud platform.* Google Patents.

ASHRAE. (2017). Ventilation and infiltration. In 2017 ASHRAE handbook fundamentals. American Society of Heating.

Ashton, K. (2009). That 'internet of things' thing. *RFID Journal, 22*(7), 97-114.

ASTM. (1985). *Standard practice for maintaining constant relative humidity by means of aqueous solutions.* American Society for Testing and Materials. https://www.astm. org/DATABASE.CART/HISTORICAL/E104-85R96.htm

Atlam, H. F., Walters, R. J., & Wills, G. B. (2018). Fog computing and the internet of things: a review. *Big Data and Cognitive Computing, 2*(2), 10.

Azab, Ahmed, Noha, Mostafa, & Park. (2016). OnTimeCargo: A Smart Transportation System Development in Logistics Management by a Design Thinking Approach. *20th Pacific Asia Conference on Information Systems (PACIS 2016),* 44.

Bacco, M., Barsocchi, P., Ferro, E., Gotta, A., & Ruggeri, M. (2019). The Digitisation of Agriculture: a Survey of Research Activities on Smart Farming. *Array, 3*(4).

Bahrin, M. A. K., Othman, M. F., Azli, N. N., & Talib, M. F. (2016). Industry 4.0: A review of industrial automation and robotics. *Jurnal Teknologi, 78*(6-13), 137-143.

Bairi, A., Nithyadevi, N., Bairi, I., Martin-Garin, A., & Millan-Garcia, J.A. (2017). *Thermal design of a sensor for building control equipped with QFN electronic devices subjected to free convection.* Academic Press.

Baker, S. B., Xiang, W., & Atkinson, I. (2017). Internet of things for smart healthcare: Technologies, challenges, and opportunities. *IEEE Access: Practical Innovations, Open Solutions, 5,* 26521–26544. doi:10.1109/ACCESS.2017.2775180

Bakhshi, Z., Balador, A., &Mustafa, J. (2018). Industrial IoT Security Threats and Concerns by Considering Cisco and Microsoft IoT reference Models. *2018 IEEE Wireless Communications and Networking Conference Workshops,* 173-178.

Balaguer, C., & Abderrahim, M. (2008). Trends in robotics and automation in construction. In *Robotics and Automation in Construction.* IntechOpen. doi:10.5772/5865

Balaji, S., Nathani, K., & Santhakumar, R. (2019). IoT technology, applications and challenges: A contemporary survey. *Wireless Personal Communications, 108*(1), 363–388. doi:10.100711277-019-06407-w

Banerjee, U., Juvekar, C., Wright, A., & Chandrakasan, A. P. (2018, February). An energy-efficient reconfigurable DTLS cryptographic engine for End-to-End security in iot applications. In *2018 IEEE International Solid-State Circuits Conference-(ISSCC)* (pp. 42-44). IEEE. 10.1109/ISSCC.2018.8310174

Baotong, Wan, Lei, Shu, Peng, Li, Mithun, Mukherjee, & Yin. (2018, September). Smart Factory of Industry 4.0: Key Technologies, Application Case, and Challenges. In Special Section on Key Technologies for Smart Factory of Industry 4.0. IEEE.

Barham, P., Dragovic, B., Fraser, K., Hand, S., Harris, T., Ho, A., Neugebauer, R., Pratt, I., & Warfield, A. (2003). Xen and the art of virtualization. *Operating Systems Review*, *37*(5), 164–177. doi:10.1145/1165389.945462

Bartholdi, J., & Hackman, S. (2016). Warehouse and Distribution Science. The Supply Chain and Logistics Institute, School of Industrial and Systems Engineering, Georgia Institute of Technology.

Bayoumi, A., & McCaslin, R. (2016). *Internet of Things – A Predictive Maintenance Tool for General Machinery, Petrochemicals and Water Treatment*. Sustainable Vital Technologies in Engineering & Informatics.

Bedhief, I., Foschini, L., Bellavista, P., Kassar, M., & Aguili, T. (2019). Toward Self-Adaptive Software Defined Fog Networking Architecture for IIoT and Industry 4.0. *2019 IEEE 24th International Workshop on Computer Aided Modeling and Design of Communication Links and Networks (CAMAD)*, 1–5.

Beetz, J., Mosenlechner, L., & Tenorth, M. (2010). Cram: A cognitive robot abstract machine for everyday manipulation in human environments. *Intl. Conf. on Intelligent Robots and Sys (IROS)*, 1012-1017. 10.1109/IROS.2010.5650146

Behrad, Bagheri, & Kao. (2014, December). A Cyber-Physical Systems architecture for Industry 4.0-based manufacturing systems. In Society of Manufacturing Engineers (SME). Elsevier Ltd.

Beier, G., Niehoff, S., & Xue, B. (2018). More sustainability in industry through industrial internet of things? *Applied Sciences (Basel, Switzerland)*, *8*(2), 219. doi:10.3390/app8020219

Besada, J. A., Bernardos, A. M., Bergesio, L., Vaquero, D., Campaña, I., & Casar, J. R. (2019). Drones-as-a-service: A management architecture to provide mission planning, resource brokerage and operation support for fleets of drones. *2019 IEEE International Conference on Pervasive Computing and Communications Workshops (PerCom Workshops)*, 931–936. 10.1109/PERCOMW.2019.8730838

Bhati, M. S. (2018). *Industrial Internet of Things (IIoT): A Literature Review*. doi:10.18231/2454-9150.2018.0340

Bhogaraju, S. D., & Korupalli, V. R. K. (2020). Design of Smart Roads-A Vision on Indian Smart Infrastructure Development. *2020 International Conference on COMmunication Systems & NETworkS (COMSNETS)*, 773–778. 10.1109/COMSNETS48256.2020.9027404

Billon, M., Lera-Lopez, F., & Marco, R. (2010). Differences in digitalization levels: A multivariate analysis studying the global digital divide. *Review of World Economics*, *146*(1), 39–73. doi:10.100710290-009-0045-y

Bittencourta, L., Immicha, R., & Sakellarioub, R. (2018). *The Internet of Things, Fog and Cloud Continuum: Integration and Challenges*. arXiv:1809.09972v1

Bligh-Wall, S. (2017). Industry 4.0: Security imperatives for IoT — converging networks, increasing risks. *Cyber Security: A Peer-Reviewed Journal*, 61-68.

Bonomi, F., Milito, R., Zhu, J., & Addepalli, S. (2012, August). Fog computing and its role in the internet of things. In *Proceedings of the first edition of the MCC workshop on Mobile cloud computing* (pp. 13-16). 10.1145/2342509.2342513

Bordes, A. Y. B., & Weston, J. (2017). Learning end-to-end goal-oriented dialog. *Intl. Conf. on Learning Representations (ICLR)*, 1-15.

Bose, R. (2009). Advanced analytics: Opportunities and challenges. *Industrial Management & Data Systems, 109*(2), 155–172. doi:10.1108/02635570910930073

Bousdekis, A., Apostolou, D., & Mentzas, G. (2019). Predictive Maintenance in the 4th Industrial Revolution: Benefits, Business Opportunities, and Managerial Implications. *IEEE Engineering Management Review, 48*(1), 57–62. doi:10.1109/EMR.2019.2958037

Boyes, H., Hallaq, B., Cunningham, J. & Watson, T. (2018). The industrial internet of things (IIoT): An analysis framework. *Computers in Industry, 101.*

Boyes, H., Hallaq, B., Cunningham, J., & Watson, T. (2018). 10). The industrial internet of things (IIoT): An analysis framework. *Computers in Industry, 101,* 1–12. doi:10.1016/j.compind.2018.04.015

Bramley, R. G. V., & Ouzman, J. (2019). Farmer attitudes to the use of sensors and automation in fertilizer decision-making: Nitrogen fertilization in the Australian grains sector. *Precision Agriculture, 20*(1), 157–175. doi:10.100711119-018-9589-y

Brandao, M., Hashimoto, K., Santos-Victor, J., & Takanishi, A. (2014, November). Gait planning for biped locomotion on slippery terrain. In *2014 IEEE-RAS International Conference on Humanoid Robots* (pp. 303-308). IEEE. 10.1109/HUMANOIDS.2014.7041376

Bravo, C. E., Saputelli, L., Rivas, F., Pérez, A. G., Nickolaou, M., Zangl, G., De Guzmán, N., Mohaghegh, S. D., & Nunez, G. (2014). State of the art of artificial intelligence and predictive analytics in the E&P industry: A technology survey. *SPE Journal, 19*(04), 547–563. doi:10.2118/150314-PA

Breivold, H. P., & Sandström, K. (2015). Internet of things for industrial automation—challenges and technical solutions. *2015 IEEE International Conference on Data Science and Data Intensive Systems,* 532–539. 10.1109/DSDIS.2015.11

Buchanan & Honey. (1993 February). *Energy and carbon dioxide implications of building construction.* University of Canterbury.

Buchli, J., Theodorou, E., Stulp, F., & Schaal, S. (2011). Variable impedance control a reinforcement learning approach. *Robotics Science and Systems: Online Proceedings, VI,* 153–160.

Buschmann, T., Favot, V., Schwienbacher, M., Ewald, A., & Ulbrich, H. (2013). Dynamics and Control of the Biped Robot Lola. In H. Gattringer & J. Gerstmayr (Eds.), *Multibody System Dynamics, Robotics and Control.* Springer. doi:10.1007/978-3-7091-1289-2_10

BusinessLine. (2019, October 1). Elgi Wins Deming Prize. *BusinessLine.* https://www.thehindubusinessline.com/companies/elgi-wins-2019-deming-prize/article29567774.ece

Butschan, J., Heidenreich, S., Weber, B., & Kraemer, T. (2019). Tackling hurdles to digital transformation— the role of competencies for successful industrial internet of things (iiot) implementation. *International Journal of Innovation Management, 23*(04), 1950036. doi:10.1142/S1363919619500361

Caiza, G., Saeteros, M., Oñate, W., & Garcia, M. V. (2020). Fog computing at an industrial level, architecture, latency, energy, and security: A review. *Heliyon, 6*(4), e03706. doi:10.1016/j.heliyon.2020.e03706 PMID:32300668

Cañadas, J., Sánchez-Molina, J. A., Francisco, R., & Águila, I. M. (2017). Improving automatic climate control with decision support techniques to minimize disease effects in greenhouse tomatoes. *Information Processing in Agriculture, 4*(1), 50–63. doi:10.1016/j.inpa.2016.12.002

Cañizares, E., & Valero, F. A. (2018). Analyzing the Effects of Applying IoT to a Metal-Mechanical Company. *Journal of Industrial Engineering and Management, 11*(2), 308–317. doi:10.3926/jiem.2526

Cannon, J. P., & Perreault, W. D. Jr. (1999). Buyer–seller relationships in business markets. *JMR, Journal of Marketing Research, 36*(4), 439–460.

Carlier, F., & Renault, V. (2016). Iot-a, embedded agents for smart internet of things: Application on a display wall. In *2016 IEEE/WIC/ACM International Conference on Web Intelligence Workshops (WIW)* (pp. 80–83). 10.1109/WIW.2016.034

Carlos, M. (2019). *Digitization in agriculture – what it means and what you need to know.* Challenge Advisory LLP.

Carlsson, C., Heikkilä, M., & Mezei, J. (2016). Fuzzy entropy used for predictive analytics. In *Fuzzy Logic in Its 50th Year* (pp. 187–209). Springer. doi:10.1007/978-3-319-31093-0_9

Cashmore, M. (2015). Rosplan: Planning in the robot operating system. *Proc. of the Intl. Conf. on Automated Planning and Scheduling (ICAPS)*, 1-9.

Castiglione, A., Santis, A. D., Castiglione, A., Palmieri, F., & Fiore, U. (2013). An energy-aware framework for reliable and secure end-to-end ubiquitous data communications. In *Proceedings of the 2013 5th International Conference on Intelligent Networking and Collaborative Systems, INCOS '13* (pp. 157–165). IEEE Computer Society. 10.1109/INCoS.2013.32

Celso, A. R. L., Domingos, F., Antonio, A. F., & Leandro, A. (2016). Fox: A Traffic Management System Of Computer-based Vehicles Fog. In *2016 IEEE Symposium on Computers and Communication (ISCC)*. IEEE.

Cerruto, E., Consoli, A., Raciti, A., & Testa, A. (1995). A robust adaptive controller for PM motor drives in robotic applications. *IEEE Transactions on Power Electronics, 10*(1), 62–71. doi:10.1109/63.368459

Challa, S., Wazid, M., Kumar Das, A., Kumar, N., Goutham Reddy, A., Yoon, E.-J., & Yoo, K.-Y. (2017). Secure Signature-Based Authenticated Key Establishment Scheme for Future IoT Applications. *IEEE Access: Practical Innovations, Open Solutions, 5*, 3028–3043. doi:10.1109/ACCESS.2017.2676119

Chandramohan, D., Vengattaraman, T., Rajaguru, D., &Dhavachelvan, P. (2016). A new privacy preserving technique for cloud service user endorsement using multi-agents. *Journal of King Saud University - Computer and Information Sciences, 28*(1), 37-54. doi:10.1016/j.jksuci.2014.06.018

Chandramohan, D., Vengattaraman, T., Rajaguru, D., Baskaran, R., & Dhavachelvan, P. (2013b). Hybrid Authentication Technique to Preserve User Privacy and Protection as an End Point Lock for the Cloud Service Digital Information. In *International Conference on Green High Performance Computing* (pp. 1-4). IEEE. 10.1109/ICGHPC.2013.6533904

Chandramohan, D., Vengattaraman, T., Rajaguru, D., & Dhavachelvan, P. (2013a). A Novel Framework to Prevent Privacy Breach in Cloud Data Storage Area Service. In *2013 International Conference on Green High Performance Computing* (pp. 1-4). IEEE. 10.1109/ICGHPC.2013.6533903

Charles, J. Kibert., (2012). Sustainable construction – Green Building Design & Delivery (3rd ed.). John Wiley & Sons, Inc.

Chehri, A., & Jeon, G. (2019). The industrial internet of things: examining how the IIoT will improve the predictive maintenance. In *Innovation in Medicine and Healthcare Systems, and Multimedia* (pp. 517–527). Springer. doi:10.1007/978-981-13-8566-7_47

Chekired, D. A., Khoukhi, L., & Mouftah, H. T. (2017). Decentralized cloud-SDN architecture in smart grid: A dynamic pricing model. *IEEE Transactions on Industrial Informatics, 14*(3), 1220–1231. doi:10.1109/TII.2017.2742147

Chen, A. &Stegner, Z. (2018). Rethinking Connectivity: Considerations for Designing Industrial IoT Networks. *IoT Now Magazine.*

Chen, D., Bovornkeeratiroj, P., Irwin, D., & Shenoy, P. (2018). Private memoirs of iot devices: Safeguarding user privacy in the IoT era. In *2018 IEEE 38th International Conference on Distributed Computing Systems (ICDCS)* (pp. 1327–1336). IEEE.

Chen, B., Wan, J., Shu, L., Li, P., Mukherjee, M., & Yin, B. (2017). Smart factory of industry 4.0: Key technologies: application case, and challenges. *IEEE Access: Practical Innovations, Open Solutions*, *6*, 6505–6519. doi:10.1109/AC-CESS.2017.2783682

Chen, J., Cao, X., Cheng, P., Xiao, Y., & Sun, Y. (2010). Distributed collaborative control for industrial automation with wireless sensor and actuator networks. *IEEE Transactions on Industrial Electronics*, *57*(12), 4219–4230. doi:10.1109/TIE.2010.2043038

Chen, J., & Zhou, J. (2018, July). Revisiting Industry 4.0 with a Case Study. In *2018 IEEE International Conference on Internet of Things (iThings) and IEEE Green Computing and Communications (GreenCom) and IEEE Cyber, Physical and Social Computing (CPSCom) and IEEE Smart Data (SmartData)* (pp. 1928-1932). IEEE. 10.1109/Cybermatics_2018.2018.00319

Chiang, M., Ha, S., Risso, F., Zhang, T., & Chih-Lin, I. (2017). Clarifying fog computing and networking: 10 questions and answers. *IEEE Communications Magazine*, *55*(4), 18–20. doi:10.1109/MCOM.2017.7901470

Chiaraviglio, L., Blefari-Melazzi, N., Liu, W., Jairo, A., Gutierrez, J. A., Beek, J., Birke, R., Chen, L., Idzikowski, F., Kilper, D., Paolo, M., Bagula, A., & Wu, J. (2017). Bringing 5G into Rural and Low-Income Areas: Is it Feasible? *IEEE Communications Magazine*.

Chiu, D. K., & Leung, H. F. (2005). Towards ubiquitous tourist service coordination and integration: A multi-agent and semantic web approach. In *Proceedings of the 7th International Conference on Electronic Commerce* (pp. 574–581). ACM. 10.1145/1089551.1089656

Chong, Li, & Li. (2019, May). IEEE 5G Ultra-Reliable Low-Latency Communications in Factory Automation Leveraging Licensed and Unlicensed Bands. IEEE Communications Magazine, 57(5).

Choudhary, G., & Jain, A. K. (2017). Internet of things: a survey on architecture, technologies, protocols and challenges. *International Conference on Recent Advances and Innovations in Engineering, ICRAIE 2016*. 10.1109/ICRAIE.2016.7939537

Cicirello, V., Peysakhov, M., Anderson, G., Naik, G., Tsang, K., Regli, W., & Kam, M. (2004). Designing dependable agent systems for mobile wireless networks. *IEEE Intelligent Systems*, *19*(5), 39–45. doi:10.1109/MIS.2004.41

CISCO-JASPER. (2017). *The hidden costs of delivering Internet of Things (IoT) Services*. White paper. Retrieved from https://www.cisco.com/c/dam/m/en_ca/never-better/manufacture/pdfs/hidden-costs-of- delivering-iiot-services-white-paper.pdf

Clapp, J. (2017). Responsibility to the rescue? Governing private financial investment in global agriculture. *Agriculture and Human Values*, *34*(1), 223–235. doi:10.100710460-015-9678-8

Colombo, A. W., Karnouskos, S., Kaynak, O., Shi, Y., & Yin, S. (2017). Industrial Cyberphysical Systems: A Backbone of the Fourth Industrial Revolution. In *Special Section on Key Technologies for Smart Factory of Industry 4.0* (pp. 6–16). IEEE. doi:10.1109/MIE.2017.2648857

Colter, T., Guan, M., Mahdavian, M., Razzaq, S., & Schneider, J. (2018, January 4). *What the future science of B2B sales growth looks like*. McKinsey & Company. https://www.mckinsey.com/business-functions/marketing-and-sales/our-insights/what-the-future-science-of-b2b-sales-growth-looks-like

Consumer Health Informatics Research Resource - Behavioral Intention. (n.d.). Retrieved October 25, 2018, from Consumer Health Informatics Research Resource: https://chirr.nlm.nih.gov/behavioral-intention.php

Conti, M., Kaliyar, P., & Lal, C. (2017). Remi: A reliable and secure multicast routing protocol for IoT networks. In *Proceedings of the 12th International Conference on Availability, Reliability and Security, ARES '17* (pp. 84:1–84:8). 10.1145/3098954.3106070

Conti, M., Kaliyar, P., Rabbani, M. M., & Ranise, S. (2018). SPLIT: A secure and scalable RPL routing protocol for Internet of Things. In *14th International Conference on Wireless and Mobile Computing, Networking and Communications (WiMob), Limassol* (pp. 1–8). 10.1109/WiMOB.2018.8589115

Cook, D. J., Duncan, G., Sprint, G., & Fritz, R. L. (2018). Using Smart City Technology to Make Healthcare Smarter. *Proceedings of the IEEE, 106*(4), 708–722. doi:10.1109/JPROC.2017.2787688 PMID:29628528

Cova, B., & Salle, R. (2008). Marketing solutions in accordance with the SD logic: Co-creating value with customer network actors. *Industrial Marketing Management, 37*(3), 270–277. doi:10.1016/j.indmarman.2007.07.005

Cristina, C.A., Coppola, M., Tregua, M., & Bifulco, F. (2017). Knowledge Sharing in Innovation Ecosystems: A Focus on Functional Food Industry. *International Journal of Innovation and Technology Management, 14*(5), 1750030-1 – 1750030-18.

Da Xu, L., He, W., & Li, S. (2014). Internet of things in industries: A survey. *IEEE Transactions on Industrial Informatics, 10*(4), 2233–2243. doi:10.1109/TII.2014.2300753

Dai, H., Valenzuela, A., & Tedrake, R. (2014). Whole-body motion planning with centroidal dynamics and full kinematics. In *2014 IEEE-RAS International Conference on Humanoid Robots* (pp. 295-302). IEEE. 10.1109/HUMANOIDS.2014.7041375

Daily, M., Medasani, S., Behringer, R., & Trivedi, M. (2017). Self-Driving Cars. *Computer, 50*(12), 18–23. doi:10.1109/MC.2017.4451204

Dakhnovich, A. D., Moskvin, D. A., & Zegzhda, D. P. (2019). An Approach to Building Cyber-Resistant Interactions in the Industrial Internet of Things. *Automatic Control and Computer Sciences, 53*(8), 948–953. doi:10.3103/S0146411619080078

Daniel, K. (2019). Big Data Analytics in Industry 4.0: Sustainable Industrial Value Creation, Manufacturing Process Innovation, and Networked Production Structures. *Journal of Self-Governance and Management Economics, 7*(3), 34. doi:10.22381/JSME7320195

Daponte, P., De Vito, L., Glielmo, L., Iannelli, L., Liuzza, D., Picariello, F., & Silano, G. (2019). A review on the use of drones for precision agriculture. *IOP Conference Series: Earth and Environmental Science, 275*(1), 12022. 10.1088/1755-1315/275/1/012022

Dastjerdi, A. V., & Buyya, R. (2016). Fog computing: Helping the Internet of Things realize its potential. *Computer, 49*(8), 112–116. doi:10.1109/MC.2016.245

Data, B., & Management, D. C. (n.d.). *Digital Prescriptive Disrupting Manufacturing Value Streams through Internet.* Academic Press.

Davenport, T., Guha, A., Grewal, D., & Bressgott, T. (2020). How artificial intelligence will change the future of marketing. *Journal of the Academy of Marketing Science, 48*(1), 24–42. doi:10.100711747-019-00696-0

Davis, F. D. (1989). Perceived Usefulness, Perceived Ease of Use and User Acceptance of Information Technology. *Management Information Systems Quarterly, 13*(3), 319–340. doi:10.2307/249008

Day, G. S. (2000). Managing market relationships. *Journal of the Academy of Marketing Science, 28*(1), 24–30. doi:10.1177/0092070300281003

De Rango, F., Potrino, G., Tropea, M., Santamaria, A. F., & Fazio, P. (2019). Scalable and ligthway bio-inspired coordination protocol for FANET in precision agriculture applications. *Computers & Electrical Engineering, 74*, 305–318. doi:10.1016/j.compeleceng.2019.01.018

Dedy Irawan, J., Adriantantri, E., & Farid, A. (2018). Rfid and IoT for attendance monitoring system. In *MATEC Web of Conferences*. 10.1051/matecconf/201816401020

Deeken, H., Wiemann, T., & Hertzberg, J. (2018). A spatio-semantic model for agricultural environments and machines. Lecture Notes in Artificial Intelligence and Lecture Notes in Bioinformatics, 10868, 589–600. doi:10.1007/978-3-319-92058-0_57

Deits, R., & Tedrake, R. (2014, November). Footstep planning on uneven terrain with mixed-integer convex optimization. In *2014 IEEE-RAS international conference on humanoid robots* (pp. 279-286). IEEE.

Depuru, S. S. S. R., Wang, L., Devabhaktuni, V., & Gudi, N. (2011, March). *Smart meters for power grid—Challenges, issues, advantages and status. In 2011 IEEE/PES Power Systems Conference and Exposition*. IEEE.

Dubey, H., Yang, J., Constant, N., Amiri, A. M., Yang, Q., & Makodiya, K. (2015, October). Fog data: Enhancing telehealth big data through fog computing. In *Proceedings of the ASE bigdata & socialinformatics 2015* (p. 14). ACM.

Eastwood, C., Klerkx, L., Ayre, M., & Rue, B. D. (2019). Challenges in the Development of Smart Farming: From a Fragmented to a Comprehensive Approach for Responsible Research and Innovation. *Journal of Agricultural & Environmental Ethics, 32*(5-6), 741–768. doi:10.100710806-017-9704-5

Elson, D. S., Cleary, K., Dupont, P., Merrifield, R., & Riviere, C. (2018). Medical Robotics. *Annals of Biomedical Engineering, 46*(10), 1433–1436. doi:10.100710439-018-02127-7 PMID:30209705

Endres, H., Indulska, M., Broser, S., Endres, H., & Indulska, M. (2019). *Association for Information Systems AIS Electronic Library (AISeL) Industrial Internet of Things (IIoT) Business Model Classification Industrial Internet of Things (IIoT)*. Business Model Classification.

Erol, K., Hendler, J., & Nau, D. (1994). Htn planning: Complexity and expressivity. AAAI, 1123-1128.

Eskola, L., Alev, U., Arumeagi, E., Jokisalo, J., Donarelli, A., & Siren, K. (2015). Airtightness, air exchange and energy performance in historic residential buildings with different structures. *International Journal of Ventilation, 14*, 11–26. . doi:10.1080/14733315.2015.11684066

Eswari, T., Sampath, P., Lavanya, S., & ... (2015). Predictive methodology for diabetic data analysis in big data. *Procedia Computer Science, 50*, 203–208. doi:10.1016/j.procs.2015.04.069

Eti, M. C., Ogaji, S. O. T., & Probert, S. D. (2006). Strategic maintenance-management in Nigerian industries. *Applied Energy, 83*(3), 211–227. doi:10.1016/j.apenergy.2005.02.004

Evans. (2011). *The Internet of Things How the Next Evolution of the Internet Is Changing Everything*. Cisco Internet Business Solutions Group (IBSG).

Fabisch, A., Petzoldt, C., Otto, M., & Kirchner, F. (2019). *A Survey of Behavior Learning Applications in Robotics--State of the Art and Perspectives*. arXiv preprint arXiv:1906.01868

Faniyi, F., Lewis, P. R., Bahsoon, R., & Yao, X. (2014). Architecting self-aware software systems. *IEEE/IFIP Conf. on Software Architecture*, 91-94.

Farhangi, H. (2010). The path of the smart grid. *Power and Energy Magazine, 8*(1), 18-28.

Feng, Y., Yang, Y., & Huang, B. (2019). Corrosion analysis and remaining useful life prediction for storage tank bottom. *International Journal of Advanced Robotic Systems, 16*(5), 1–9. doi:10.1177/1729881419877051

Field, U. S. A., & Interim, M. (2008, Dec.). For Official Use Only for Official Use Only. *Distribution.*

Fielke, S., Taylor, B., & Emma, J. (2020). Digitalisation of agricultural knowledge and advice networks state-of-the art review. *Agricultural Systems Elsevier, 18*, 120763. doi:10.1016/j.agsy.2019.102763

Fletcher, S. R., Johnson, T. L., & Larreina, J. (2019). Putting people and robots together in manufacturing: are we ready? In *Robotics and Well-Being* (pp. 135–147). Springer. doi:10.1007/978-3-030-12524-0_12

Floreano, D., & Wood, R. J. (2015). Science, technology and the future of small autonomous drones. *Nature, 521*(7553), 460–466. doi:10.1038/nature14542 PMID:26017445

Freimuth, H. & Keonig, M. (2018). Planning and executing construction inspections with unmanned aerial vehicles. *Automation in Construction, 96*, 540-553. . doi:10.1016/j.autcon.2018.10.016

Gavin, R., Harrison, L., Plotkin, C., Spillecke, D., & Stanley, J. (2020, April 30). *The B2B digital inflection point: How sales have changed during COVID-19.* McKinsey & Company. https://www.mckinsey.com/business-functions/marketing-and-sales/our-insights/the-b2b-digital-inflection-point-how-sales-have-changed-during-covid-19

Gawron-Deutsch, T., Diwold, K., Cejka, S., Matschnig, M., & Einfalt, A. (2018). Industrial IoT für Smart Grid-Anwendungenim Feld. *E&I Elektrotechnik und Informationstechnik, 135*(3), 256–263. doi:10.100700502-018-0617-4

Gazis, V., Leonardi, A., Mathioudakis, K., Sasloglou, K., Kikiras, P., & Sudhaakar, R. (2015, June). Components of fog computing in an industrial internet of things context. In *2015 12th Annual IEEE International Conference on Sensing, Communication, and Networking-Workshops (SECON Workshops)* (pp. 1-6). IEEE. 10.1109/SECONW.2015.7328144

Geetha, P., & Wahida Banu, R. S. D. (2014). A compact modelling of a double-walled gate wrap around nanotube array field effect transistors. *Journal of Computational Electronics, 13*(4), 900–916. doi:10.100710825-014-0607-7

Ghodrati, Hoseinie, & Garmabaki. (2015). Reliability considerations in automated mining Systems. *International Journal of Mining Reclamation and Environment, 29*(15), 404-418.

Giesler, S. (2018). *Digitisation in agriculture - from precision farming to farming 4.0, Dossier.* BIOPRO Baden-Württemberg GmbH.

Gochhayat, S. P., Kaliyar, P., Conti, M., Tiwari, P., Prasath, V. B. S., Gupta, D., & Khanna, A. (2019). LISA: Lightweight context-aware IoT service architecture. *Journal of Cleaner Production, 212*, 1345–1356. doi:10.1016/j.jclepro.2018.12.096

Golestan, S., Mahmoudi-Nejad, A., & Moradi, H. (2019). A Framework for Easier Designs: Augmented Intelligence in Serious Games for Cognitive Development. *IEEE Consumer Electronics Magazine, 8*(1), 19–24. doi:10.1109/MCE.2018.2867970

Golnabi, H., & Asadpour, A. (2007). Design and application of industrial machine vision systems. *Robotics and Computer-integrated Manufacturing, 23*(6), 630–637. doi:10.1016/j.rcim.2007.02.005

Gowtham, Banga, & Patil. (2019, July). Secure Internet of Things: Assessing Challenges and Scopes for NextGen Communication. In *2nd IEEE International Conference on Intelligent computing, Instrumentation and Control Technologies (ICICICT-2019).* IEEE. 10.1109/ICICICT46008.2019.8993327

Gowtham, Banga, & Patil. (2020, April). Cyber-Physical Systems and Industry 4.0 Practical Applications and Security Management an Intelligent Traffic Management System. In *An Intelligent Traffic Management System.* CRC Press.

Goyal, M., Hancock, M. Q., & Hatami, H. (2012). Selling into micromarkets. *Harvard Business Review, 90*(7-8), 79–86.

Graboyes, R. F., & Skorup, B. (2020). *Medical Drones in the United States and a Survey of Technical and Policy Challenges.* Mercatus Center Policy Brief. doi:10.2139srn.3565463

Greeneltch, K. M., Haudenschild, C. C., Keegan, A. D., & Shi, Y. (2004). The opioid antagonist naltrexone blocks acute endotoxic shock by inhibiting tumor necrosis factor-α production. *Brain, Behavior, and Immunity, 18*(5), 476–484. doi:10.1016/j.bbi.2003.12.001 PMID:15265541

Greengard, S. (2015). *The Internet of Things.* MIT Press. doi:10.7551/mitpress/10277.001.0001

Grimm, S., Hitzler, P., & Abecker, A. (2007). *Knowledge representation and ontologies.* Springer Semantic Web Services.

Groom, F. M., & Jones, S. S. (Eds.). (2018). *Enterprise cloud computing for non-engineers.* CRC Press. doi:10.1201/9781351049221

Groover, M. P. (2016). *Automation, production systems, and computer-integrated manufacturing.* Pearson Education India.

Gudlur, V. V. R., Shanmugan, V. A., Perumal, S., & Mohammed, R. M. S. R. (2020). Industrial Internet of Things (IIoT) of Forensic and Vulnerabilities. *International Journal of Recent Technology and Engineering, 8*(5).

Gunasekaran, A., Subramanian, N., & Ngai, W. T. E. (2019). *Quality management in the 21st century enterprises: Research pathway towards Industry 4.0.* Academic Press.

Gungor, V. C., Sahin, D., Kocak, T., Ergut, S., Buccella, C., Cecati, C., & Hancke, G. P. (2011). Smart grid technologies: Communication technologies and standards. *IEEE Transactions on Industrial Informatics, 7*(4), 529–539. doi:10.1109/TII.2011.2166794

Guo, H., Liu, J., & Zhang, J. (2018). Computation Offloading for Multi-Access Mobile Edge Computing in Ultra-Dense Networks. *IEEE Communications Magazine, 56*(8), 14–19. doi:10.1109/MCOM.2018.1701069

Gurtov, A., Liyanage, M., & Korzun, D. (2016). Secure Communication and Data Processing Challenges in the Industrial Internet. *Baltic Journal of Modern Computing*, 1058-1073.

Hallaq, Cunningham, & Watson. (2018, October). The industrial internet of things (IIoT): An analysis framework. Computers in Industry, 101, 1-12.

Hamdi, Abouabdellah, & Oudani. (2018, December). Disposition of Moroccan SME Manufacturers to Industry 4.0 with the Implementation of ERP as A First Step. In *Sixth International Conference on Enterprise Systems (ES).* IEEE.

Han, G., Que, W., Jia, G., & Zhang, W. (2018). Resource-utilization-aware energy efficient server consolidation algorithm for green computing in IIOT. *Journal of Network and Computer Applications, 103*, 205–214. doi:10.1016/j.jnca.2017.07.011

Harish, Nagaraju, Harish, & Shaik. (2019). A Review on Fog Computing and its Applications. *International Journal of Innovative Technology and Exploring Engineering, 8*(6C2).

Hartmann, M. H., & Halecker, B. (2015). Management of Innovation in the Industrial Internet of Things. *XXVI ISPIM Conference – Shaping the Frontiers of Innovation Management.*

Hartmann, M., & Halecker, B. (2015). Management of Innovation in the Industrial Internet of Things. *XXVI ISPIM Conference – Shaping the Frontiers of Innovation Management In*, 1–17.

Hasan, H. (2017). Secure lightweight ECC-based protocol for multi-agent IoT systems. In *IEEE 13th International Conference on Wireless and Mobile Computing, Networking and Communications (WiMob), Rome* (pp. 1–8). IEEE.

Hassanalian, M., & Abdelkefi, A. (2017). Classifications, applications, and design challenges of drones: A review. *Progress in Aerospace Sciences, 91*, 99–131. doi:10.1016/j.paerosci.2017.04.003

Hatzivasilis, G., Askoxylakis, I., Alexandris, G., & Anicic, D. (2018). The Interoperability of Things: Interoperable solutions as an enabler for IoT and Web 3.0. *Conference Paper*. 10.1109/CAMAD.2018.8514952

Hatzivasilis, G., Fysarakis, K., Soultatos, O., Askoxylakis, I., Papaefstathiou, I., & Demetriou, G. (2018). The Industrial Internet of Things as an enabler for a Circular Economy Hy-LP: A novel IIoT protocol, evaluated on a wind park's SDN/NFV-enabled 5G industrial network. *Computer Communications, 119*, 127–137. doi:10.1016/j.comcom.2018.02.007

Hein, A., Weking, J., Schreieck, M., Wiesche, M., Böhm, M., & Krcmar, H. (2019). Value co-creation practices in business-to-business platform ecosystems. *Electronic Markets, 29*(3), 503–518. doi:10.100712525-019-00337-y

Hermann, M., Pentek, T., & Otto, B. (2016). Design principles for industry 4.0 scenarios. In *Proceedings of the 49th Hawaii International Conference on System Sciences (HICSS)*, (pp. 3928–3937). 10.1109/HICSS.2016.488

He, Y., Gu, C., Chen, Z., & Han, X. (2017). Integrated predictive maintenance strategy for manufacturing systems by combining quality control and mission reliability analysis. *International Journal of Production Research, 55*(19), 5841–5862. doi:10.1080/00207543.2017.1346843

Hindriks, K., & Dix, J. (2014). Goal: A multi-agent programming language applied to an exploration game. *Springer Agent-Oriented Software Engineering*, 112-136.

Hossain, M. S., & Muhammad, G. (2016). Cloud-assisted Industrial Internet of Things (IIoT) – Enabled framework for health monitoring. *Computer Networks, 101*, 192–202. doi:10.1016/j.comnet.2016.01.009

Huang, W., Kim, J., & Atkeson, C. G. (2013, May). Energy-based optimal step planning for humanoids. In *2013 IEEE International Conference on Robotics and Automation* (pp. 3124-3129). IEEE. 10.1109/ICRA.2013.6631011

Huebscher, M., & McCann, J. (2008). A survey of autonomic computing – degrees, models, and applications. *ACM Computing Surveys, 40*(3), 1–28. doi:10.1145/1380584.1380585

Iorga, M., Feldman, L., Barton, R., Martin, M. J., Goren, N. S., & Mahmoudi, C. (2018). *Fog computing conceptual model* (No. Special Publication (NIST SP)-500-325).

Iqbal, K. (2014). Drones under UN scrutiny. *Defence Journal, 17*(6), 68.

Ivan Stojmenovic, S. I. T., & Wen, S. (2014)The Fog Computing Paradigm: Scenarios and Security Issues. *Proceedings of the 2014 Federated Conference on Computer Science and Information Systems*. 10.15439/2014F503

Jaidka, H., Sharma, N., & Singh, R. (2020). *Evolution of IoT to IIoT: Applications & Challenges*. Available at SSRN 3603739

Jalasri, M., & Lakshmanan, D. L. (2018). A Survey: Integration of IoT and Fog Computing. *Second International Conference on Green Computing and Internet of Things (ICGCIoT)*. 10.1109/ICGCIoT.2018.8753010

Janssen, C. P., Donker, S. F., Brumby, D. P., & Kun, A. L. (2019). History and future of human-automation interaction. *International Journal of Human-Computer Studies, 131*(January), 99–107. doi:10.1016/j.ijhcs.2019.05.006

Jazdi, N. (2014, May). *Cyber physical systems in the context of Industry 4.0. In 2014 IEEE international conference on automation, quality and testing, robotics*. IEEE.

Jehoon & Sung. (2018, March). The Fourth Industrial Revolution and Precision Agriculture. In *Automation in Agriculture - Securing Food Supplies for Future Generations*. INTECH.

Jelali, M. (2012). *Control performance management in industrial automation: assessment, diagnosis and improvement of control loop performance.* Springer Science & Business Media.

Jeschke, S., Brecher, C., Meisen, T., Özdemir, D., & Eschert, T. (2017). Industrial internet of things and cyber manufacturing systems. In *Industrial Internet of Things* (pp. 3–19). Springer. doi:10.1007/978-3-319-42559-7_1

Jha, S. B., Babiceanu, R. F., & Seker, R. (2019). Formal modeling of cyber-physical resource scheduling in IIoT cloud environments. *Journal of Intelligent Manufacturing, 31*(5), 1149–1164. doi:10.100710845-019-01503-x

Jo, H. S., & Mir-Nasiri, N. (2013). Development of minimalist bipedal walking robot with flexible ankle and split-mass balancing systems. *International Journal of Automation and Computing, 10*(5), 425–437. doi:10.100711633-013-0739-4

Jokinen, K., Nishimura, S., Watanabe, K., & Nishimura, T. (2019). *Human-Robot Dialogues for Explaining Activities.* Academic Press.

Jones, V., & Jo, J. H. (2004). Ubiquitous learning environment: An adaptive teaching system using ubiquitous technology. In *Beyond the comfort zone: Proceedings of the 21st ASCILITE Conference* (*vol. 468*, p. 474). Academic Press.

Judit, Olah, Erdei, & Mate. (2018, September). The Role and Impact of Industry 4.0 and the Internet of Things on the Business Strategy of the Value Chain—the Case of Hungary. Sustainability.

Kagermann, H., Helbig, J., Hellinger, A., &Wahlster, W. (2013). *Recommendations for implementing the strategic initiative INDUSTRIES 4.0.* Securing the future of German manufacturing industry; final report of the Industries 4.0 Working Group, Forschungsunion.

Kagermann, H., Wahlster, W., & Helbig, J. (2013, April). Recommendations for Implementing the Strategic Initiative INDUSTRIE 4.0 -- Securing the Future of German Manufacturing Industry. National Academy of Science and Engineering, Munchen.

Kai, K., Cong, W., & Tao, L. (2016). Fog computing for vehicular ad-hoc networks: paradigms, scenarios, and issues. *The Journal of China Universities of Posts and Telecommunications, 23*(2), 56-96.

Kamble, S. S., Gunasekaran, A., Parekh, H., & Joshi, S. (2019). Modeling the internet of things adoption barriers in food retail supply chains. *Journal of Retailing and Consumer Services, 48*, 154–168. doi:10.1016/j.jretconser.2019.02.020

Kamei, K., & Arai, T. (2018). Optimization for Line of Cars Manufacturing Plant using Constrained Genetic Algorithm. Journal of Robotics. *Networking and Artificial Life, 5*(2), 131–134. doi:10.2991/jrnal.2018.5.2.13

Kamel, S. O., & Hegazi, N. H. (2018). A Proposed Model of IoT Security Management System Based on A study of Internet of Things (IoT) Security. *International Journal of Scientific & Engineering Research, 9*(9).

Kamilaris, A., & Gao, F. (2016). A semantic framework for internet of things-enabled smart farming applications. In *Proceedings of the 2016 IEEE 3rd World Forum on Internet of Things (WF-IoT)*, (pp. 442–447). 10.1109/WF-IoT.2016.7845467

Kan, C., Yang, H., & Kumara, S. (2018). Parallel computing and network analytics for fast Industrial Internet-of-Things (IIoT) machine information processing and condition monitoring. *Journal of Manufacturing Systems, 46*, 282–293. doi:10.1016/j.jmsy.2018.01.010

Kaneko, K., Kanehiro, F., Morisawa, M., Akachi, K., Miyamori, G., Hayashi, A., & Kanehira, N. (2011, September). Humanoid robot hrp-4-humanoid robotics platform with lightweight and slim body. In *2011 IEEE/RSJ International Conference on Intelligent Robots and Systems* (pp. 4400-4407). IEEE. 10.1109/IROS.2011.6094465

Karati, A., Islam, S. K. H., & Karuppiah, M. (2018). Provably secure and lightweight certificateless signature scheme for IIoT environments. *IEEE Transactions on Industrial Informatics, 14*(8), 3701–3711. doi:10.1109/TII.2018.2794991

Karmakar, A., Dey, N., Baral, T., Chowdhury, M., & Rehan, M. (2019). Industrial Internet of Things: A Review. *International Conference on Opto-Electronics and Applied Optics (Optronix)*. 10.1109/OPTRONIX.2019.8862436

Kasina, H., Bahubalendruni, M. R., & Botcha, R. (2017). Robots in medicine: Past, present and future. *International Journal of Manufacturing, Materials, and Mechanical Engineering*, 7(4), 44–64. doi:10.4018/IJMMME.2017100104

Kattepur, A., & Balamuralidhar, P. (2019). Robo-planner: Autonomous robotic action planning via knowledge graph queries. *Proc. of the 34th ACM/SIGAPP Symposium on Applied Computing*, 953-956. 10.1145/3297280.3297568

Kaufmann, D., Ruaux, X. & Jacob, M., (2018). *Digitalization of the Construction Industry: The Revolution is Underway*. Academic Press.

Kaur, K., Garg, S., Aujla, G. S., Kumar, N., Rodrigues, J. J., & Guizani, M. (2018). Edge computing in the industrial internet of things environment: Software-defined-networks-based edge-cloud interplay. *IEEE Communications Magazine*, 56(2), 44–51. doi:10.1109/MCOM.2018.1700622

Kawa, A. (2012). *SMART logistics chain. Intelligent Information and Database Systems*. ACIIDS.

Kettunen, K., & Salmela, E. (2017). Internet of Things as a Digital Transformation Driver in the Finnish Manufacturing Technology Industry. *Journal of Innovation & Business Best Practice*.

Khanna, A., & Kaur, S. (2019). Evolution of Internet of Things (IoT) and its significant impact in the field of Precision Agriculture. *Computers and Electronics in Agriculture*, 157, 218–231. doi:10.1016/j.compag.2018.12.039

Khan, W. Z., Rehman, M. H., Zangoti, H. M., Afzal, M. K., Armi, N., & Salah, K. (2020). Industrial internet of things: Recent advances, enabling technologies and open challenges. *Computers & Electrical Engineering*, 81, 106522. doi:10.1016/j.compeleceng.2019.106522

Kharb, S., & Singhrova, A. (2019). Fuzzy based priority aware scheduling technique for dense industrial IoT networks. *Journal of Network and Computer Applications*, 125, 17–27. doi:10.1016/j.jnca.2018.10.004

Khusainov, R., Shimchik, I., Afanasyev, I., & Magid, E. (2015, July). Toward a human-like locomotion: modelling dynamically stable locomotion of an anthropomorphic robot in simulink environment. In *2015 12th International Conference on Informatics in Control, Automation and Robotics (ICINCO)* (Vol. 2, pp. 141-148). IEEE. 10.5220/0005576001410148

Kim, D. S., & Tran-Dang, H. (2019). An Overview on Industrial Internet of Things. In *Industrial Sensors and Controls in Communication Networks* (pp. 207–216). Springer. doi:10.1007/978-3-030-04927-0_16

Kim, J. Y., Park, I. W., & Oh, J. H. (2007). Walking control algorithm of biped humanoid robot on uneven and inclined floor. *Journal of Intelligent & Robotic Systems*, 48(4), 457–484. doi:10.100710846-006-9107-8

Kim, S., Cho, J., Lee, C., & Shon, T. (2020). Smart seed selection-based effective black box fuzzing for IIoT protocol. *The Journal of Supercomputing*, 76(12), 10140–10154. Advance online publication. doi:10.100711227-020-03245-7

King, P. J., & Mamdani, E. H. (1977). The application of fuzzy control systems to industrial processes. *Automatica*, 13(3), 235–242. doi:10.1016/0005-1098(77)90050-4

Kitchin, D., Quark, A., Cook, W., & Misra, J. (2009). The orc programming language. *Proc. of FMOODS/FORTE*, 1-25.

Kotseruba, I., & Tsotsos, J. (2018). 40 years of cognitive architectures: Core cognitive abilities and practical applications. *Springer Artificial Intelligence Review*, 53(1), 17–94. doi:10.100710462-018-9646-y

Koubâa, A., Qureshi, B., Sriti, M.-F., Allouch, A., Javed, Y., Alajlan, M., Cheikhrouhou, O., Khalgui, M., & Tovar, E. (2019). Dronemap planner: A service-oriented cloud-based management system for the internet-of-drones. *Ad Hoc Networks*, 86, 46–62. doi:10.1016/j.adhoc.2018.09.013

Krieg, J. G., Jakllari, G., & Beylot, A. L. (2018, May). InPReSS: INdoor Plan REconstruction Using the Smartphone's Five Senses. In *2018 IEEE International Conference on Communications (ICC)* (pp. 1-6). IEEE. 10.1109/ICC.2018.8422975

Kumaran, B. S., & Kirubakaran, S. J. (2018). Implementation of 6-DOF Biped Footstep Planning Under Different Terrain Conditions. *Dimension, 18*(13).

Kumar, K. V. R., Kumar, K. D., Poluru, R. K., Basha, S. M., & Reddy, M. P. K. (2020). Internet of things and fog computing applications in intelligent transportation systems. In *Architecture and Security Issues in Fog Computing Applications* (pp. 131–150). IGI Global. doi:10.4018/978-1-7998-0194-8.ch008

Kusters, D., Praß, N., &Gloy, Y. S. (2017). Textile Learning Factory 4.0–Preparing Germany's Textile Industry for the Digital Future. *Procedia Manufacturing, 9,* 214-221. doi:10.1016/j.promfg.2017.04.035

Kychkin, A., Deryabin, A., Neganova, E., & Markvirer, V. (2019). IoT-Based Energy Management Assistant Architecture Design. *IEEE 21st Conference on Business Informatics (CBI).* DOI:10.1109/cbi.2019.00067

Lackner, G., & Tscheliesnig, P. (2007). *Acoustic emission (AE) field application: Testing of aboveground storage tanks (AST) for corrosion and leakage.* Academic Press.

Laird, J., Kinkade, K., Mohan, S., & Xu, J. (2012). Cognitive robotics using the soar cognitive architecture. *AAAI Tech. Report,* 46-54.

Lam, H. Y. K., Kim, P. M., Mok, J., Tonikian, R., Sidhu, S. S., Turk, B. E., Snyder, M., & Gerstein, M. B. (2010). MOTIPS: Automated motif analysis for predicting targets of modular protein domains. *BMC Bioinformatics, 11*(1), 243. doi:10.1186/1471-2105-11-243 PMID:20459839

Langmann, R., & Stiller, M. (2019). The PLC as a Smart Service in Industry 4.0 Production Systems. *Applied Sciences (Basel, Switzerland), 9*(18), 1–20. doi:10.3390/app9183815

Lasi, H., Fettke, P., Kemper, H. G., Feld, T., & Hoffmann, M. (2014). Industry 4.0. *Business & Information Systems Engineering, 6*(4), 239–242. doi:10.100712599-014-0334-4

Lask, J. (2020). Alphabeet – the green-fingered smartphone. *Digitisation in agriculture – from precision farming to farming 4.0.*

Laudien, S. M., &Daxböck, B. (2019). The Influence of the Industrial Internet of Things on Business Model Design: A Qualitative-Empirical Analysis. *Digital Disruptive Innovation Series on Technology Management,* 271-303. doi:10.1142/9781786347602_0010

Lavanya, R. (2019). Fog Computing and Its Role in the Internet of Things. In *Advancing Consumer-Centric Fog Computing Architectures* (pp. 63–71). IGI Global.

Lay, P., Hewlin, T., & Moore, G. (2009). In a downturn, provoke your customers. *Harvard Business Review, 87*(3), 48–56.

Lee, W., Nam, K., Roh, H. G., & Kim, S. H. (2016, January). A gateway based fog computing architecture for wireless sensors and actuator networks. In *2016 18th International Conference on Advanced Communication Technology (ICACT)* (pp. 210-213). IEEE.

Lee, I., & Lee, K. (2015). The Internet of Things (IoT): Applications, investments, and challenges for enterprises. *Business Horizons, 58*(4), 431–440. doi:10.1016/j.bushor.2015.03.008

Lee, Y., Lee, K. M., & Lee, S. H. (2019). Blockchain-based reputation management for custom manufacturing service in the peer-to-peer networking environment. *Peer-to-Peer Networking and Applications, 13*(2), 671–683. doi:10.100712083-019-00730-6

Lehmann, H. (2020). Digitisation in agriculture – from precision farming to farming 4.0. *Sensors for the Bioeconomy.*

Lemaignan, S., Ros, R., Mosenlechner, L., Alami, R., & Beetz, M. (2010). Oro, a knowledge management platform for cognitive architectures in robotics. *Intl. Conf. on Intelligent Robots and Systems*, 3548-3553. 10.1109/IROS.2010.5649547

Leminen, S., Rajahonka, M., Wendelin, R., & Westerlund, M. (2020). Industrial internet of things business models in the machine-to-machine context. *Industrial Marketing Management, 84*, 298–311. doi:10.1016/j.indmarman.2019.08.008

Levesque, H., & Lakemeyer, G. (2010). Cognitive robotics. *Dagstuhl Seminar Proc., 10081*, 1-19.

Liao, Y., Shen, X., Sun, G., Dai, X., & Wan, S. (2019). EKF/UKF-based channel estimation for robust and reliable communications in V2V and IIoT. *EURASIP Journal on Wireless Communications and Networking, 2019*(1). doi:10.118613638-019-1424-2

Liao, Y., Loures, E., & Deschamps, F. (2018). Industrial Internet of Things: A Systematic Literature Review and Insights. *IEEE Internet of Things Journal, 5*(6), 4515–4525. doi:10.1109/JIOT.2018.2834151

Li, G., Wu, J., Li, J., Wang, K., & Ye, T. (2018). Service popularity-based smart resources partitioning for fog computing-enabled industrial Internet of Things. *IEEE Transactions on Industrial Informatics, 14*(10), 4702–4711. doi:10.1109/TII.2018.2845844

Lima, F., de Carvalho, C. N., Acardi, M. B., dos Santos, E. G., de Miranda, G. B., Maia, R. F., & Massote, A. A. (2019). Digital Manufacturing Tools in the Simulation of Collaborative Robots: Towards Industry 4.0. *Brazilian Journal of Operations & Production Management, 16*(2), 261–280. doi:10.14488/BJOPM.2019.v16.n2.a8

Lin, C. Y., Tseng, C. K., Teng, W. C., Lee, W. C., Kuo, C. H., Gu, H. Y., ... Fahn, C. S. (2009, June). The realization of robot theater: Humanoid robots and theatric performance. In *2009 International Conference on Advanced Robotics* (pp. 1-6). IEEE.

Li, Q. Q., Prasad Gochhayat, S., Conti, M., & Liu, F. A. (2017). EnergIoT: A solution to improve network lifetime of IoT devices. *Pervasive and Mobile Computing, 42*, 124–133. doi:10.1016/j.pmcj.2017.10.005

Li, Q., Yue, Y., & Wang, Z. (2020). Deep Robust Cramer Shoup Delay Optimized Fully Homomorphic For IIOT secured transmission in cloud computing. *Computer Communications, 161*, 10–18. doi:10.1016/j.comcom.2020.06.017

Li, R., Asaeda, H., & Li, J. (2017). A distributed publisher-driven secure data sharing scheme for information-centric IoT. *IEEE Internet of Things Journal, 4*(3), 791–803. doi:10.1109/JIOT.2017.2666799

Li, S., Da Xu, L., & Zhao, S. (2018). 5G Internet of Things: A survey. *Journal of Industrial Information Integration, 10*, 1–9. doi:10.1016/j.jii.2018.01.005

Liu, X., Leon-Garcia, A., & Zhu, P. (2017). A distributed software-defined multi-agent architecture for unifying IoT applications. In *8th IEEE Annual Information Technology, Electronics and Mobile Communication Conference (IEMCON)* (pp. 49–55). 10.1109/IEMCON.2017.8117142

Liu, W., Huang, G., Zheng, A., & Liu, J. (2020). Research on the optimization of IIoT data processing latency. *Computer Communications, 151*, 290–298. doi:10.1016/j.comcom.2020.01.007

Liu, Z., Li, Z., Liu, B., Fu, X., Raptis, I., & Ren, K. (2015). Rise of mini-drones: Applications and issues. *Proceedings of the 2015 Workshop on Privacy-Aware Mobile Computing*, 7–12. 10.1145/2757302.2757303

Li, Y., Hou, M., Liu, H., & Liu, Y. (2012). Towards a theoretical framework of strategic decision, supporting capability and information sharing under the context of Internet of Things. *Information Technology Management, 13*(4), 205–216. doi:10.100710799-012-0121-1

Lopes, I. S., Figueiredo, M. C., & Sá, V. (2020). Criticality evaluation to support maintenance management of manufacturing systems. *International Journal of Industrial Engineering and Management, 11*(1), 3–18. doi:10.24867/IJIEM-2020-1-248

Loupos, K., Caglayan, B., Papageorgiou, A., Starynkevitch, B., Vedrine, F., Skoufis, C., ... Boulougouris, G. (2019). Cognition Enabled IoT Platform for Industrial IoT Safety, Security, and Privacy — The CHARIOT Project. *IEEE 24th International Workshop on Computer-Aided Modeling and Design of Communication Links and Networks (CAMAD).* DOI:10.1109/camad.2019.8858488

Lu, S., Xu, C., Zhong, R. Y., & Wang, L. (2017). A RFID-enabled positioning system in automated guided vehicle for smart factories. *Journal of Manufacturing Systems, 44*, 179–190. doi:10.1016/j.jmsy.2017.03.009

Luvisotto, M., Tramarin, F., Vangelista, L., & Vitturi, S. (2018). On the use of LoRaWAN for indoor industrial IoT applications. *Wireless Communications and Mobile Computing, 2018*, 2018. doi:10.1155/2018/3982646

Maheshwari, N., & Dagale, H. (2018). Secure communication and firewall architecture for IoT applications. In *2018 10th International Conference on Communication Systems Networks (COMSNETS)* (pp. 328–335). 10.1109/COMS-NETS.2018.8328215

Mahmud, R., Kotagiri, R., & Buyya, R. (2018). Fog computing: A taxonomy, survey and future directions. In *Internet of everything* (pp. 103–130). Springer. doi:10.1007/978-981-10-5861-5_5

Maiorino, A., & Muscolo, G. G. (2020). Biped robots with compliant joints for walking and running performance growing. *Frontiers of Mechanical Engineering, 6*, 11. doi:10.3389/fmech.2020.00011

Management, A. (2014). *Using the Internet of Things for preventive maintenance.* Academic Press.

Manikanthan, S. V., & Padmapriya, T. (2017). Relay Based Architecture For Energy Perceptive For Mobile Adhoc Networks. *Advances and Applications in Mathematical Sciences, 17*(1), 165–179.

Manyika, J., Ramaswamy, S., Khanna, S., Sarrazin, H., Pinkus, G., & Sethupathy, G. (2015). *Digital America: a tale of the haves and have-mores.* McKinsey Global Institute. https://www.mckinsey.com/industries/high-tech/our-insights/digital-america-a-tale-of-the-haves-and-have-mores

Marakakis, I., & Kalimeri, T. (2019) Remote sensing and multi-criteria evaluation techniques with GIS application for the update of Greek Land Parcel Identification System. *Scandinavian Journal of Information Systems, 44*, 103–106.

Marcu, I., Suciu, G., Bălăceanu, C., Vulpe, A., & Drăgulinescu, A.-M. (2020). Arrowhead Technology for Digitalization and Automation Solution: Smart Cities and Smart Agriculture. *Sensors (Basel), 20*(5), 1464. doi:10.3390200051464 PMID:32155934

Marín-Tordera, E., Masip-Bruin, X., García-Almiñana, J., Jukan, A., Ren, G. J., & Zhu, J. (2017). Do we all really know what a fog node is? Current trends towards an open definition. *Computer Communications, 109*, 117–130. doi:10.1016/j.comcom.2017.05.013

Markets and Markets. (2020, June). *Predictive Maintenance Market by Component (Solutions and Services), Deployment Mode, Organization Size, Vertical (Government and Defense, Manufacturing, Energy and Utilities, Transportation and Logistics), and Region - Global Forecast to 2025.* MarketsandMarkets. https://www.marketsandmarkets.com/Market-Reports/operational-predictive-maintenance-market-8656856.html

MarkitI. (2017). https://news.ihsmarkit.com/prviewer/release_only/slug/number-connected-iot-devices-will-surge-125-billion-2030-ihs-markit-says

Maro, M., Valentino, M. A. R., & Origlia, A. (2017). Graph databases for designing high-performance speech recognition grammars. *Proc. of the 12th Intl. Conf. on Computational Semantics*, 1-9.

Maroua, Fourati, Fourati., & Chouaya. (2017, November). Internet of Things in Industry 4.0 Case Study: Fluid Distribution Monitoring System. *9th International Conference on Networks & Communications.*

Martin, G. (2013). Acoustic emission for tank bottom monitoring. *Key Engineering Materials, 558*(December), 445–455. doi:10.4028/www.scientific.net/KEM.558.445

Martins, B. O., & Küsters, C. (2019). Hidden security: EU public research funds and the development of European drones. *Journal of Common Market Studies, 57*(2), 278–297. doi:10.1111/jcms.12787

Matthyssens, P., & Vandenbempt, K. (1998). Creating competitive advantage in industrial services. *Journal of Business and Industrial Marketing, 13*(4/5), 339–355. doi:10.1108/08858629810226654

Maximilian, L., Markl, E., & Mohamed, A. (2018). Cyber security Management for (Industrial) Internet of Things: Challenges and Opportunities. *Inform Tech Software Eng, 2018*(8), 5. doi:10.4172/2165-7866.1000250

McCue, C. (2014). *Data mining and predictive analysis: Intelligence gathering and crime analysis.* Butterworth-Heinemann.

Mcninch, M., Parks, D., Jacksha, R., & Miller, A. (2019). Leveraging IIoT to Improve Machine Safety in the Mining Industry. *Mining. Metallurgy & Exploration, 36*(4), 675–681. doi:10.100742461-019-0067-5 PMID:33005876

M-CORD. (2020, July 30). http://opencord.org/

Meghana, B., Kumari, S., & Pushphavathi, T. (2017). Comprehensive traffic management system: Real-time traffic data analysis using RFID. In *2017 International conference of Electronics, Communication and Aerospace Technology (ICECA)* (pp. 168-171). Coimbatore: IEEE. 10.1109/ICECA.2017.8212787

Mekki, K., Bajic, E., Chaxel, F., & Meyer, F., (2019). A comparative study of LPWAN technologies for large-scale IoT deployment. *ICT Express, 5,* 1-7. . doi:10.1016/j.icte.2017.12.005

Melich, G., Pai, A., Shoela, R., Kochar, K., Patel, S., Park, J., Prasad, L., & Marecik, S. (2018). Rectal Dissection Simulator for da Vinci Surgery: Details of Simulator Manufacturing With Evidence of Construct, Face, and Content Validity. *Diseases of the Colon and Rectum, 61*(4), 514–519. doi:10.1097/DCR.0000000000001044 PMID:29521834

Meonghun, Lee, & Shin. (2018). IoT-Based Strawberry Disease Prediction System for Smart Farming. Sensors 2018.

Mercer, R. (2019). *Global Connected and IoT Device Forecast.* Academic Press.

Merino, R., Bediaga, I., Iglesias, A., & Munoa, J. (2019). Hybrid Edge–Cloud-Based Smart System for Chatter Suppression in Train Wheel Repair. *Applied Sciences (Basel, Switzerland), 9*(20), 1–18. doi:10.3390/app9204283

Mezentsev, O., & Collin, J. (2019, April). Design and Performance of Wheel-mounted MEMS IMU for Vehicular Navigation. In *2019 IEEE International Symposium on Inertial Sensors and Systems (INERTIAL)* (pp. 1-4). IEEE. 10.1109/ISISS.2019.8739733

Mgonja, C. T. (2017). *Evaluation on Use of Industrial Radiography for Weld Joints Inspection in Tanzania.* Academic Press.

Migiro, S. O., & Magangi, B. A. (2011). Mixed methods: A review of literature and the future of the new research paradigm. *African Journal of Business Management, 5*(10), 3757–3764. doi:10.5897/AJBM09.082

Misra, S., & Vaish, A. (2011). Reputation-based role assignment for rolebasedaccess control in wireless sensor networks. *Computer Communications, 34*(3), 281–294. doi:10.1016/j.comcom.2010.02.013

Mitchell, S., Weersink, A., & Erickson, B. (2018). Adoption of precision agriculture technologies in Ontario crop production. *Canadian Journal of Plant Science, 98*(6), 1384–1388. doi:10.1139/cjps-2017-0342

Mittal, S., Negi, N., & Chauhan, R. (2017). Integration of edge computing with cloud computing. *International Conference on Emerging Trends in Computing and Communication Technologies (ICETCCT)*. DOI:10.1109/icetcct.2017.8280340

Mittal, V., Sarkees, M., & Murshed, F. (2008). The right way to manage unprofitable customers. *Harvard Business Review*, *86*(4), 94–103.

Mobley, R. K. (2002). *An Introduction to Predictive Maintenance* (2nd ed.). Butterworth-Heinemann., doi:10.1016/B978-0-7506-7531-4.X5000-3

Mora, H., Gil, D., Terol, R. M., Azorín, J., & Szymanski, J. (2017). An IoT-based computational framework for healthcare monitoring in mobile environments. Sensors (Basel). doi:10.339017102302

Moser, A., & Korstjens, I. (2018). Series: Practical guidance to qualitative research. Part 3: Sampling, data collection and analysis. *The European Journal of General Practice*, *24*(1), 9–18. doi:10.1080/13814788.2017.1375091 PMID:29199486

Moura, R. L., Ceotto, L. L. F., Gonzalez, A., & Toledo, R. A. (2018). Industrial Internet of Things (IIoT) Platforms: An Evaluation Model. *International Conference on Computational Science and Computational Intelligence (CSCI)*. DOI 10.1109/CSCI46756.2018.00194

Mourtzis, D., Vlachou, E., & Milas, N. (2016). Industrial Big Data as a result of IoT adoption in Manufacturing. *Procedia CIRP*, *55*, 290–295. doi:10.1016/j.procir.2016.07.038

Muhonen, T. (2015). *Standardization of Industrial internet and IoT (IoT – Internet of Things) – Perspective on condition-based maintenance*. Academic Press.

Mukherjee, A., Goswami, P., Yang, L., Tyagi, S. K., Samal, U. C., & Mohapatra, S. K. (2020). Deep neural network-based clustering technique for secure IIoT. *Neural Computing & Applications*, *32*(20), 16109–16117. Advance online publication. doi:10.100700521-020-04763-4

Muylle, S., Dawar, N., & Rangarajan, D. (2012). B2B brand architecture. *California Management Review*, *54*(2), 58–71. doi:10.1525/cmr.2012.54.2.58

Myles, G., Friday, A., & Davies, N. (2003). Preserving privacy in environments with location-based applications. *IEEE Pervasive Computing*, *2*(1), 56–64. doi:10.1109/MPRV.2003.1186726

N., M., Srinivasan, S., Ramkumar, K., Pal, D., Vain, J., & Ramaswamy, S. (2019). A model-based approach for design and verification of Industrial Internet of Things. *Future Generation Computer Systems, 95*, 354-363. doi:10.1016/j.future.2018.12.012

Nagy, J., Olah, J., Erdei, E., Mate, D., & Popp, J. (2018). The Role and Impact of Industry 4.0 and the Internet of Things on the Business Strategy of the Value Chain—The Case of Hungary. *Sustainability*, *2018*(10), 3491. doi:10.3390u10103491

Nallapaneni, Kumar, Archana, & Dash. (2017 November). The Internet of Things: An Opportunity for Transportation and Logistics. In *IEEE International Conference on Inventive Computing and Informatics (ICICI)*. IEEE.

Nallappan, K., Guerboukha, H., Nerguizian, C., & Skorobogatiy, M. (2018). Live Streaming of Uncompressed HD and 4K Videos Using Terahertz Wireless Links. *IEEE Access: Practical Innovations, Open Solutions*, *6*, 58030–58042. doi:10.1109/ACCESS.2018.2873986

Narayandas, D. (2005). Building loyalty in business markets. *Harvard Business Review*, *83*(9), 131–139. PMID:16171217

Nayak, J.K., & Prajapati, J.A. (2006). *Handbook On Energy Conscious Buildings*. Prepared under the interactive R & D project no. 3/4(03)/99-SEC between Indian Institute of Technology, Bombay and Solar Energy Centre, Ministry of Non-conventional Energy Sources.

Neeraj & Singh, A. (2016). Internet of Things and Trust Management in IoT – Review. *International Research Journal of Engineering and Technology, 3*(6).

Nettlea, R., Crawforda, A., & Brightling, P. (2018). How private-sector farm advisors change their practices: An Australian case study. *Journal of Rural Studies, 58*, 20–27. doi:10.1016/j.jrurstud.2017.12.027

Neumann, P. (2007). Communication in industrial automation—What is going on? *Control Engineering Practice, 15*(11), 1332–1347. doi:10.1016/j.conengprac.2006.10.004

Neville, S., Adams, J., & Cook, C. (2016). Using internet-based approaches to collect qualitative data from vulnerable groups: Reflections from the field. *Contemporary Nurse, 52*(6), 657–668. doi:10.1080/10376178.2015.1095056 PMID:26394073

Ni, J., Zhang, K., Lin, X., & Shen, X. S. (2017). Securing fog computing for internet of things applications: Challenges and solutions. *IEEE Communications Surveys and Tutorials, 20*(1), 601–628. doi:10.1109/COMST.2017.2762345

Nikolić, M., Branislav, B., & Raković, M. (2014). Walking on slippery surfaces: Generalized task-prioritization framework approach. In *Advances on Theory and Practice of Robots and Manipulators* (pp. 189–196). Springer. doi:10.1007/978-3-319-07058-2_22

Niranjan, M., Madhukar, N., Ashwini, A., Muddsar, J., & Saish, M. (2017). IOT Based Industrial Automation. *IOSR Journal of Computer Engineering (IOSR-JCE)*, 36-40.

Niu, D., Xu, H., Li, B., & Zhao, S. (2012). *Quality-assured cloud bandwidth auto-scaling for video-on-demand applications. In 2012 Proceedings IEEE INFOCOM*. IEEE. doi:10.1109/INFCOM.2012.6195785

Noha, Hamdy, Hisham, & Alawady. (2019 March). Impacts of Internet of Things on Supply Chains: A Framework for Warehousing. Social Sciences.

Noura, M., Atiquzzaman, M., & Gaedke, M. (2019). Interoperability in internet of things: taxon- omies and open challenges. *Mobile Networks and Applications, 24*, 796-809. . doi:10.100711036-018-1089-9

Nyasimi, M., Kimeli, P., Sayula, G., Radeny, M., Kinyangi, J., & Mungai, C. (2017). Adoption and Dissemination Pathways for Climate-Smart Agriculture Technologies and Practices for Climate-Resilient Livelihoods in Lushoto, Northeast Tanzania. *Climate (Basel), 5*(3), 63. doi:10.3390/cli5030063

O'flaherty, K. W. (2005). *Building predictive models within interactive business analysis processes*. Google Patents.

Oberländer, A. M., Übelhör, J., & Häckel, B. (2019). IIoT-basierteGeschäftsmodellinnovationimIndustrie-Kontext: Archetypen und praktischeEinblicke. *HMD Praxis Der Wirtschaftsinformatik, 56*(6), 1113–1125. doi:10.136540702-019-00570-1

Oh, S., & Kim, Y. (2017). Security requirements analysis for the IoT. In *2017 International Conference on Platform Technology and Service (PlatCon)* (pp. 1–6). Academic Press.

Open Fog Consortium. (2017). *OpenFog Reference Architecture for Fog Computing*. Retrieved from https://www. openfogconsortium.org/wp-content/uploads/OpenFog_Reference_Architecture_2_09_17-FINAL.pdf

Open Fog Consortium. (2018). *Top 10 Myths of Fog Computing*. Retrieved from https://www.openfogconsortium.org/top-10-myths-of-fog-computing/

Ott, C., Baumgärtner, C., Mayr, J., Fuchs, M., Burger, R., Lee, D., . . . Hirzinger, G. (2010, December). Development of a biped robot with torque controlled joints. In *2010 10th IEEE-RAS International Conference on Humanoid Robots* (pp. 167-173). IEEE. 10.1109/ICHR.2010.5686340

Otto, A., Agatz, N., Campbell, J., Golden, B., & Pesch, E. (2018). Optimization approaches for civil applications of unmanned aerial vehicles (UAVs) or aerial drones: A survey. *Networks*, *72*(4), 411–458. doi:10.1002/net.21818

Overgoor, G., Chica, M., Rand, W., & Weishampel, A. (2019). Letting the computers take over: Using AI to solve marketing problems. *California Management Review*, *61*(4), 156–185. doi:10.1177/0008125619859318

Pacis, D. M., Subido, E. D. Jr, & Bugtai, N. T. (2017). *Research on the Application of Internet of Things (IoT) Technology towards a Green Manufacturing Industry: A Literature Review. DLSU Research Congress*, Manila.

Pang, Z., Yang, G., Khedri, R., & Zhang, Y. T. (2018). Introduction to the special section: Convergence of automation technology, biomedical engineering, and health informatics toward the healthcare 4.0. *IEEE Reviews in Biomedical Engineering*, *11*, 249–259. doi:10.1109/RBME.2018.2848518

Parker, L. E., & Draper, J. V. (1998). Robotics applications in maintenance and repair. Handbook of Industrial Robotics, 1378.

Park, S. Y. (2009). An Analysis of the Technology Acceptance Model in Understanding University Students' Behavioral Intention to Use e-Learning. *Journal of Educational Technology & Society*, 150–162.

Parsa, Najafabadi, & Salmasi. (2017). Implementation of smart optimal and automatic control of electrical home appliances (IoT). In *IEEE Proceedings 2017 Smart Grid Conference. SGC* (pp. 1–6). IEEE.

Pearson, N. R., Mason, J. S. D., & Priewald, R. H. (n.d.). *The influence of maintenance on the life cycle of above ground storage tanks*. Academic Press.

Pei Breivold, H. (2019). Towards factories of the future: Migration of industrial legacy automation systems in the cloud computing and Internet-of-things context. *Enterprise Information Systems*, 1–21.

Peng, M., Yan, S., Zhang, K., & Wang, C. (2016). Fog-computing-based radio access networks: Issues and challenges. *IEEE Network*, *30*(4), 46–53. doi:10.1109/MNET.2016.7513863

Pesce, M. D., & Standards, R. (2011). *NDT of Welded Steel Tanks*. Academic Press.

Peter, N. (2015). Fog Computing and It's Real-Time Applications. *International Journal of Emerging Technology and Advanced Engineering*, *5*(6).

Phan, L., & Kim, T. (2020, May 14). Breaking down the Compatibility Problem in Smart Homes: A Dynamically Updatable Gateway Platform. *Sensors (Basel)*, *20*(10), 2783. doi:10.339020102783 PMID:32422946

Popescu, D., Dragana, C., Stoican, F., Ichim, L., & Stamatescu, G. (2018). A Collaborative UAV-WSN Network for Monitoring Large Areas. *Sensors (Basel)*, *18*(12), 1–25. doi:10.339018124202 PMID:30513655

Pouryousefzadeh, S., & Akbarzadeh, R. (2019). Internet of Things (IoT) systems in future Cultural Heritage. *International Conference on Internet of Things and Applications (IoT)*. DOI:10.1109/iicita.2019.8808838

Puiu, Bischof, Serbanescu, Nechifor, Parreira, & Schreiner. (2017). A public transportation journey planner enabled by IoT data analytics. In *20th Conference on Innovations in Clouds, Internet and Networks (ICIN), Paris, 2017* (pp. 355-359). Academic Press.

Puliafito, C., Mingozzi, E., Longo, F., Puliafito, A., & Rana, O. (2019). Fog computing for the internet of things: A Survey. *ACM Transactions on Internet Technology*, *19*(2), 1–41. doi:10.1145/3301443

Puri, V., Nayyar, A., & Raja, L. (2017). Agriculture drones: A modern breakthrough in precision agriculture. *Journal of Statistics and Management Systems*, *20*(4), 507–518. doi:10.1080/09720510.2017.1395171

Pye, A. (2014). The internet of things: Connecting the unconnected. *Engineering & Technology, 9*(11), 64–64. doi:10.1049/et.2014.1109

Rabuzin, K., Sestak, M., & Konecki, M. (2016). Implementing unique integrity constraint in graph databases. *Intl Multi-Conf. on Computing in the Global Information Technology.*

Radanliev, P., De Roure, D. C., Nurse, J. R., Montalvo, R. M., & Burnap, P. (2019). The Industrial Internet-of-Things in the Industry 4.0 supply chains of small and medium sized enterprises. University of Oxford.

Rahimi, H., Zibaeenejad, A., & Safavi, A. A. (2018). A Novel IoT Architecture based on 5G-IoT and Next Generation Technologies. *2018 IEEE 9th Annual Information Technology, Electronics and Mobile Communication Conference (IEMCON)*, 81–88.

Rajab, H., & Cinkelr, T. (2018). IoT based Smart Cities. In *2018 International Symposium on Networks, Computers and Communications (ISNCC)* (pp. 1-4). Rome: IEEE. doi:10.1109/ISNCC.2018.8530997

Ramalingam, M., Puviarasi, R., Chinnavan, E., & Foong, H. K. (2019). Self-monitoring framework for patients in IoT-based healthcare system. International Journal Innovation Technology and Engineering, 3641-3645.

Ramli, M. R., Bhardwaj, S., & Kim, D. S. (2019). *Toward Reliable Fog Computing Architecture for Industrial Internet of Things.* Academic Press.

Ramya, Krishnan, Renuka, Swetha, & Ramakrishnan. (2016). Effective Automatic Attendance Marking System Using Face Recognition with RFID. *International Journal of Scientific Research in Science and Technology, 2*(2).

Rashid, K. M., & Louis, J. (2019). Times-series data augmentation and deep learning for construction equipment activity recognition. *Advanced Engineering Informatics, 42*, 100944. doi:10.1016/j.aei.2019.100944

Rathee, G., Balasaraswathi, M., Chandran, K. P., Gupta, S. D., & Boopathi, C. S. (2020). A secure IoT sensors communication in industry 4.0 using blockchain technology. *Journal of Ambient Intelligence and Humanized Computing.* Advance online publication. doi:10.100712652-020-02017-8

Ravichandran, R. (2019, November 12). *Elgi Equipments to focus more on global markets, eyes No 2 slot.* Financial Express. https://www.financialexpress.com/industry/elgi-equipments-to-focus-more-on-global-markets-eyes-no-2-slot/1761786/

Ray, P. D. (2019). *Pervasive, domain and situational-aware, adaptive, automated, and coordinated big data analysis, contextual learning and predictive control of business and operational risks and security.* Google Patents.

Reddy, K., Hari Priya, K., & Neelima, N. (2015). Object Detection and Tracking — A Survey. In *2015 International Conference on Computational Intelligence and Communication Networks (CICN)* (pp. 418-421). Jabalpur: IEEE. doi:10.1109/CICN.2015.317

Rehman, M. H., Yaqoob, I., Salah, K., Imran, M., Jayaraman, P. P., & Perera, C. (2019). The role of big data analytics in industrial Internet of Things. *Future Generation Computer Systems, 99*, 247–259. doi:10.1016/j.future.2019.04.020

Reinartz, W., & Ulaga, W. (2008). How to sell services more profitably. *Harvard Business Review, 86*(5), 90–96. PMID:18543811

Reinsel, D., Gantz, J., & Rydning, J. (2018). T*he Digitization of the World from Edge to Core, Seagate.* https://www.seagate.com/files/www-content/our-story/trends/files/idc-seagate-dataage-whitepaper.pdf

Riasanow, T., Jäntgen, L., Hermes, S., Böhm, M., & Krcmar, H. (2020). Core, intertwined, and ecosystem-specific clusters in platform ecosystems: Analyzing similarities in the digital transformation of the automotive, blockchain, financial, insurance and IIoT industry. *Electronic Markets.* Advance online publication. doi:10.100712525-020-00407-6

Richards, K. A., & Jones, E. (2008). Customer relationship management: Finding value drivers. *Industrial Marketing Management, 37*(2), 120–130. doi:10.1016/j.indmarman.2006.08.005

Rittinghouse, J. W., & Ransome, J. F. (2017). *Cloud computing: implementation, management, and security.* CRC Press. doi:10.1201/9781439806814

Riva, G., & Riva, E. (2019). SARAFun: Interactive Robots Meet Manufacturing Industry. *Cyberpsychology, Behavior, and Social Networking, 22*(4), 295–296. doi:10.1089/cyber.2019.29148.ceu PMID:30958039

Rizvi, S. S., Zubair, M., Ahmad, J., Hashmani, M., & Khan, M. W. (2019). Wireless Communication as a Reshaping Tool for Internet of Things (IoT) and Internet of Underwater Things (IoUT) Business in Pakistan: A Technical and Financial Review. *Wireless Personal Communications.* Advance online publication. doi:10.100711277-019-06937-3

Rizwan, P., Suresh, K., & Rajasekhara Babu, M. (2017). Real-time smart traffic management system for smart cities by using Internet of Things and big data. *Proceedings of IEEE International Conference on Emerging Technological Trends in Computing, Communications and Electrical Engineering, ICETT 2016.* 10.1109/ICETT.2016.7873660

Robertson, M., Keating, B. A., Daniel, W., Bonnett, G., & Hall, A. J. (2016). Five Ways to Improve the Agricultural Innovation System in Australia. *Farm Policy Journal., 13*, 1–13.

Rogers, E. M. (2010). *Diffusion of Innovations.* The Free Press.

Roman, R., Lopez, J., & Mambo, M. (2018). Mobile edge computing, Fog et al.: A survey and analysis of security threats and challenges. *Future Generation Computer Systems, 78*, 680–698. doi:10.1016/j.future.2016.11.009

Rose, D. C., & Jason, C. (2018). Agriculture 4.0: Broadening Responsible Innovation in an Era of Smart Farming. *Frontiers in Sustainable Food Systems, 2*, 87. doi:10.3389/fsufs.2018.00087

Rosser, J. C. Jr, Vignesh, V., Terwilliger, B. A., & Parker, B. C. (2018). Surgical and medical applications of drones: A comprehensive review. *JSLS: Journal of the Society of Laparoendoscopic Surgeons, 22*(3), e2018.00018. doi:10.4293/JSLS.2018.00018 PMID:30356360

Roth, J. (2002). Context-aware web applications using the pinpoint infrastructure. In *IADIS International Conference WWW/Internet 2002* (pp. 13–15). IADIS Press.

Rubio, F., Valero, F., & Llopis-Albert, C. (2019). A review of mobile robots: Concepts, methods, theoretical framework, and applications. *International Journal of Advanced Robotic Systems, 16*(2), 1729881419839596. doi:10.1177/1729881419839596

Rudra Kumar. (2019). Energy Efficient Scheduling of Cloud Data Center Servers. *International Journal of Innovative Technology and Exploring Engineering, 8*(11), 1769-1772.

Ruiz-Sarmiento, J. R., Monroy, J., Moreno, F. A., Galindo, C., Bonelo, J. M., & Gonzalez-Jimenez, J. (2020). A predictive model for the maintenance of industrial machinery in the context of industry 4.0. *Engineering Applications of Artificial Intelligence, 87*, 103289. doi:10.1016/j.engappai.2019.103289

Rupnik, R., Kukar, M., Petar, V., Košir, D., Darko, P., & Bosnic, Z. (2018). AgroDSS: A decision support system for agriculture and farming. *Computers and Electronics in Agriculture.*

Russell, S., & Norvig, P. (2015). *Artificial Intelligence: A Modern Approach* (3rd ed.). Pearson.

Rüßmann, M., Lorenz, M., Gerbert, P., Waldner, M., Justus, J., Engel, P., & Harnisch, M. (2015). Industry 4.0: The future of productivity and growth in manufacturing industries. *Boston Consulting Group, 9*(1), 54-89.

Ryoo, I., Sun, K., Lee, J., & Kim, S. (2018). A 3-dimensional group management MAC scheme for mobile IoT devices in wireless sensor networks. *Journal of Ambient Intelligence and Humanized Computing, 9*(4), 1223–1234. doi:10.100712652-017-0557-6

Saadaoui, S., Khalil, A., Tabaa, M., Chehaitly, M., Monteiro, F., & Dandache, A. (2020). Improved many-to-one architecture based on discrete wavelet packet transform for industrial IoT applications using channel coding. *Journal of Ambient Intelligence and Humanized Computing*. Advance online publication. doi:10.100712652-020-01972-6

Sabella, D., Vaillant, A., Kuure, P., Rauschenbach, U., & Giust, F. (2016). Mobile-Edge Computing Architecture: The role of MEC in the Internet of Things. *IEEE Consumer Electronics Magazine, 5*(4), 84–91. doi:10.1109/MCE.2016.2590118

Sadeghi, A. R., Wachsmann, C., & Waidner, M. (2015, June). Security and privacy challenges in industrial internet of things. In *2015 52nd ACM/EDAC/IEEE Design Automation Conference (DAC)* (pp. 1-6). IEEE. 10.1145/2744769.2747942

Sahal, R., Breslin, J. G., & Ali, M. I. (2020). Big data and stream processing platforms for Industry 4.0 requirements mapping for a predictive maintenance use case. *Journal of Manufacturing Systems, 54*, 138–151. doi:10.1016/j.jmsy.2019.11.004

Saha, R. K., Tsukamoto, Y., Nanba, S., Nishimura, K., & Yamazaki, K. (2018). *Novel M-CORD Based Multi-Functional Split Enabled Virtualized Cloud RAN Testbed with Ideal Fronthaul*. IEEE. doi:10.1109/GLOCOMW.2018.8644390

Saleem, J., Hammoudeh, M., Raza, U., Adebisi, B., & Ande, R. (2018, June). IoT standardisation: Challenges, perspectives and solution. *Proceedings of the 2nd International Conference on Future Networks and Distributed Systems (ICFNDS'18)*, 26–27.

Salman, O., Elhajj, I., Kayssi, A., & Chehab, A. (2015, December). Edge computing enabling the Internet of Things. In *2015 IEEE 2nd World Forum on Internet of Things (WF-IoT)* (pp. 603-608). IEEE. 10.1109/WF-IoT.2015.7389122

Sandström, K., Vulgarakis, A., Lindgren, M., & Nolte, T. (2013): Virtualization technologies in embedded real-time systems. In *2013 IEEE 18th conference on emerging technologies and factory automation (ETFA)* (pp. 1–8). Los Alamitos: IEEE Press. 10.1109/ETFA.2013.6648012

Saroa, M. K., & Aron, R. (2018). Fog Computing and Its Role in the Development of Smart Applications. *2018 IEEE Intl Conf on Parallel & Distributed Processing with Applications, Ubiquitous Computing & Communications, Big Data & Cloud Computing, Social Computing & Networking, Sustainable Computing & Communications, (ISPA/IUCC/BDCloud/SocialCom/SustainCom)*. DOI:10.1109/bdcloud.2018.00166

Satyanarayanan, M. (2017). The emergence of edge computing. *Computer, 50*(1), 30–39. doi:10.1109/MC.2017.9

Schiavone, F., & Simoni, M. (2019). Strategic marketing approaches for the diffusion of innovation in highly regulated industrial markets: The value of market access. *Journal of Business and Industrial Marketing, 34*(7), 1606–1618. doi:10.1108/JBIM-08-2018-0232

Schleicher, E., Graffi, K., & Rabaya, A. (2019). Fog Computing with P2P: Enhancing Fog Computing Bandwidth for IoT Scenarios. *International Conference on the Internet of Things (iThings) and IEEE Green Computing and Communications (GreenCom) and IEEE Cyber, Physical and Social Computing (CPSC), and IEEE Smart Data (SmartData)*. DOI: com/cpscom/smartdata.2019.0003610.1109/ithings/green

Schmidt, D. R., & Trenta, L. (2018). Changes in the law of self-defence? Drones, imminence, and international norm dynamics. *Journal on the Use of Force and International Law, 5*(2), 201–245. doi:10.1080/20531702.2018.1496706

Schmitt, F., Piccin, O., Barbé, L., & Bayle, B. (2018). Soft robots manufacturing: A review. *Frontiers in Robotics and AI, 5*, 84. doi:10.3389/frobt.2018.00084

Schober, K. S., Hoff, P., & Sold, K. (2016). *Digitization in the construction industry: building Europe's road to "Construction 4.0". Think Act.* https://www.rolandberger.com/en/ Publications/Digitization-of-the-construction industry.html

Schroder, C. (2016). *The challenges of industry 4.0 for small and medium-sized enterprises.* Friedrich-Ebert-Stiftung.

Sengupta, J., Ruj, S., & Das Bit, S. (2020). A Comprehensive survey on attacks, security issues and blockchain solutions for IoT and IIoT. *Journal of Network and Computer Applications, 149,* 102481. doi:10.1016/j.jnca.2019.102481

Sensirion. (2011). *Datasheet SHT21 Humidity and temperature sensor IC.* https://www.sensirion.com/fileadmin/user_upload/customers/sensirion/Dokumente/0_Datashets/Humidity/Sensirion_Humidity_Sensors_SHT21_Datasheet.pdf

Seo, J., Han, S., Lee, S., & Kim, H., (2015). Computer vision techniques for construction safety and health monitoring. *Advanced Engineering Informatics, 29,* 239-251. doi:10.1016/j.aei.2015.02.001

Seraji, H. (1998). A new class of nonlinear PID controllers with robotic applications. *Journal of Robotic Systems, 15*(3), 161–181. doi:10.1002/(SICI)1097-4563(199803)15:3<161::AID-ROB4>3.0.CO;2-O

Sethi, P., & Sarangi, S. R. (2017). Internet of Things: Architectures, Protocols, and Applications. *Journal of Electrical and Computer Engineering, 2017,* 9324035. Advance online publication. doi:10.1155/2017/9324035

Sha, L., Xiao, F., Chen, W., & Sun, J. (2017). IIoT-SIDefender: Detecting and defense against the sensitive information leakage in industry IoT. *World Wide Web (Bussum), 21*(1), 59–88. doi:10.100711280-017-0459-8

Shankar, V., Berry, L. L., & Dotzel, T. (2009). A practical guide to combining products services. *Harvard Business Review, 87*(11), 94–99.

Sharkey, N. (2011). Automating warfare: Lessons learned from the drones. *Journal of Library and Information Science, 21*(2), 140. doi:10.5778/JLIS.2011.21.Sharkey.1

Sharma, N. K., Singh, R. J., Mandal, D., Kumar, A., Alam, N. M., & Keesstra, S. (2017). Increasing farmer's income and reducing soil erosion using intercropping in rainfed maize-wheat rotation of Himalaya, India. *Agriculture, Ecosystems & Environment, 247,* 43–53. doi:10.1016/j.agee.2017.06.026

Sharma, S., & Saini, H. (2020). Fog assisted task allocation and secure deduplication using 2FBO2 and MoWo in cluster-based industrial IoT (IIoT). *Computer Communications, 152,* 187–199. doi:10.1016/j.comcom.2020.01.042

Sherman, M. H., & Grimsrud, D. T. (1980). Infiltration-pressurization Correlation. Simplified physical modeling, California Univ.

Shi, Y., Ding, G., Wang, H., Roman, H. E., & Lu, S. (2015, May). The fog computing service for healthcare. In *2015 2nd International Symposium on Future Information and Communication Technologies for Ubiquitous HealthCare (Ubi-HealthTech)* (pp. 1-5). IEEE. 10.1109/Ubi-HealthTech.2015.7203325

Shijie, W., & Yingfeng, Z. (2020). A credit-based dynamical evaluation method for the smart configuration of manufacturing services under Industrial Internet of Things. *Journal of Intelligent Manufacturing.* Advance online publication. doi:10.100710845-020-01604-y

Shimanuki, Y. (1999). OLE for process control (OPC) for new industrial automation systems. *IEEE SMC'99 Conference Proceedings. 1999 IEEE International Conference on Systems, Man, and Cybernetics (Cat. No. 99CH37028), 6,* 1048–1050.

Shi, W., Pallis, G., & Xu, Z. (2019). Edge Computing. *Proceedings of the IEEE, 107*(8), 1474–1481. doi:10.1109/JPROC.2019.2928287

Shrouf, F., Ordieres, J., & Miragliotta, G. (2014). Smart factories in Industry 4.0: A review of the concept and of energy management approached in production based on the Internet of Things paradigm. *2014 IEEE International Conference on Industrial Engineering and Engineering Management.* 10.1109/IEEM.2014.7058728

Silva, B. N., Khan, M., & Han, K. (2017). Internet of Things : A Comprehensive Review of Enabling Technologies, Architecture, and Challenges. *IETE Technical Review, 0*(0), 1–16. doi:10.1080/02564602.2016.1276416

Silva, M. F., & Machado, J. T. (2012). A literature review on the optimization of legged robots. *Journal of Vibration and Control, 18*(12), 1753–1767. doi:10.1177/1077546311403180

Singh, D., Tripathi, G., & Jara, A. J. (2014, March). A survey of Internet-of-Things: Future vision, architecture, challenges and services. In 2014 IEEE world forum on Internet of Things (WF-IoT) (pp. 287-292). IEEE.

Sinha, R.S., Wei, Y., & Hwang, S. (2017). A survey on LPWA technology: LoRa and NB-IoT. *ICT Express, 3*, 14-21. . doi:10.1016/j.icte.2017.03.004

Sinnott, D., & Dyer, M., (2012). Air-tightness field data for dwellings in Ireland. *Building and Environment, 51*, 269-275. . doi:10.1016/j.buildenv.2011.11.016

Sishi, M. N., & Telukdarie, A. (2017, December). Implementation of Industry 4.0 Technologies in the Mining Industry: A Case Study. In *IEEE International Conference on Industrial Engineering and Engineering Management (IEEM).* IEEE. 10.1109/IEEM.2017.8289880

Sisinni, E., Han, S., Jennehag, U., & Gidlund, M. (2018). *Industrial Internet of Things : Challenges, Opportunities, and Directions.* (doi:10.1109/TII.2018.2852491

Sivashanmugam, K., Verma, K., Sheth, A. P., & Miller, J. (2003). *Adding semantics to web services standards.* Academic Press.

Ślusarczyk, B. (2018). Industry 4.0 – Are we ready? *Polish Journal of Management Studies, 17*(1), 232–248. doi:10.17512/pjms.2018.17.1.19

Srinivasan, K., & Agrawal, N. K. (2018). A study on M-CORD based architecture in traffic offloading for 5G-enabled multiaccess edge computing networks. In *2018 IEEE International Conference on Applied System Invention (ICASI)* (pp. 303-307). Chiba: IEEE. 10.1109/ICASI.2018.8394593

Srinivasan, K., Agrawal, N. K., Cherukuri, A. K., & Pounjeba, J. (2018). An M-CORD Architecture for Multi-Access Edge Computing: A Review. In *2018 IEEE International Conference on Consumer Electronics-Taiwan (ICCE-TW)* (pp. 1-2). Taichung: IEEE. doi:10.1109/ICCE-China.2018.8448950

Statista. (2020). *Industrial Internet of Things market size worldwide from 2017 to 2025.* Statista. https://www.statista.com/statistics/611004/global-industrial-internet-of-things-market-size/

Steiner & Poledna.(2016). Fog computing as an enabler for the Industrial Internet of Things. *e &iElektrotechnik und Informationstechnik, 133*(7), 310-314.

Stenerson, J. (2002). *Industrail Automation and Process Control.* Prentice Hall Professional Technical Reference.

Stojmenovic, I., & Wen, S. (2014, September). *The fog computing paradigm: Scenarios and security issues. In 2014 federated conference on computer science and information systems.* IEEE.

Strauß, P., Schmitz, M., Wöstmann, R., & Deuse, J. (2018). Enabling of Predictive Maintenance in the Brownfield through Low-Cost Sensors, an IIoT-Architecture and Machine Learning. *2018 IEEE International Conference on Big Data (Big Data)*, 1474–1483. 10.1109/BigData.2018.8622076

Suliman, W., Albitar, C., & Hassan, L. (2020). Optimization of Central Pattern Generator-Based Torque-Stiffness-Controlled Dynamic Bipedal Walking. *Journal of Robotics, 2020*, 2020. doi:10.1155/2020/1947061

Sulimin, V. V., Shvedov, V. V., & Lvova, M. I. (2019) Digitization of agriculture: innovative technologies and development models. *Proceedings of IOP Conf. Ser.: Earth Environ. Sci.*, 34, 012215. 10.1088/1755-1315/341/1/012215

Sutton, J., & Austin, Z. (2015). Qualitative reserch: Data collection,analysis,and managment. *The Canadian Journal of Hospital Pharmacy*, 68(3), 226–231. doi:10.4212/cjhp.v68i3.1456 PMID:26157184

Tang, B., Chen, Z., Hefferman, G., Pei, S., Wei, T., He, H., & Yang, Q. (2017). Incorporating intelligence in fog computing for big data analysis in smart cities. *IEEE Transactions on Industrial Informatics*, 13(5), 2140–2150. doi:10.1109/TII.2017.2679740

Taylor, P., Liu, F., Guo, X., Hu, D., & Guo, W. (n.d.). *Nondestructive Testing and Evaluation Comprehensive inspection and evaluation technique for atmospheric storage tanks*. doi:10.1080/10589750902795358

Taylor, K., Griffith, C., Lefort, L., Gaire, R., Compton, M., Wark, T., Lamb, D., Falzon, G., & Trotter, M. (2013). Farming the Web of Things. *IEEE Intelligent Systems*, 28(6), 12–19. doi:10.1109/MIS.2013.102

Tenorth, M., & Beetz, M. (2013). Knowrob – a knowledge processing infrastructure for cognition-enabled robots. *The International Journal of Robotics Research*, 32(5), 566–590. doi:10.1177/0278364913481635

Teslya, N., & Ryabchikov, I. (2018). Blockchain-based platform architecture for industrial IoT. *Conference of Open Innovation Association, FRUCT*, 321–329. doi:10.23919/FRUCT.2017.8250199

Thareja, C., & Singh, N. P. (2019). Role of Fog Computing in IoT-Based Applications. In *Emerging Research in Computing, Information, Communication and Applications* (pp. 99–112). Springer. doi:10.1007/978-981-13-5953-8_9

Thramboulidis, K., & Frey, G. (2011). Towards a model-driven IEC 61131-based development process in industrial automation. *Journal of Software Engineering and Applications*, 4(04), 217–226. doi:10.4236/jsea.2011.44024

Torre, I., Koceva, F., Sanchez, O. R., & Adorni, G. (2016). A framework for personal data protection in the IoT. In *2016 11th International Conference for Internet Technology and Secured Transactions (ICITST)* (pp. 384–391). 10.1109/ICITST.2016.7856735

Trafton, G., Hiatt, L., Harrison, A., Tamborello, F., Khemlani, S., & Schultz, A. (2013). Act-r/e: An embodied cognitive architecture for human-robot interaction. *J. of Human Robot Interaction*, 2(1), 30–55. doi:10.5898/JHRI.2.1.Trafton

Trendov, N. M., Varas, S., & Zeng, M. (2019). *Digital technologies in agriculture and rural areas briefing paper*. Food and Agriculture Organization of the United Nations Rome.

Trindade, E. P., Hinnig, M., Costa, E., Marques, J., Bastos, R., & Yigitcanlar, T. (2017). Sustainable development of smart cities: A systematic review of the literature. *Journal of Open Innovation*, 3(1), 1–14. doi:10.118640852-017-0063-2

Umar, B., Hejazi, H., Lengyel, L., & Farkas, K. (2018). Evaluation of IoT Device Management Tools. IARIA, 2018.

Vaquero, L. M., & Rodero-Merino, L. (2014). Finding your way in the fog: Towards a comprehensive definition of fog computing. *Computer Communication Review*, 44(5), 27–32. doi:10.1145/2677046.2677052

Varney, M. (2018). *Why Machine Intelligence Is the Key to Solving the Data Integration Problem for the IIOT?* Bit Stew Systems Inc.

Velmurugan, R. S., & Dhingra, T. (2014). Maintenance strategy selection theory & practices in natural gas industry: A case study of an Indian gas utility company. *International Gas Research Conference Proceedings, 3*, 2223–2250.

Verma, A. L., Mendelsohn, R., & Bernstein, H. J. (1974). Resonance Raman spectra of the nickel, cobalt, and copper chelates of mesoporphyrin IX dimethyl ester. *The Journal of Chemical Physics*, *61*(1), 383–390. doi:10.1063/1.1681648

Vijayalakshmi, A. (2007). *Climate responsive building envelope to design energy efficient buildings for moderate climate.* Conference on sustainable building South East Asia, Malaysia.

Vimal, S., Khari, M., Dey, N., Crespo, R. G., & Robinson, Y. H. (2020). Enhanced resource allocation in mobile edge computing using reinforcement learning based MOACO algorithm for IIOT. *Computer Communications*, *151*, 355–364. doi:10.1016/j.comcom.2020.01.018

Vinther, K. S., & Müller, S. D. (2018). The imbrication of technologies and work practices: The case of Google Glass in Danish agriculture. *Scandinavian Journal of Information Systems*, *30*, 32–46.

Visvizi, A., & Lytras, M. (2018). It's Not a Fad: Smart Cities and Smart Villages Research in European and Global Contexts. *Sustainability*, *10*(8), 1–10. doi:10.3390u10082727

Vitturi, S., Zunino, C., & Sauter, T. (2019). Industrial communication systems and their future challenges: Next-generation Ethernet, IIoT, and 5G. *Proceedings of the IEEE*, *107*(6), 944–961. doi:10.1109/JPROC.2019.2913443

Walker, M. J. (2018). *Hype Cycle for Emerging Technologies*. Gartner Inc. https://www.gartner.com/en/documents/3885468/hype-cycle-for-emerging-technologies-2018

Waller, M. A., & Fawcett, S. E. (2013). Data science, predictive analytics, and big data: A revolution that will transform supply chain design and management. *Journal of Business Logistics*, *34*(2), 77–84. doi:10.1111/jbl.12010

Walter, A., Finger, R., Huber, R., & Buchmann, N. (2017). Opinion: Smart farming is key to developing sustainable agriculture. *Proceedings of the National Academy of Sciences of the United States of America*, *114*(24), 6148–6150. doi:10.1073/pnas.1707462114 PMID:28611194

Wang, H., Li, L., Chen, H., Li, Y., Qiu, S., & Gravina, R. (n.d.). *Motion Recognition for Smart Sports Based on Wearable Inertial Sensors*. Bodynets.

Wang, K., Wang, Y., Sun, Y., Guo, S., & Wu, J. (2016, December 16th). Green Industrial Internet of Things Architecture: An Energy-Efficient Perspective. *Article in IEEE Communications Magazine*, *54*(12), 48–54. doi:10.1109/MCOM.2016.1600399CM

Wang, S., Wan, J., Li, D., & Zhang, C. (2016). Implementing smart factory of industries 4.0: An outlook. *International Journal of Distributed Sensor Networks*, *12*(1), 3159805. doi:10.1155/2016/3159805

Wang, T., Gao, H., & Qiu, J. (2015). A combined adaptive neural network and nonlinear model predictive control for multirate networked industrial process control. *IEEE Transactions on Neural Networks and Learning Systems*, *27*(2), 416–425. doi:10.1109/TNNLS.2015.2411671 PMID:25898246

Wang, W., Capitaneanu, S. L., Marinca, D., & Lohan, E.-S. (2019). Comparative Analysis of Channel Models for Industrial IoT Wireless Communication. *IEEE Access: Practical Innovations, Open Solutions*, *7*, 91627–91640. doi:10.1109/ACCESS.2019.2927217

Wazid, M., Das, A. K., Odelu, V., Kumar, N., Conti, M., & Jo, M. (2018). Design of secure user authenticated key management protocol for generic iot networks. *IEEE Internet of Things Journal*, *5*(1), 269–282. doi:10.1109/JIOT.2017.2780232

Wei, Mao-Jiun, & Wang. (2004). A comprehensive framework for selecting an ERP system. International Journal of Project Management.

Whitmore, A., Agarwal, A., & Da Xu, L. (2015). The Internet of Things—A survey of topics and trends. *Information Systems Frontiers*, *17*(2), 261–274. doi:10.100710796-014-9489-2

Wilson, A. D. (2013). Diverse Applications of Electronic-Nose Technologies in Agriculture and Forestry. *Sensors (Basel)*, *13*(2), 2295–2348. doi:10.3390130202295 PMID:23396191

Wise, R., & Morrison, D. (2000). Beyond the exchange—the future of B2B. *Harvard Business Review*, *78*(6), 86–96. PMID:11184979

Wollschlaeger, M., Sauter, T., & Jasperneite, J. (2017). The future of industrial communication: Automation networks in the era of the internet of things and industry 4.0. *IEEE Industrial Electronics Magazine*, *11*(1), 17–27. doi:10.1109/MIE.2017.2649104

Wu, M., Lu, T. J., Ling, F. Y., Sun, J., & Du, H. Y. (2010, August). Research on the architecture of Internet of Things. In *2010 3rd International Conference on Advanced Computer Theory and Engineering (ICACTE)* (Vol. 5, pp. V5-484). IEEE.

Wurman, P., D'Andrea, R., & Mountz, M. (2008). Coordinating hundreds of cooperative, autonomous vehicles in warehouses. *AAAI Artificial Intelligence Mag.*, *29*, 9–19.

Wurm, J., Hoang, K., Arias, O., Sadeghi, A., & Jin, Y. (2016).Security analysis on consumer and industrial IoT devices. *21st Asia and South Pacific Design Automation Conference (ASP-DAC)*, 519–524. 10.1109/ASPDAC.2016.7428064

Xiong, M., Kang, Y., Lin, M., & Sun, Y. (2009). *Study on In-service Inspection Methods for the Above-ground Oil Tanks Floors Master, Petroleum Engineer - PetroChina pipeline R & D center Doctor, Petroleum Engineer - PetroChina pipeline R & D center Doctor, Petroleum Engineer - PetroChina pipeline R*. Academic Press.

Xu, H., Yu, W., Griffith, D., & Golmie, N. (2018). A Survey on Industrial Internet of Things: A Cyber-Physical Systems Perspective. *IEEE Access: Practical Innovations, Open Solutions*, 1–1. doi:10.1109/ACCESS.2018.2889501

Xu, L. D., Xu, E. L., & Li, L. (2018). Industry 4.0: State of the art and future trends. *International Journal of Production Research*, *56*(8), 2941–2962. doi:10.1080/00207543.2018.1444806

Xun, Xu, Lu, Aristizabal, Velásquez, Yesid, & Valencia. (2020 January). IoT-enabled smart appliances under industry 4.0: A case study. Advanced Engineering Informatics.

Xu, X. (2012). From cloud computing to cloud manufacturing. *Robotics and Computer-integrated Manufacturing*, *28*(1), 75–86. doi:10.1016/j.rcim.2011.07.002

Yan, C., Deng, W., Jin, L., Yang, T., Wang, Z., Chu, X., Su, H., Chen, J., & Yang, W. (2018). Epidermis-Inspired Ultra-thin 3D Cellular Sensor Array for Self-Powered Biomedical Monitoring. *ACS Applied Materials & Interfaces*, *10*(48), 41070–41075. doi:10.1021/acsami.8b14514 PMID:30398047

Yang, H., & Kim, Y. (2019). Design and Implementation of High-Availability Architecture for IoT-Cloud Services. *Sensors (Basel)*, *2019*(19), 327. doi:10.339019153276 PMID:31349629

Yang, Y., Song, Y., Bo, X., Min, J., Pak, O. S., Zhu, L., Wang, M., Tu, J., Kogan, A., Zhang, H., Hsiai, T. K., Li, Z., & Gao, W. (2020). A laser-engraved wearable sensor for sensitive detection of uric acid and tyrosine in sweat. *Nature Biotechnology*, *38*(2), 217–224. doi:10.103841587-019-0321-x PMID:31768044

Yannuzzi, M., Milito, R., Serral-Gracià, R., Montero, D., & Nemirovsky, M. (2014, December). Key ingredients in an IoT recipe: Fog Computing, Cloud computing, and more Fog Computing. In *2014 IEEE 19th International Workshop on Computer Aided Modeling and Design of Communication Links and Networks (CAMAD)* (pp. 325-329). IEEE.

Yin, R. K. (2017). *Case study research and applications: Design and methods*. Sage Publications.

Yi, S., Li, C., & Li, Q. (2015, June). A survey of fog computing: concepts, applications and issues. In *Proceedings of the 2015 workshop on mobile big data* (pp. 37-42). 10.1145/2757384.2757397

Yokoyama, K., Handa, H., Isozumi, T., Fukase, Y., Kaneko, K., Kanehiro, F., . . . Hirukawa, H. (2003, September). Co-operative works by a human and a humanoid robot. In *2003 IEEE International Conference on Robotics and Automation* (Cat. No. 03CH37422) (Vol. 3, pp. 2985-2991). IEEE. 10.1109/ROBOT.2003.1242049

Younan, M., Houssein, E. H., Elhoseny, M., & Ali, A. A. (2020). Challenges and recommended technologies for the industrial internet of things: A comprehensive review. *Measurement, 151*, 107198. doi:10.1016/j.measurement.2019.107198

Yun, J., Jeong, E., Lee, Y., & Kim, K. (2018). The effect of open innovation on technology value and technology transfer: A comparative analysis of the automotive, robotics, and aviation industries of Korea. *Sustainability, 10*(7), 2459. doi:10.3390u10072459

Yu, W., Dillon, T., Mostafa, F., Rahayu, W., & Liu, Y. (2019). A global manufacturing big data ecosystem for fault detection in predictive maintenance. *IEEE Transactions on Industrial Informatics, 16*(1), 183–192. doi:10.1109/TII.2019.2915846

Yu, X., & Guo, H. (2019). A Survey on IIoT Security. *IEEE VTS Asia Pacific Wireless Communications Symposium (APWCS).* 10.1109/VTS-APWCS.2019.8851679

Zaier, R. (Ed.). (2012). *The Future of Humanoid Robots: Research and Applications.* BoD–Books on Demand. doi:10.5772/1407

Zamora-Izquierdo, M. A., Santa, J., Martínez, J. A., Martínez, V., & Skarmeta, A. F. (2019). Smart farming IoT platform based on edge and cloud computing. *Biosystems Engineering, 177*, 4–17. doi:10.1016/j.biosystemseng.2018.10.014

Zehner, N., Umstätter, C., Niederhauser, J.J., & Schick, M. (2017). System specification and validation of a noseband pressure sensor for measurement of ruminating and eating behavior in stable-fed cows. *Computers and Electronics in Agriculture, 136*, 31–41.

Zeiid, A., Sundaam, S., Moghaddam, M., Kamarthi, S., & Marion, T. (2019). Interoperability in Smart Manufacturing: Research Challenges. *Machines, 2019*(7), 21. doi:10.3390/machines7020021

Zhang, H., Yan, X., Li, H., Jin, R., & Fu, H. (2019). Real-time alarming, monitoring, and locating for non-hard-hat use in construction. *Journal of Construction Engineering and Management, 145.* ,1943-7862.0001629 doi:10.1061/(ASCE)CO

Zhang, X., & Hassanein, H. (2010). Video on-demand streaming on the Internet — A survey. In *2010 25th Biennial Symposium on Communications* (pp. 88-91). Kingston: IEEE. doi:10.1109/BSC.2010.5472998

Zhang, H., & . . . . (2016). Dorapicker: An autonomous picking system for general objects. *IEEE Intl. Conf. on Automation Science and Engineering (CASE)*, 721-726. 10.1109/COASE.2016.7743473

Zhang, K., Yang, K., Liang, X., Su, Z., Shen, X., & Luo, H. H. (2015). Security and privacy for mobile healthcare networks: From a quality of protection perspective. *IEEE Wireless Communications, 22*(4), 104–112. doi:10.1109/MWC.2015.7224734

Zhang, W., Wu, Z., Han, G., Feng, Y., & Shu, L. (2020). LDC: A lightweight dada consensus algorithm based on the blockchain for the industrial Internet of Things for smart city applications. *Future Generation Computer Systems, 108*, 574–582. doi:10.1016/j.future.2020.03.009

Zhang, Y., Jonsson, M., & Li, M. (2017). Guest Editorial Special Issue on Industrial IoT Systems and Applications. *IEEE Systems Journal, 11*(3), 1337–1339. doi:10.1109/JSYST.2017.2702940

Zhong, C., Zhu, Z., & Huang, R. (2017). *Study on the IOT Architecture and Access Technology*. doi:10.1109/DCABES.2017.32

Zhong, R. Y., Xu, X., Klotz, E., & Newman, S. T. (2017). Intelligent Manufacturing in the Context of Industry 4.0: A Review. *Engineering*, *3*(5), 616–630. doi:10.1016/J.ENG.2017.05.015

Zhong, Z., & Hu, W. (2020). Error detection and control of IIoT network based on CRC algorithm. *Computer Communications*, *153*, 390–396. doi:10.1016/j.comcom.2020.02.035

Zhou, K., Liu, T., & Zhou, L. (2015): Industry 4.0: Towards future industrial opportunities and challenges. In *12th International Conference on Fuzzy Systems and Knowledge Discovery (FSKD)* (pp. 2147-2152). IEEE.

Zhou, K., Liu, T., & Zhou, L. (2016). Industry 4.0: Towards future industrial opportunities and challenges. *2015 12th International Conference on Fuzzy Systems and Knowledge Discovery, FSKD 2015*, 2147–2152. 10.1109/FSKD.2015.7382284

Zhou, L., Guo, H., & Deng, G. (2019). A fog computing based approach to DDoS mitigation in IIoT systems. *Computers & Security*, *85*, 51–62. doi:10.1016/j.cose.2019.04.017

Zhou, W., Jia, Y., Peng, A., Zhang, Y., & Liu, P. (2018). The effect of iot new features on security and privacy: New threats, existing solutions, and challenges yet to be solved. *IEEE Internet of Things Journal*, *6*(2), 1606–1616. doi:10.1109/JIOT.2018.2847733

Zhou, X., Guan, Y., Zhu, H., Wu, W., Chen, X., Zhang, H., & Fu, Y. (2014). Bibot-u6: A novel 6-dof biped active walking robot-modeling, planning and control. *International Journal of Humanoid Robotics*, *11*(02), 1450014. doi:10.1142/S0219843614500145

Zhu, C., Rodrigues, J. J., Leung, V. C., Shu, L., & Yang, L. T. (2018). Trust-based communication for the industrial Internet of Things. *IEEE Communications Magazine*, *56*(2), 16–22. doi:10.1109/MCOM.2018.1700592

Zhu, C., Sheng, W., & Liu, M. (2015). Wearable Sensor-based Behavioral Anomaly Detection in Smart Assisted Living Systems. *IEEE Transactions on Automation Science and Engineering*, *4*(4), 1225–1234. doi:10.1109/TASE.2015.2474743

About the Contributors

Sam Goundar is an Editor-in-Chief of the International Journal of Blockchains and Cryptocurrencies (IJFC) – Inderscience Publishers, Editor-in-Chief of the International Journal of Fog Computing (IJFC) – IGI Publishers, Section Editor of the Journal of Education and Information Technologies (EAIT) – Springer and Editor-in-Chief (Emeritus) of the International Journal of Cloud Applications and Computing (IJCAC) – IGI Publishers. He is also on the Editorial Review Board of more than 20 high impact factor journals.As a researcher, apart from Blockchains, Cryptocurrencies, Fog Computing, Mobile Cloud Computing and Cloud Computing, Dr. Sam Goundar also researches in Educational Technology, MOOCs, Artificial Intelligence, ICT in Climate Change, ICT Devices in the Classroom, Using Mobile Devices in Education, e-Government, and Disaster Management. He has published on all these topics. He was a Research Fellow with the United Nations University.He is a Senior Lecturer in IS at The University of the South Pacific, Adjunct Lecturer in IS at Victoria University of Wellington and an Affiliate Professor of Information Technology at Pontificia Universidad Catolica Del Peru.

J. Avanija is Associate Professor, Dept. of Computer Science & Engineering Sree Vidyanikethan Engineering College, Tirupati, India: B.Sc. (Physics), MCA, ME (CSE), PhD., in Information and Communication Engineering Teaching Experience:17 Years in Engineering Colleges at India Research publications in International/National journals/Conferences.

Gurram Sunitha is a professor of CSE at Sree Vidyanikethan Engineering College, Tirupati, India. Her research interests include Data Mining, Spatio-Temporal Analytics, Internet of Things and Artificial Intelligence. She has 7 patents and 3 books published to her credit. She has published around 30 papers in reputed journals and conferences. She has been serving as reviewer for several journals; and has served on the program committees and co-chaired for various international conferences.

S. Bharath Bhushan has received his Ph.D. in Computer Science from VIT University, Vellore, India. He has pursued his M.Tech in Computer Networks and Information Security from JNTUA, India. He has received his B.Tech in Information Technology from JNTUA, India. Currently he is working as an Assistant Professor in school of computing science and engineering, VIT Bhopal University, India. He has authored many national and international journal papers and one book. Also, he has published many chapters in different books published by International publishers. He is holding membership in many professional bodies like CSTA, MCDM and IAENG. He is in the editorial board of international journals like IJFC, IJGR and is a reviewer of over 06 international journals. His areas of research include Cloud Computing, Multi Criteria Decision Making, Data Analytics, Networks, Internet of Things.

* * *

Vijayalakshmi Akella is currently working as Professor and Head, Dept of Civil Engineering, K S School of Engineering and Management, Visvesvaraya Technological University, PhD in Civil Engineering, JNT University, Hyderabad, August 2007, M.E in Structural Engineering, Andhra University, Vizag, 1988, B.Tech in Civil Engineering, JNT University, Hyderabad, June 1986. Guided 1 PhD student and presently guiding 6 students in VTU.

Abhilash B. L. is currently working as Assistant Professor in Department of Civil Engineering, Vidyavardhaka College of Engineering, Mysuru. He as qualified Indian Green Building Council - Accredited Professional (IGBC-AP) from CII Hyderabad, worked as Green Building Engineer in ECO360 Holistic Sustainable Solutions. Projects done - Green Rating of Chennai Metro Rail Limited (CMRL) Chennai (underground & Elevated) which attained IGBC Platinum rated from IGBC MRTS Rating system Hyderabad. CMRL L&T package U004 & U005. Pursuing Environmental Law from National Law School of India University, Bengaluru.

Senthil Kumar Babu is an Associate Professor in Sree Vidyanikethan Engineering College, Tirupathi, India. He obtained his Doctoral degree in the area of Cognitive Spectrum Sensing from the St. Peters University, Chennai, India. He obtained his Master in Applied Electronics from the M.G.R. University, Chennai, India. He obtained his Bachelor's degree from the VIT University, Vellore, India. His research interests include cognitive spectrum sensing and mm-wave communications.

Akashdeep Bhardwaj achieved his PhD from University of Petroleum & Energy Studies (UPES), Post Graduate Diploma in Management (PGDM), Engineering graduate in Computer Science. He has worked as Head of Cyber Security Operations and currently is a Professor in a leading university in India. He has over 24 year experience working as an Enterprise Risk and Resilience and Information Security and Technology professional for various global multinationals.

C. Sasikala Chinthakunta completed Ph.D. form JNT University Anatapur in the year 2019. Her research interests are in the fields of Cloud Computing, Edge Computing and Network Security.

Vishwas D. B. is currently working as Assistant Professor, Department of Computer Science and Engineering, NIE Institute of Technology, Mysuru, India. His research interests include Block Chain Technology, Cyber Security, Wireless Senor Network, Ad-hoc networks, IOT, Data Mining, Cloud Computing and Machine Learning.He has published more than 5 research papers.

Tapan Kumar Das, currently working as Associate Professor in School of Information Technology and Engineering, Vellore Institute of Technology, Vellore. He has graduated from Berhampur University, received M. Tech. in computer science from Utkal University in 2003 and Ph.D. degree in Information Technology from VIT University, India in 2015. He has about 17 years of experience in various capacity in IT industry and academics. He has authored more than 30 journal & conference research articles, book chapters with various reputed publishers. His research interests include Artificial Intelligence, Machine Learning and Data Analytics, Medical informatics and cognitive computing. He is a member of IEEE, life Member of Computer Society of India and Indian Science Congress Association.

Chandramohan Dhasarathan is currently Senior Assistant Professor, Department of Computer Science and Engineering, Madanapalle Institute of Technology & Science, Madanapalle, Andhara Pradesh, India. His area of interest includes Distributed Web Service, Web Service (Evaluation) Testbed, Software Metrics, GVANET and Cloud Computing, Opportunistic Computing, Evolutionary Computing, Service Computing, Software Engineering, Multi-Agent, Pervasive & Ubiquitous Computing, Fog & Edge Computing, Underwater Communication, Privacy and Security. Currently he is working on E-Waste Management, Disaster Management, Bio-Inspired Algorithms and Privacy Preserving Generic Framework for Cloud Data Storage, Optimization approach for minimizing Agro-crops. He is having 9-Years of academic and research expertise and 3-years of industrial experience. He is serving as the Guest Editorial Member of International Journal of Handheld Computing and Research, IGI Global (IJHCR). Acting as an editorial Board Member of International Journal of Information Technology, Modeling and Computing (IJITMC). Review member of International Journal JOHN WILEY-Concurrency and Computation: Practice and Experience, Security and Communication Networks, IEEE- ACCESS International Journal and ACM Computing Survey-International journal.

Seeja G. is a blooming researcher in the field of robotics at Vellore Institute of Technology- Chennai with her masters in robotics and automation from Amritha University. She has three years of research experience along with two years of teaching experience at VIT itself with accomplishments of 3 paper publications.

A. Peda Gopi is currently working as Asst.Professor in CSE department in Vignans Nirula Institute of Technology and Science for Women, peda palakaluru, Guntur. He has good number of publications in scopus and SCIE indexed journals. His areas of interests include Big data, machine learning, neural networks and Blockchain.

Praveena H. D. obtained her B.Tech (ECE) from Nagarjuna University, Guntur in 1995 and M.Tech (Digital Electronics and Communication Systems) from Sree Vidyanikethan Engineering College, Tirupati affiliated to JNTUA, Anantapur in 2010. She has published and presented 35 technical papers in various International Journals & conferences and registered Ph.D at JNTUA, Ananthapuramu. Her areas of research include Communications, Signal and image processing. She has 15 years of teaching experience. Currently she is working as Assistant Professor at Sree Vidyanikethan Engineering College, Sree Sainath Nagar, Tirupati, A.P.

Gururaj H. L. is currently working as Associate Professor, Department of Computer Science and Engineering, Vidyavardhaka College of Engineering, Mysuru, India. He holds a Ph.D. Degree in Computer Science and Engineering from Visweswaraya Technological University, Belagavi, India in 2019. He is a professional member of ACM and working as ACM Distinguish Speaker from 2018. He is the founder of Wireless Internetworking Group(WiNG). He is a Senior member of IEEE and lifetime member of ISTE and CSI. Dr. Gururaj received young scientist award from SERB, DST, Government of India in December 2016. He has 9 years of teaching experience at both UG and PG level. His research interests include Block Chain Technology, Cyber Security, Wireless Senor Network, Ad-hoc networks, IOT, Data Mining, Cloud Computing and Machine Learning. He is an Editorial Board member of the International Journal of Block chains and Cryptocurrencies (Inderscience Publishers) and Special Editor of EAI publishers. He has published more than 75 research papers including 2 SCI publications in various international journals such in IEEE Access, Springer Book Chapter, WoS, Scopus, and UGC referred journals. He has presented 30 papers at various international Conferences. He has authored 1 Book on Network Simulators. He worked as reviewer for various journals and conferences. He also received Best paper awards at various National and International Conferences. He was honored as Chief Guest, Resource Person, Session chair, Keynote Speaker, TPC member, Advisory committee member at National and International Seminars, Workshops and Conferences.

Mohamamd Farukh Hashmi is a Faculty at NIT Warangal.

Shaik Jaffar Hussain received the B.Tech and M.Tech from Madina Engineering College (affiliated to JNTU), Kadapa, in 2005 and 2010, respectively. In June 2006, he became an Assistant Professor at the Department of Computer Science and Engineering, Sri Sai Institute of Technology, Rayachoty, Kadapa(District), where he became an Associate Professor in June 2011. He is the author or coauthor of more than 22 papers in international refereed journals. He has given several invited/plenary talks at different Engineering Colleges. Now currently he is working for KSRM College of Engineering, Kadapa.

Ajay Kattepur is a Senior Researcher at Ericsson Research. He was previously a scientist at the Tata Consultancy Service Research division. His research interests are in autonomous systems, distributed systems and performance evaluation. He received his PhD from Inria, France and an M.Eng. and B.Eng. from NTU, Singapore.

Shailesh Khapre is currently working in ASET-CSE, Amity University, Noida, India. Having 10+ years of research and academic experience. His research area focusing on Information retrieval, Cloud computing, Automata Theory, Blockchain technology, IoT, Web Service.

E. Sudheer Kumar, pursuing a Ph.D. from JNTUA University, Anantapur, received B.Tech and M.Tech in Information Technology from JNTUA, Ananthapuramu, India. His research interests include Data Science, Software Engineering, Software Architecture, and other latest trends in technology.

Korupalli V. Rajesh Kumar is currently a Research Associate in the School of Electronics Engineering of VIT University Chennai. He received his Bachelor's degree in Electronics and Communication Engineering from Institution of Engineers India(IEI), Kolkata. Later he obtained his Master's degree in specialization of Embedded Systems from Jawaharlal Nehru Technological University Kakinada (JNTUK). He has two years of teaching experience as Asst.Prof in Pragati Engineering College Kakinada, he taught subjects like Embedded Systems, IoT and VLSI Design prior to taking up full time research. His expertise in Embedded Systems and IoT based research and product development further helped him in achieving mentor certification from Tata Communication Services - for mentoring students in a project "Self-organizing Networks using ZigBee" under remote internship programme. He has hands-on experience in developing projects on hardware boards like Raspberry -pi, MSP- Texas, Node MCU, Simblee, Arduino, Atmel controllers and Xilinx FPGA's programmed in Assembly, Embedded C, RTOS, VHDL & Verilog(hardware based coding). He started his full time research in Biomechanical and kinesiology related simulation using OpenSim and predicting muscles activation systems and levels for injury prediction using Machine learning and Deep learning methods in 2016. He achieved certificate of merit in concept presentation from the unit FIPS Physicon-2017, DRDO Delhi. He is now working with Machine learning and Deep Learning Frame works using Python and R – languages.

Shonal Kumar is a masters student in Information Systems at The University of the South Pacific.

Gowtham M. is currently working as Assistant Professor, Department of Computer Science and Engineering, NIE Institute of Technology, Mysuru, India. He is doing his research (Ph.D. Degree) in Computer Science and Engineering from Dayananda Sagar University, Bengaluru, India. He has 6 years of teaching experience at both UG and PG level. His research interests include Block Chain Technology, Cyber Security, Wireless Senor Network, Ad-hoc networks, IOT, Data Mining, Cloud Computing and Social IoT. He has published more than 30 research papers including 1 SCI publications in various international journals, Springer, Elsevier, Taylor and Francis and Wiley Book Chapter, WoS, Scopus, and UGC referred journals. He worked as reviewer for various journals and conferences.

Shanmugam M. is currently working as Associate professor in Department of Computer Science and Engineering, Vignan's Foundation for Science, Technology & Research, Guntur, AP, India, having nearly 12+ Years of Academic and research experience. His main research area focusing on VANET, Wireless Networking, Cloud Computing, IoT and Blockchain technologies.

Tawanda Mushiri is a Corporate Member Engineer of the Zimbabwe Institute of Engineers. His research interests are in Artificial Intelligence and Robotics in Machines and the Health sector. Tawanda is in the process of finalizing his patent for Robotic First Aid in passenger vehicle, which will reduce deaths in the global society as a whole at scenes where accidents occur. This has motivated him to do a second Masters in Medical Physics, which is underway to have an appreciation of the Medical Physics and Equipment. For his PhD, he did artificial intelligence for reducing machinery failure in industries and was obtained at the University of Johannesburg in 2017. He possesses a Masters degree in Manufacturing Systems and Operations Management, which he completed in 2012, and a BSc in Mechanical Engineering in 2008 all from the University of Zimbabwe. He is currently reading for an MSc in Medical Physics at National University of Science and Technology in Zimbabwe (2019 – 2021).

D. Narendarsingh was born in Nizamabad, India, in 1981. He received the B.Tech degree in Electronics and Communication engineering from the JNTUH University, Hyderabad, India, in 2003, and the M.Tech.in Embedded Systems from Bharath University, Chennai in 2006 and also he received Post Graduation Diploma from West Minister college of Computing,London,UK. In 2006. Registered for Ph.D. at JNU university. In 2003, he joined XL-Telecoms Pvt. Ltd, Hyderabad, as a Jr. Mobile Testing Engineer for One year and After completion of PG. Diploma, Under Graduate Placement Scheme he worked as Network Trainee Engineer at UTILITY WARE HOUSE, Birmingham, UK. Since November 2009, He has been with the Department of Electronics and Communication, as a Assistant Professor, became an Associate Professor in 2013. His current research interests include IOT, AI & ML Embedded Systems, Embedded Networking, Wireless Communication& Networks, Mobile Computing and IIOT. He is a Student Member of the IEEE. He has Published 34 Papers in various National and International Journals by Supervising M.Tech students and also attended 2 National Conferences 6 International conference.

Safiya Nur is a postgraduate student at The University of the South Pacific.

B. Pavitra was born in Nellore, Andhra Pradesh, India, in 1986. She received the B.Tech degree with distinction in Electronics and Communication engineering from ANNA University Chennai(TN), India, in 2007, and the M.Tech.in Embedded Systems with Distinction from JNTUH University, Hyderabad, India in 2014. Registered for Ph.D. at JNU, Jaipur in the year 2019, worked as Lecturer at AVS College of Engineering from July 2007 to March 2009. Worked as Assistant professor at Avanthi College of Engineering from July 2009 to March 2010. Since 2010 Working as Assistant professor at Anurag Group of Institutions from Till Date. Her current research interests include Embedded Systems, Embedded & IOT, Machine Learning, and Deep Learning. She is a Member of the ISTE. She has Published Papers in various National Journals by Supervising M.Tech students and also Published 2 patents and 8 papers, presented a papers in 2 (two) International Conferences.) Received educational Professional membership qualification certificate from the innovative scientific research professional Malaysia. Received educational Life membership registration card from SIESRP (The society of innovative educationalist and scientific research Professional.

Geetha Prahalathan is working as Associate Professor in Sree Vidyanikethan Engineering College, Tirupathi, India. She got her Doctorate from Anna University in the Field of Nano Electronics. Her Postgraduate and undergraduate are from the University of Madras, Chennai, India. Her interested areas of research are sensors, device modeling, signal modeling, and processing.

G. Rama Subba Reddy is an Associate Professor and Head of the Department at Computer Science & Engineering, Mother Theresa Institute of Engineering & Technology, Palamaner, Andhra Pradesh. He is the author of over 50 articles in various fields like cloud computing, IoT, WSN, Network protocols & Security. His current research focus is on Machine Learning, Data mining, and Cloud computing.

K. Rangaswamy is currently working as Assistant professor and pursuing part time PhD in the area of Networking. Published various Scopus, Web of Science, and UGC journals. Participated various national and international conference, attended workshops, organised technical event's and Ratified as a Assistant Professor by JNTUA Ananthapuramu.

Subhasis Ray is Professor at Xavier Institute of Management Bhubaneswar, India. He researches and writes on marketing and sustainability.

R. Obulakonda Reddy received Bachelor's Degree in Computer Science and Engineering from Sri Krishnadevaraya University, Anantapur, Master's Degree in Computer Science and Engineering from JNTUH, Hyderabad and Ph.D from JNTUA, Anantapur. He is having 12 years of teaching and 2 research experience. He published and presented research papers in national and international conferences and reputed journals. His area of interest includes Pattern Recognition, Image Analysis, Cloud Computing and Network Systems.

Senthil Kumaran S. is currently working as Associate Professor, Department of Manufacturing Engineering, School of Mechanical Engineering, Vellore Institute of Technology-VIT University, Vellore, Tamil Nadu, India. He has been working for more than 10 years in various Engineering Institutions. He has guided various under-graduate, post-graduate students and 3 Ph.D. research scholars supervised in the field of quality improvement of the process in a different field such as friction welding, Tribology,

Non-Traditional Machining Process, and composite materials. He has completed his research work in Friction welding process improvement at the National Institute of Technology, Tiruchirappalli, and Tamil Nadu. His research interests in Advance solid-state welding process, Materials and Metallurgy, Composite materials and Quality Management, Unconventional machining process, Optimization, Material characterization and Mechanical Behavior of Material. He has published more than 175 'International' reputed Journals and conferences. His contribution to the research work such as Genetic Algorithm, Acoustic Emission, Artificial Neural networks, fuzzy logic and optimization techniques in quality improvement of Manufacturing Process, etc. He received Young Scientist Award from Department of Science and Technology (DST), Science and Engineering Research Board (SERB), New Delhi in the year 2014 and he has completed a research project in the year 2017 under the topic of friction welding of SA 213 tube to SA 387 tube plate using an external tool. Also, he received an outstanding reviewer award and recognized the reviewer award from Elsevier journals. He was a lifetime member of Indian Society of Technical Education (ISTE), Indian Welding Society (IWS), International Association of Engineers (IAEng) and International Research Engineers and Doctors (IRED). Also, looking forward to guiding many research scholars, often developing his own interests in the field which he was expertized.

Achyut Shankar is working as an assistant professor in ASET-CSE, Amity University, Noida, India. Having 2.5 Years of experience in academic and research. His research area focuses on Soft computing, Cloud computing, & Blockchain.

Alok Shukla is working as Associate Professor in the Department of Computer Science and Engineering, G L Bajaj Institute of Technology and Management Greater Noida India. Having an experience of 7+ Years in research and academics. His research area focuses on optimization techniques, soft computing, IoT, Cloud Computing, Video analysis, Blockchain, AI, and ML.

V. Shunmughavel received his PhD in Computer Science and Engineering from Anna University Chennai, Tamil Nadu, India in 2015. He is currently serving as Professor & Head, Computer Science, SSM Institute of Engineering & technology. He also served as Dean, Head of the Department and Professor in various Engineering colleges. He received funds from All India Council for Technical Education for Entrepreneur Development Programme. He won many awards for project contest and academics. He is representing Internal Quality Assurance Cell (NAAC- National Assessment and Accreditation Council) of SSM Institute of Engineering & Technology as the Chief Coordinator.

Kathiravan Srinivasan is presently working as an Associate Professor in the School of Information Technology and Engineering at Vellore Institute of Technology (VIT), India. He was previously working as a faculty in the Department of Computer Science and Information Engineering and also as the Deputy Director - Office of International Affairs at National Ilan University, Taiwan.

M. Sudhakar, currently working as Research Associate in Vellore Institute of technology, Chennai Campus. He finished master's degree in the stream of Computer Science and Engineering in the year of 2012 from JNTU University, Anantapur. He finished graduation from the same university in the year of 2010. He started his teaching career in Modugula Kalavathamma Engineering College located in Rajampet and he also worked as Assistant Professor in the department of C.S.E in Sri Sai Institute of Technology and Science around 3.5 years. He started his research work in 2015 in the school of Computing Science and Engineering in VIT Chennai Campus. He published several UGC and Scopus Indexed journals. His research interests include Image Analytics, Image Processing and Deep Learning in computer vision.

Lakshman Vejendla completed his Ph.D in Vel Tech Rangarajan Dr. Sagunthala R&D Institute of Science and Technology, Avadi, Chennai, Tamil Nadu. He is working as Assoc. Professor & HOD in IT Department in Vignan's Nirula Institute of Technology & Science for Women, Guntur, Andhra Pradesh. He has good number of publications in Scopus and SCIE indexed journals. His areas of interests include cryptography, network security, machine learning and Blockchain.

Index

Ensure Quality Research is Introduced to the Academic Community

Become an IGI Global Reviewer for Authored Book Projects

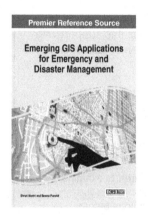
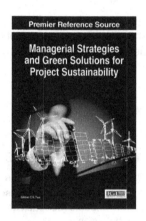

The overall success of an authored book project is dependent on quality and timely reviews.

In this competitive age of scholarly publishing, constructive and timely feedback significantly expedites the turnaround time of manuscripts from submission to acceptance, allowing the publication and discovery of forward-thinking research at a much more expeditious rate. Several IGI Global authored book projects are currently seeking highly-qualified experts in the field to fill vacancies on their respective editorial review boards:

Applications and Inquiries may be sent to:
development@igi-global.com

Applicants must have a doctorate (or an equivalent degree) as well as publishing and reviewing experience. Reviewers are asked to complete the open-ended evaluation questions with as much detail as possible in a timely, collegial, and constructive manner. All reviewers' tenures run for one-year terms on the editorial review boards and are expected to complete at least three reviews per term. Upon successful completion of this term, reviewers can be considered for an additional term.

If you have a colleague that may be interested in this opportunity, we encourage you to share this information with them.

Printed in the United States
By Bookmasters